Cartography

Cartography

The Ideal and Its History

Matthew H. Edney

THE UNIVERSITY OF CHICAGO PRESS • CHICAGO AND LONDON

Publication of this book has been aided by a grant from the Neil Harris Endowment Fund, which honors the innovative scholarship of Neil Harris, the Preston and Sterling Morton Professor Emeritus of History at the University of Chicago. The Fund is supported by contributions from the students, colleagues, and friends of Neil Harris.

The University of Chicago Press, Chicago 60637
The University of Chicago Press, Ltd., London
© 2019 by The University of Chicago
Published 2019
Printed in the United States of America

28 27 26 25 24 23 22 21 20 19   1 2 3 4 5

ISBN-13: 978-0-226-60554-8 (cloth)
ISBN-13: 978-0-226-60568-5 (paper)
ISBN-13: 978-0-226-60571-5 (e-book)
DOI: https://doi.org/10.7208/chicago/9780226605715.001.0001

LIBRARY OF CONGRESS CATALOGING-IN-PUBLICATION DATA

Names: Edney, Matthew H., author.
Title: Cartography : the ideal and its history / Matthew H. Edney.
Description: Chicago ; London : The University of Chicago Press, 2019. | Includes bibliographical references and index.
Identifiers: LCCN 2018025842 | ISBN 9780226605548 (cloth : alk. paper) | ISBN 9780226605685 (pbk. : alk. paper) | ISBN 9780226605715 (e-book)
Subjects: LCSH: Cartography—Philosophy. | Cartography—Philosophy—History.
Classification: LCC GA102.3 .E36 2019 | DDC 526—dc23
LC record available at https://lccn.loc.gov/2018025842

To
ART AND JAN HOLZHEIMER
and
HAROLD AND PEGGY OSHER
in gratitude

Contents

Acknowledgments

This book is the product of my entire career as a map historian (so far) and I have many people to thank. My first scholarly debts are to Hugh Prince and Arthur Allan at University College London: Hugh introduced me to historical geography and inadvertently pointed me toward early maps; Arthur grounded me in surveying practices. David Woodward, my graduate advisor, and Brian Harley then introduced me to the intricacies and complexities of map history. Since 1999, I have been extremely privileged to spend much time in close partnership with Mary Pedley, as we have together designed and edited *Cartography in the European Enlightenment*, an experience that has tested and tempered many of my ideas. And my wife Kathryn has long been a wonderful and loving support (and occasional sounding board).

I am honored to dedicate this book to two couples who have materially promoted the field of map history with their philanthropic support and who, in the process, have advanced my own career. Art and Jan Holzheimer provided the funding that permitted me to complete my dissertation, but that was just one small drop in the ocean of support they have given to fellowships, lectures, and collections in the field. Harold and Peggy Osher gave me their friendship and support, but again only as part of a much larger pattern of remarkable generosity to the University of Southern Maine and the study of map history.

My special thanks go to those who read this book in manuscript, in all or part, especially Max Edelson, Ian Fowler, Carla Lois, Katie Parker, and the Press's readers. Max, in particular, helped me reframe what had become a much larger project to make it more cogent; Ian and Katie independently suggested the same reorganization. Any failings in how I have implemented these changes are, of course, my own.

Many other people have—willingly or not—helped me or contributed to my ideas over the last four decades. I apologize unreservedly to anyone whom I may have unintentionally omitted from the following list. My two families, of course: Sarah Edney, Philip Edney, Rebecca and Graham Moss, Caroline and Paul Lang, and all my aunts, uncles, cousins, nieces, and nephews; Mel and Irene Tremper and the whole Mazurek–Hylan–Tremper–Allman clan. My third family, the History of Cartography Project: Roz Woodward, Beth Freundlich, and Jude Leimer; Mark Monmonier and Roger Kain; Sarah Tyacke, Bob Karrow, and Dennis Reinhartz; and all the project and office assistants, particularly Dana Freiburger, Jen Martin, and Jed Wadworth. Also in or from Madison: Jim Burt; Susan Cook; Bill Gartner, Michelle Szabo, and clan; Susan Cook and Lea Jacobs; Paul, Ann, and Emmon Hedrich Rogers; Jennie and Justin Woodward and clans; and, never last, John Taylor. In the Osher Map Library: Adinah Barnett, Stuart Hunter, Renee Keul, Heather McGaw, David Neikirk, David Nutty, Roberta Ransley-Rideau, Bob Spencer, and especially Yolanda Theunissen; Richard d'Abate, Bill Browder, Glen Parkinson, and Harry Pringle. At or from USM more generally: Lydia Savage and Greg Willette, Rayne Carroll and Joe Medley, Donna Cassidy, Joe Conforti, Nathan Hamilton, Adrienne Hess, Wim Klooster, Christi Mitchell and Drew Sell, Damir Porobic, Tracy and Tom Quimby, Conor Quinn, Kent Ryden, and Joe Wood; Angela Connolly and David Jones; and all my students who have prompted me to look in odd corners of the field and to think about their, and my own, preconceptions.

And my professional friends and colleagues over more than three decades: Jim Akerman; Tom Baerwald; Peter Barber; Chris Baruth; Gordon Bleach; Paul van den Brink; Martin Brückner; David Buisseret; Charles Burroughs; Tony Campbell; George Carhart; Susan Cimburek; David Cobb; Paul Cohen; Tom Conley; Andrew Cook; Karen Cook; Denis Cosgrove; Jeremy Crampton; Jack Crowley; Ed Dahl; Chris Dando; Susan Danforth; Stephen Daniels; Surekha Davies; Catherine Delano Smith; Imre Demhardt; Felix Driver; Jordana Dym; Ralph Ehrenberg; Norman Fiering; Chris Fleet; Nadina Gardner; Joe Garver; Anne Godlewska; Ron Grim; Tom Harper; John Hébert; Mike Heffernan; Markus Heinz; Christie Henry; Francis Herbert; John Hessler; Catherine Hoffman; Stephen Hornsby; Christian Jacob; Bert Johnson; Richard Kagan; Penny Kaiserlian; Art Kelly; Anne Knowles; Charlie Kolb; Peter van der Krogt; Fred Kronz; John Krygier; Daniel Lawrence; Paul Laxton; Malcolm Lewis; Nick Millea; Salim Ghouse Mohammed; Carme Montaner; Jossy and Ken Nebenzahl; Pete Nekola; Nick Noyes; Barbara Paulson; Margaret Pearce; Alexey Postnikov; Don Quataert; Sumathi Ramaswamy; Bill Rankin; Ed Redmond; Jamie Kingman Rice; Penny Richards; Jonathan Rosenwasser; Barry Ruderman; David Rumsey; Neil Safier; Tom

Sander; Paul Schellinger; Susan Schulten; Martijn Storms; Bud Sweet and Julie Camilo; John Tagg; Richard Talbert; Henry Taliaferro; Sinharaja Tammita Delgoda; Zsolt Török; Tim Wallace; David Weimer; Charlie Withers; Denis Wood; Joel Wurl; and Kees Zandvliet. Mary Laur has been a wonderful and supportive editor at Chicago and it has been a pleasure to work with her on the production of this tome. Finally, I must thank all the librarians, bar staff, and baristas who have put up with me over these many years, most recently, as I've actually been writing this book, at Gritty's in Freeport and at the Freeport and Diamond Street branches of Coffee by Design.

Conventions Used in This Book

Chronology

early modern: The era from ca. 1450 to ca. 1800, the subject of Woodward (2007b) and Edney and Pedley (2019)

modern: The era from ca. 1800 to the present, as covered by both Kain (forthcoming) and Monmonier (2015)

Terminology

Quotation marks are used for a generic concept, e.g., "map."

Italics are used for a specific linguistic incarnation of a generic concept, e.g., *map*, *carte*, or *Karte*. When discussing specific terms from different languages, I generally present them in triplets of French/German/English terms, e.g., *carte/Karte/map*.

Coordinate Systems

(ϕ, λ): The cosmographical coordinate pair, respectively latitude (ϕ) and longitude (λ)

(x, y): The two-dimensional Cartesian coordinate pair, respectively abscissa (x) and ordinate (y)

$(x[\phi,\lambda], y[\phi,\lambda])$: The projected coordinate pair, in which the abscissa and ordinate are each a function of latitude and longitude. I express projective coordinates in this manner to emphasize that they are not really the same as two-dimensional Cartesian coordinates when circumscribed by mapping practices within territorial discourses.

1

Introducing the Ideal of Cartography

There is no such thing as cartography, and this is a book about it

The study of maps and mapping is bedeviled by a profound act of cultural misdirection. Modern culture deploys an idealized conception of mapping that obscures the myriad ways in which people actually go about producing, circulating, and consuming maps, whether in the past or the present. The actual behavior, what people do, is mapping. The idealized behavior, what people *think* they do, is cartography. As the apparently singular endeavor by which the world is reduced to paper, or nowadays to digital screens, cartography stands as a fundamental element of human society and culture. Cartography seems coterminous with civilization. It appears to be universal and timeless. And this singular endeavor has only ever had just one product, "the map."

The image—or desideratum or perhaps simulacrum—of cartography is the product of a complex belief system that permeates modern culture: the "ideal of cartography." The ideal normalizes "the map," requiring that it be construed only in certain confined and quite unrealistic ways; in particular, the map is understood to be the product of a restricted set of specific practices. The ideal smothers the actual messiness of mapping with a thick, fluffy, warm, and comforting blanket of cartographic uniformity and transcultural universality. The *ideal of cartography* is the entire belief system, while *cartography* is the fiction generated by the ideal.

According to the ideal, cartography is simply the making of maps. In more academic formulations, it also includes the study of maps and map making. From this perspective, it seems axiomatic that "'maps' define the domain of cartography" (Vasiliev et al. 1990, 119), and maps, not cartography and certainly not the ideal of cartography, have been the focus of scholarly attention. Conceptual studies bear titles such as *The Nature of Maps* (Robinson and Petchenik 1976); "Deconstructing the Map" (Harley 1989); *The Power of Maps* (Wood 1992b); *How Maps Work* (MacEachren 1995); *Maps and Politics* (Black 1997); and *Rethinking Maps* (Dodge, Kitchin, and Perkins 2009b). None has a title like "the nature of cartography," "deconstructing cartography," or "the power of cartography." Those scholars who have taken a broader viewpoint—especially those who have concerned themselves with the intellectual identity of the academic discipline of cartography—might have referenced "cartography" in the titles to their essays or books, yet they too consistently focus their analyses on "the map" and on how "the map" is made and used.

Scholarly assessments and reconsiderations of the nature of maps have flourished since the mid-1960s, when academic cartographers began to reflect seriously on the nature of their field and when historians of cartography began to consider how early maps might be studied as part of the humanities (Edney 2016). Their disciplinary concerns and historiographical reflections subsequently merged with much broader intellectual movements, notably the postmodernist reevaluation of the nature of representation and political concerns for access to and the shaping of knowledge. The result has been two kinds of critical reflection about "the map": the normative and the sociocultural. The normative critique has sought to validate and uphold the normative map and, with it, all the other idealizations promoted by the ideal of cartography; such critics are relatively few and have largely been limited to academic cartographers and geographers (e.g., Robinson and Petchenik 1976; MacEachren 1995), although some scholars in other disciplines have made notable contributions (e.g., Black 1997). The wide-ranging and interdisciplinary sociocultural critique has sought to replace the normative map with new interpretations that take into account not only the cultural and social significance of maps but also those kinds of maps that the ideal dismisses as unorthodox or abnormal (e.g., Harley 1989; Wood 1992b; Dodge, Kitchin, and Perkins 2009b). Yet, after fifty years, the debate between normative and sociocultural critics is still inconclusive, and scholars continue to discuss, ponder, and debate the nature of maps.

Many scholars are unpersuaded by the sociocultural critique. They are intellectually unmoved by unfamiliar concepts adopted by sociocultural critics from other disciplines. They find politically unpalatable the radicalism of many sociocultural critics. And they find the new interpretations to have little, if anything, to contribute to their primary concerns for the technicalities of map making and for the instrumentality of map use. Some scholars accordingly dismiss the sociocultural critique out of hand, but most just ignore it as irrelevant. The few normative critics who have engaged with the sociocultural critique have reconciled the conflicting positions by supplementing the still-dominant technical concerns with greatly watered-down social and cultural elements.

At the same time, the sociocultural critique has centered on refuting the normative map, not the ideal as a whole, and as such has been both incomplete and ineffective. Sociocultural critics have been able to address some of modern society's deeply rooted cartographic norms, but have been unable to dispel them completely, while leaving others barely touched. Despite the wealth of new insights, basic issues that should have been settled long ago must continue to be revisited. Sociocultural critics still have to explain to their colleagues that, yes, maps are in fact rich, human-made documents and not simply normative statements of spatial fact (Edney 2015a). Unnoticed and unacknowledged, flawed concepts continue to infect even the most intellectually radical and carefully constructed positions.

Both sides have been hindered in this debate by their mutual misunderstanding of just what it is they criticize and defend. The debate hinges on definitions of "the map" as a generic phenomenon. Normative critics seek to sustain the normative map; sociocultural critics seek an alternative, yet equally all-encompassing, conception of "the map." The problem is that such generic categories are utterly unstable and can be explained and understood only through historical analysis (Lois 2015).

When "maps," "charts," and "plans" are placed in their social contexts—why they are commissioned, how they are used, and who uses them—then it becomes apparent that what separates them is much greater and of more import than what they have in common. "Map," "chart," and "plan" acquired semantic stability in the seventeenth and eighteenth centuries not only in English but also in French and German. Each term refers to different kinds of images:

- *Cartes géographiques / Landkarten / maps* delineate regions or the whole world beyond the ability of one individual to observe and survey directly.
- *Plans / Pläne / plans* delineate parts of the world observed and measured by one surveyor or organized teams of surveyors.
- *Cartes marines / Seekarten / charts* delineate the hydrosphere, for the use of navigators.

Each set of imagery entails a particular conception of the world, which it depicts with different strategies and techniques, in order to support specific functions; each set is produced and consumed within certain social institutions and contexts. Overall, the common terms suggest not a single and coherent category but a diffuse and conceptually extensive array of artifacts (Jacob 2006, 18–21). These same pragmatic distinctions are as evident today as at any time in the past (Rankin 2016, 16).

Why should "maps," "charts," and "plans" be thought of as things that must possess some common character? Why do scholars and lay commentators obscure the clear differences between them by assuming that there exists some essential commonality, some *je ne sais quoi*, whose elucidation is the proper subject of scholarly inquiry? Why do they all hypostatize a Platonic form and insist that it is legitimate to construe a generic category as having significance in the real world?

3

What is the evidence for any generic map, other than some ingrained cultural commitment? In short, why do scholars continue to ask, what is a map?

My succinct answer to these questions is that the conviction that there exists a single and unambiguous category of things called "maps" is in fact the visible tip of the culturally pervasive and as-yet-unexamined belief system that is the ideal of cartography. The ideal regulates and controls all conceptions of the nature of maps and mapping. Under the ideal's cultural hegemony, maps are all the same. Regardless of their form and function, they can be described with the same vocabulary and evaluated by the same standards. A map of the whole earth in an atlas or a detailed plan of one or two city blocks, a medieval chorographical map of Palestine or one sheet of a modern topographical map of Israel, an eighteenth-century chart of Europe's coasts and an eighteenth-century map of Europe: all of the disparate cultural, social, historical, and functional significances of these and other maps are ignored, and each map is normalized as nothing more than a depiction of spatial fact. Scholars and lay commentators alike have worked hard to erect and maintain a barricade around the normative map by refusing to accept as maps any works that reveal even a hint of the unorthodox. "The map" is just one of many normative concepts that are sustained by the ideal and that are effaced and obscured by the conviction that cartography is just map making.

It is not that maps somehow hide their interestedness and partiality behind a naturalized veneer of objectivity, as Denis Wood has argued (especially Wood 1992b, 2010). Rather, it is intrinsic to the ideal of cartography that maps can only be statements of spatial fact. Furthermore, the emphasis on "the map" tends to grant an "efficacy and agency to cartographic documents which no one involved" in their production and consumption "ever held to be the case" (Barford 2016, 10; also Edney 2015a). It is not that "maps" are capable of independent action and do things, but rather that people do things with maps, to maps, and without maps. These complaints are not merely quibbles over academic shorthand: neither normative nor sociocultural map scholars have appreciated that the maps they study are only the visible part of the conceptual iceberg of the ideal of cartography. Any critique and reinterpretation of maps must address the whole ideal and the structure of its constituent beliefs, not its surficial features. In other words, we should abandon the critique of maps, whether normative or sociocultural, and engage instead in a critique of the ideal of cartography.

The ideal of cartography consists of an interlocking and resilient web of mutually reinforcing preconceptions, each of which sustains basic convictions that seem to be common-sense propositions about the nature of maps. These preconceptions and convictions together construe cartography to be the apparently transcultural endeavor of translating the world to paper or screen, with the shared goal of advancing civilization by perfecting a singular archive of spatial knowledge through the use of universal techniques of observation and communication. The foundation of the ideal was laid by the widespread adoption by European states after 1790 of systematic territorial surveys that held out the promise of a singular methodology for all map making. Further idealizations were developed, augmented, and popularized by a variety of factors, such as Europe's imperialistic engagement with the rest of the world, the active writing of histories of cartography, the formulation of set theory in mathematics, and the rise of personal mobility. Each new factor supplied the ideal's preconceptions with further intellectual burdens, progressively naturalizing cartography and normalizing "the map."

Like any other historically emergent phenomenon, the ideal of cartography is messy. It is not formed as a series of corollaries that logically and inexorably derive from one or two core axioms. Rather, its logic derives from the manner in which preconceptions both complement and contradict one another, their interconnections obscuring significant paradoxes and inconsistencies.

For example, the ideal holds that cartography is the necessarily timeless and universal endeavor of map making: whenever people have made maps, they have engaged in cartography. Scholars have therefore looked for cartography's origins in the ancient worlds of Asia and Europe, and even in prehistory. Yet the way the ideal developed—especially its integration with modern imperialism and its reliance on triumphal historical narratives and other apologias—has led scholars to identify cartography as a particular phenomenon of the Renaissance. This was supposedly the era when the mentality of Europeans acquired a new rationality that was manifested in, or caused by, a new and universal geometry of measurement and perspective vision. This transformation was explicit in early nineteenth-century map histories (e.g., Humboldt 1836–39) and has since been implicit (e.g., Cortesão 1969–71, 1: 4). More recently, the Renaissance has been explicitly identified as the moment when modern cartography came into being (e.g., Harvey 1989, 204, 244; Buisseret 1992, 1; Biggs 1999, 377–78; Wood 2010, 21–27; Gehring and Weibel 2014; Farinelli 2015; Silverberg 2015). In the nineteenth and twentieth centuries, Westerners used cartography's geometrical essence to distinguish themselves from the Asians and Africans whom they colonized: the geometrical nature of Western cartography marked Westerners as innately rational, while the apparently nongeometrical maps of colonized peoples marked them as innately irrational and therefore properly subject to Western rule. Cartography thus appears to be at once a practice found in all socially complex cultures and a particular historical formation associated with Western imperialism (Edney 2009, 42). Yet the ideal is so flexible that it does not admit of any contradiction. Indeed, the ability of the ideal to contain potentially conflicting positions that can be variously deployed as needed accounts for much of its resilience. Conversely, there are no axioms whose negation will bring down the whole edifice; revealing the flaws inherent to some of the ideal's preconceptions has no effect on the others.

The normative critique of maps accepts the ideal and does not directly challenge

5

its preconceptions. For their part, sociocultural critics have been concerned with developing new approaches to "the map" and so have only incidentally addressed some of the ideal's preconceptions. And no sooner have sociocultural critics happened to reveal the flaws of one preconception than other preconceptions pop right back up and reaffirm the ideal. The sociocultural critique has thus been a game of intellectual whack-a-mole. The ideal's preconceptions remorselessly corrupt even the most carefully argued scholarly analyses. Its idealizations continue to provide the default intellectual framework for any consideration of maps and mapping in the past and in the present, whether by well-established map scholars or by newcomers to the field.

The ideal is so big, so multifaceted, and so thoroughly naturalized within modern culture that no one has yet identified it, let alone sought to address it in its entirety. Until scholars appreciate all of the ideal's constituent norms and their implications *in toto*, they will continue to focus on maps and their reforms will fail. To move forward with an intellectually valid understanding of maps and mapping, we must first expose and eradicate the ideal's misconceptions. And that, in turn, requires that we appreciate the sheer size and complexity of this web of beliefs.

The goal of the present book is to promote just such an appreciation by exposing the ideal of cartography in as comprehensive a manner as possible. I explore the many ways in which the ideal has been expressed and the traps it lays for the unwary. I do this both by revealing the ideal's preconceptions and by telling the history of its creation and development. With cartography denaturalized, scholars of all stripes will be able to move forward intellectually without repeatedly tripping and falling over the ideal.

This book effectively challenges everyone who studies any aspect of mapping to examine and reevaluate what they think they know about their subject. Everyone, including map scholars who have been actively pursuing sociocultural approaches, will have to unlearn at least a few core concepts. It has taken me many years to recognize the complex system of beliefs hiding behind cartography's seeming innocence and innocuousness, and in doing so I have had to confront and reassess my own ideas and arguments. We all need to give serious thought to our own assumptions and attitudes, and correct them as necessary.

THE BOOK'S ARRANGEMENT AND SOME TERMINOLOGICAL IMPLICATIONS
The book proceeds in four main chapters together with a conclusion. Note that chapters 5 and 6 argue for new terminology that is necessarily used in earlier chapters, as briefly noted here.

Chapter 2, "Seeing, and Seeing Past, the Ideal," explains why the sociocultural critique has been a game of intellectual whack-a-mole when it comes to the ideal of cartography, because it is difficult to see the ideal from that perspective. The chapter then lays out the conceptual foundation of a processual approach by which it is possible to recognize the ideal as a historically emergent conception, together with its many negative influences on map studies. Furthermore, a processual approach provides an intellectual framework to pursue map studies without running afoul of the ideal's flaws and misconceptions. This is because it encourages scholars to address issues about why and how maps are produced and consumed—issues that the ideal obscures with normative concepts—in an explicit and empirically founded manner. From the perspective of a processual approach, nothing can be taken for granted. The explication of a processual approach is necessarily succinct; a further book will provide a more comprehensive explication.*

Chapter 3, "Cartography's Idealized Preconceptions," exposes the ideal by describing its constituent preconceptions—ontology, pictorialness, individuality, materiality, observation, efficacy, discipline, publicity, and morality—which together have construed cartography to be a singular and universal endeavor and have directed scholarly attention to "the map." The chapter identifies the contradictions, intellectual limitations, outright flaws, and mutual reinforcement of these preconceptions. Separating them out for individual treatment helps to reveal their inadequacies and weakens the hold of the ideal as a whole. I cannot record every manifestation I have noticed of the flaws engendered by each preconception, so I instead variously draw representative examples from three sets of scholarship and commentary: popular works, to reveal how deeply rooted the ideal has been in modern culture; scholarship that is completely indebted to the ideal, which is mostly older in origin although much has been produced after 1980; and post-1980 scholarship that might be critical of the normative map yet remains bound in some degree to the ideal of cartography. I have inevitably relied heavily on the Anglophone literature, although I have sought to engage with the French- and German-language literatures to the best of my ability.

Chapter 4, "The Ideal of Cartography Emerges," explores how the ideal emerged and developed in the nineteenth and twentieth centuries. Some of the ideal's factors predate 1800, such as the desire of many early modern princes and governors to have their territories mapped in as much detail as possible. But these factors were couched in terms that were decidedly *not* modern and the emergent ideal would adapt and reconfigure them extensively. Many more factors arose only after 1800. Chapter 4 presents all of the factors in thematic groups relating to systematic mapping, including the coinage of the word "cartography"; the interrelated themes of empire and a rational identity for Westerners; vision and observation; the development of new mapping professions; and the rise of mass consumption of maps. I identify the preconceptions promoted by the different factors, and the

---

*I discuss ideas related to this further work, tentatively entitled *Mapping as Process*, at mappingasprocess.net. I also intend to expand upon the conceptual criticisms levied here in a third book, which will explore how map history has long been written in support of the ideal of cartography and how and why map historians have been in the vanguard of the sociocultural critique of maps.

convictions they bolstered, to indicate how, working in concert, they forged a web of persistent beliefs. In other words, cartography is manifestly not a universal human phenomenon. It is not even a phenomenon characteristic of early modern Europe. It is an idealization created only in the modern age.

The threads of these chapters come together in chapter 5, "Map Scale and Cartography's Idealized Geometry." This chapter implements a processual approach to explore the specific question of the geometric foundations of different kinds of mapping in both the early modern and modern eras, in order to understand how, after 1800, the concept of "map scale" served to reduce all different mapping practices to cartography. Put simply, the map-to-world proportionality inherent to certain modes of mapping was progressively applied after 1800 to *all* kinds of maps, even when it was clearly invalid to do so. This particular idealization was enabled and reinforced by the adoption of the numerical ratio (1:x) and by the improper and inappropriate conception of the relative categories of "large scale" versus "small scale." In this respect, I offer an alternative conception of fine and coarse "resolution" as a means to compare maps without perpetuating the ideal, and I use the terms throughout the book. Chapter 5 thus adds some flesh to the previous chapter's bare-boned historical narrative of the ideal's development. It reveals a principal way in which the ideal of cartography has obscured actual and ongoing differences between different sets of mapping practices.

Overall, I argue that cartography exists only as an idealization produced by modern culture and that it cannot provide a valid conception of mapping in the past, present, or future. Cartography is not an endeavor common to all societies and cultures. Rather, it is very much a creation of the modern West and is permeated by many of the myths that Westerners tell themselves about their rationality and superiority. Scholars of all stripes and concerns need to reject the concept of the normative map and, more generally, the entire ideal and all of its preconceptions. Indeed, they must abandon "cartography" entirely except when referring to the ideal and the idealized endeavor. It makes sense to use "mapping" instead, but this word is somewhat ambiguous. For example, the obvious term for the history of this subject, "mapping history," means both the act of making sense of past spatial relationships and the history of the different ways in which people have produced, circulated, and consumed maps. Moreover, most people are drawn to the subject of mapping by the appealing aesthetics and intellectual power of the maps themselves. Maps are the primary, often the only surviving, manifestation of mapping processes. I therefore use "map studies," "map history," and "map scholars" throughout this book as the most appropriate terms for the subject, its history, and its researchers. As defined in chapter 2, this usage of "map" is underspecified and does not imply any normative idealization.

2

Seeing, and Seeing Past, the Ideal

Cartography can be unveiled

The ideal of cartography does not actively draw a curtain around its machinations to hide them from view. After all, it comprises incorporeal beliefs that have agency only to the extent that people adhere to those beliefs and act in accordance with them. In this respect, the ideal's persistence stems entirely from the failure by scholars and lay commentators to dispel the misplaced beliefs, and that failure is a direct result of their inability to see the ideal in its entirety and to appreciate its character and hegemonic status.

At best, sociocultural questioning of the normative map has challenged some of the ideal's preconceptions. But the ideal's other idealizations persist and continue to distort even the most carefully thought-out conceptual statements. In particular, its overall construction of cartography as a transcultural endeavor, as something that can be pursued by any mentally competent member of *Homo sapiens*, effectively obscures the ideal behind a veil of inconsequentiality: if anyone can make maps, then significance lies not in the act of mapping but in the maps produced. Time and again, when scholars have set out to discuss and analyze "cartography," they have immediately redirected their attention to "the map." The ideal persists.

We can see the development of this conceptual diversion in the history of graphic and written satires on the nature of cartography itself. I refer not to the long-established manipulation of maps in order to parody or allegorize personal and political relationships—from Opicinus de Canistris's fourteenth-century reconfigurations of countries and continents as male and female figures (Whittington 2014) to nineteenth-century moralistic maps that laid out the contrasting paths to heaven and hell (Reitinger 2008)—but to a small body of late-nineteenth-century satires that mocked the ideal of cartography. These early satires, by Mark Twain and Lewis Carroll, actively played the ideal's preconceptions off one another, relying for their humor on the ideal's internal paradoxes and contradictions. They referred as much to the conditions and circumstances of cartography as they did to the normative maps. They not only prove that the historically emergent ideal had indeed cohered and was starting to permeate modern culture—how else would their humor work?—they also help reveal some of its preconceptions. But while the nineteenth-century satires, and one later one by Jorge Luis Borges, remarked on the idealization of the endeavor of cartography, once the ideal had been more fully accepted in modern culture, the focus of satires in the twentieth century shifted to the normative map. Later commentaries, especially in works of literature, emphasized the mismatch apparent between the normative map and the inaccuracies and incomplete coverage of actual maps. Finally, when map scholars first discovered the satires of cartography in the 1970s, they became a significant ingredient in the questioning not of cartography but of the normative map. The increasing academic attention given to the satires since then marks the intensification and spread of the sociocultural critique of maps. In this respect, if the appearance of the satires in public discourse marked the attainment of hegemonic status for the ideal, then their appearance in academic discourse marked the inadvertent beginning of the end of that hegemony.

A history of the satires and of their importance for the understanding of the normative map thus permits us to see why a sociocultural approach to map studies has failed to come to terms with the ideal of cartography. Ultimately, the focus on "the map" has led scholars to ignore key details in the patterns of map circulation and consumption. Even as appreciation of the functions of maps has grown, there remains a firm sense that map "users" or "readers" can still be anyone in a society.

To perceive and overcome the ideal, we need a new approach that explicitly addresses everything the ideal prompts us to take for granted. After a discussion of the satires on maps and cartography, the second part of this chapter accordingly lays out a processual approach to map studies that requires the empirical analysis of why and how maps are produced, circulated, and consumed without any *a priori* presumptions about their nature. Indeed, a processual approach understands maps to be simply the products of mapping, itself a simple concept with multiple incarnations. Each and every instance of mapping, and each and every instance of map, are thus determined not by some preconception of what they should be, but by specific circumstances of human action. More generally, a processual approach constitutes an explicit ontology for mapping that actively counteracts the ontology ostensibly offered by cartography. It requires a complete reformulation of how mapping and maps are understood, with the result that scholarship can proceed without further infection and corruption by the flawed ideal.

Satire, Critique, and a Persistent Ideal

Perhaps the first satire on cartography was Mark Twain's humorous account of the "Fortifications of Paris," with its surreal and wrong-reading map printed from a woodblock that Twain himself had cut (fig. 2.1). While Paris and the River Seine are prominent in the center, the rest of the geography is confused, and it is overlain with several places from the United States: Jersey City, New Jersey; St. Cloud, Minnesota; and Omaha, Nebraska. The inclusion, at right, of a town labeled Podunk is a joke all by itself: "Podunk" had been used in the United States since at least the 1840s to refer to any insignificant, out-of-the-way place, precisely the kind of place that is commonly left off maps. Twain first published the satire on 17 September 1870 in the *Buffalo Express*, a newspaper that he co-owned and edited (Twain 1870a). The brief essay and map ridiculed the many pretty but uninformative maps generated by the American press to show the Prussians' rapid advance on Paris and the anticipated siege of that great city. The account began with the bold, mocking statement that "the accompanying map explains itself," and it ended with a series of nine fictional commendations, including these:

> It is the only map of the kind I ever saw. U. S. GRANT.

> It places the situation in an entirely new light. BISMARCK.

> My wife was for years afflicted with freckles, and though everything was done for her relief that could be done, all was in vain. But, sir, since her first glance at your map, they have entirely left her. She has nothing but convulsions now. J. SMITH.

Twain later reminisced that distractions in his personal life had led him to "heedlessly" cut the wood so that it printed wrong-reading; it does seem that a crucial element of the satire had been unintended (Twain 1995, 199–200n1). Several copies of the image and its accompanying text were soon reprinted, and there was also a proposal to have it published as a chromolithograph in Boston (Twain 1995, 203–4n2, 204n5, 205–6; Edney 2018).

Twain intensified the satire when he republished the map together with an expanded essay in a New York monthly magazine, *The Galaxy*, in November 1870. Since the work's original appearance, Twain now wrote,

> strangers to me keep insisting that this map does *not* "explain itself." One person came to me with bloodshot eyes and a harassed look about him, and shook the map in my face and said he believed I was some new kind of idiot. (Twain 1870b, original emphasis)

11

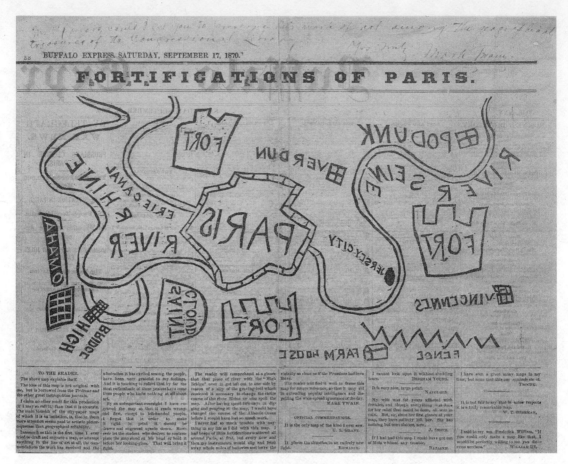

FIGURE 2.1. Mark Twain sent this impression of his "Fortifications of Paris" (Twain 1870a) to the Librarian of Congress, with an annotation requesting him to "preserve this work of art among the geographical treasures of the Congressional Library" (see Twain 1995, 207-8). Twain soon republished the work, with an enlarged commentary and a right-reading title added ("Mark Twain's Map of Paris"; Twain 1870b). Woodcut, 23 × 32 cm, plus letterpress. Courtesy of the Library of Congress, Department of Manuscripts (Samuel Clemens Papers).

The explanations stood by themselves and were further reprinted without the map in several newspapers in the eastern United States. In some of these mapless reprints, such as that in *the Plain Dealer* of Cleveland for 17 October 1870, the missing image was referred to as Twain's "burlesque map" (burlesque being crude and barely literate parody). In all these incarnations, Twain's humor challenged the expectation—the conviction—that maps are self-contained works that can be read by themselves without reference to explanatory texts (Edney 2018).

Twain undertook a quite different satire at the very end of the century, when he queried the manner in which maps were supposed, even before the advent of fixed-wing aviation, to manifest an almost divine view from above. In *Tom Sawyer*

Abroad, Twain imagines Tom and Huck Finn, the narrator, aboard a balloon, and Huck has doubts about how fast they are actually traveling:

> "Because if we was going so fast we ought to be past Illinois, oughtn't we?"
>
> "Certainly."
>
> "Well, we ain't."
>
> "What's the reason we ain't?"
>
> "I know by the color. We're right over Illinois yet. And you can see for yourself that Indiana ain't in sight."
>
> "I wonder what's the matter with you, Huck. You know by the *color*?"
>
> "Yes—of course I do."
>
> "What's the color got to do with it?"
>
> "It's got everything to do with it. Illinois is green, Indiana is pink. You show me any pink down here if you can. No, sir; it's green."
>
> "Indiana *pink*? Why, what a lie!"
>
> "It ain't no lie; I've seen it on the map, and it's pink."
>
> You never see a person so aggravated and disgusted. He says:
>
> "Well, if I was such a numskull as you, Huck Finn, I would jump over. Seen it on the map! Huck Finn, did you reckon the States was the same color out of doors that they are on the map?"
>
> "Tom Sawyer, what's a map for? Ain't it to learn you facts?"
>
> "Of course."
>
> "Well, then, how is it going to do that if it tells lies? That's what I want to know."
>
> "Shucks, you muggings! It don't tell lies."
>
> "It don't, don't it?"
>
> "No, it don't."
>
> "All right, then; if it don't, there ain't no two States the same color. You git around that, if you can, Tom Sawyer." (Twain 1894, 42–43)

Twain's satire would later be echoed in Stanislaw Lem's (1985, 31) fantasy of a planet that, seen from orbit, reveals "one continent only, down the middle of which ran a bright red line: everything on one side was yellow, everything on the other, pink." Geoff King (1996, 4) also recorded that "in the Walter Abish novel *Alphabetical Africa* (1974) . . . the population of an imaginary Tanzania is kept in employment in the Sisyphean task of painting the entire country orange to conform with its color on the map." All these satires rely on the incompatibility of several of the ideal's preconceptions—specifically, the ontological and observational preconceptions, which together hold that *all* maps are made from direct observation of the world, preferably from overhead, and the preconceptions of efficacy and pictorialness,

13

which together suggest that the world can, and perhaps should, be made over to look like the map (see also fig. 4.20)—all combined with the pragmatic conventions of using printed color (still relatively new when Twain wrote) to delineate political divisions on coarse-resolution, regional maps.

Perhaps the most famous cartographic satire is that provided by Lewis Carroll in his 1893 novel, *Sylvie and Bruno Concluded*. Carroll imagined a conversation between an English boy and his German host. While ostensibly poking fun at the period's nationalistic rivalries, the passage questions the ideal of cartography by mocking the perfect map:

> "That's another thing we've learned from *your* Nation," said Mein Herr, "map-making. But we've carried it much further than you. What do you consider the *largest* map that would be really useful?"
>
> "About six inches to the mile."
>
> "Only *six inches*!" exclaimed Mein Herr. "We very soon got to six *yards* to the mile. Then we tried a *hundred* yards to the mile. And then came the grandest idea of all! We actually made a map of the country, on the scale of *a mile to the mile*!"
>
> "Have you used it much?" I enquired.
>
> "It has never been spread out, yet," said Mein Herr: "the farmers objected: they said it would cover the whole country, and shut out the sunlight! So we now use the country itself, as its own map, and I assure you it does nearly as well." (Carroll 1893, 169, original emphasis)

Carroll's humor stemmed from contradictions among three of the ideal's several constituent preconceptions. First, the ontological preconception, which construes cartography to be an inherently geometrical and mathematical endeavor with the potential for perfection. If one could observe and measure everything with sufficient precision and accuracy, then one might indeed perfect the archive of spatial knowledge, and so replicate the world completely. Second, the pictorial preconception, which holds that maps are tools of visualization and comprehension. From this perspective, the map's purpose is not to replicate the world, but to simplify it in order to promote readers' comprehension; by stripping away the clutter of extraneous detail, the map presents a simple, but never simplistic, image of the world that allows readers to imagine the world and to readily identify relationships between geographical features. And, third, the preconception of efficacy, which presents maps as instrumental tools that are meant to be used in navigating and modifying the landscape. As Carroll suggested, when taken together, these three elements of the ideal are logically incompatible. Cartography's functionalities and pictorial nature are conflicting, and they are both negated by the ontologically perfect cartographic archive.

The tensions in Carroll's fantasy were echoed in a short fiction in which Jorge Luis Borges (1964 [1946]) imagined a passage from a fictitious early collection of travel accounts:

On Rigor in Science

... In that Empire, the Art of Cartography reached such Perfection that the map of one Province alone took up the whole of a City, and the map of the empire, the whole of a Province. In time, those Unconscionable Maps did not satisfy and the Colleges of Cartographers set up a Map of the Empire which had the size of the Empire itself and coincided with it point by point. Less Addicted to the Study of Cartography, Succeeding Generations understood that this Widespread Map was Useless and not without Impiety they abandoned it to the Inclemencies of the Sun and of the Winters. In the deserts of the West some mangled Ruins of the Map lasted on, inhabited by Animals and Beggars; in the whole Country there are no other relics of the Disciplines of Geography.

Suarez Miranda,
Viajes de Varones Prudentes,
Book Four, Chapter XLV,
Lérida, 1658.

As in Carroll's satire, Borges's unconscionable, widespread maps proved economically damaging and so fell into ruin, along with the empire itself (see also Eco 1985, 1994; Self 2013). Borges differed from Carroll in setting his satire on modern science not in the modern era but rather in some distant past, well before the mid-seventeenth century, when this passage was supposedly written. Given the satirical nature of the passage, it would be overly pedantic to complain that in choosing this temporal setting Borges inappropriately imposed the ideal of cartography onto nonmodern societies. Yet in doing so, Borges nonetheless reified the ideal's claim that there has only ever been one cartographic practice, so that it is indeed valid to interpret nonmodern mapping practices through the modern lens of cartography. Furthermore, in placing his fantasy in an empire, Borges succumbed to the ideal's association of cartography with territorial control, an association that is itself a product of modern European imperialism. (Just to be clear: much mapping *has* been concerned with territorial control, just not all mapping.) Neil Gaiman (2006, xix–xxii; 2012) offered a politically more realistic variant of the fantasy that was, once again, set in a distant, exotic, non-European past empire, in this case ancient or medieval China.

But even as the satires poked fun at the ideal, they neither challenged nor sought to overthrow it. Consider the example of Lewis Carroll's other famous cartographic passage, from "Fit the Second" of *The Hunting of the Snark*:

15

The Bellman himself they all praised to the skies—
    Such a carriage, such ease and such grace!
Such solemnity, too! One could see he was wise,
    The moment one looked in his face!

He had bought a large map representing the sea,
    Without the least vestige of land:
And the crew were much pleased when they found it to be
    A map they could all understand.

"What's the good of Mercator's North Poles and Equators,
    Tropics, Zones, and Meridian Lines?"
So the Bellman would cry: and the crew would reply
    "They are merely conventional signs!"

"Other maps are such shapes, with their islands and capes!
    But we've got our brave Captain to thank"
(So the crew would protest) "that he's bought *us* the best—
    A perfect and absolute blank!" (Carroll 1876, 15–16)

The surrealism of the passage extends to the accompanying image—a vision of blankness—whose marginal annotations of poles, equator, other circles, and so forth make no geographical sense (fig. 2.2). But then, the very next line, which map scholars never quote, undermines the whole passage:

This was charming, no doubt; . . .

The Bellman's perfectly blank chart stands as a metaphor for personal crisis, for the loss both of one's self and of the means to relocate it—losses that literary scholars have identified as the overall poem's principal concern (Lennon 1962, 243)—and this crucial line further recasts the satirical stanzas as a statement of what cartography *should* be about: locational accuracy, navigational functionality, an esoteric science, and a well-defined and constant earth reduced to paper through the consistent application of conventional representational strategies.

A PERSISTENT IDEAL (PART 1)

Such graphic and literary satires might have reduced cartography to logical absurdity, but they had little effect on the ideal's cultural sway. The ideal was sustained by a general acceptance that cartography had achieved a workable balance between the coordinate geometry of the spatial archive and the visual structure of the map. This pragmatic balance informed modernist commentaries about the nature of

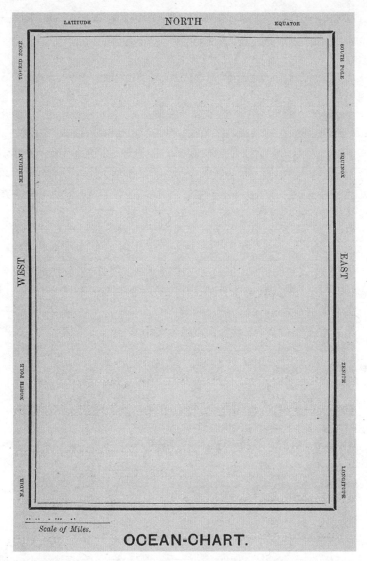

FIGURE 2.2. Henry Holiday, *Ocean-Chart* (Carroll 1876, 17). Lithograph, 19 × 13 cm (page). Courtesy of P. J. Mode Collection of Persuasive Cartography, Cornell University (1079); digital.library.cornell.edu /catalog/ss:19343175.

representation. Thus, just a few years after René Magritte first painted a pipe with the caption "Ceci n'est pas une pipe" (This is not a pipe) in 1928, with the intention of undermining naturalistic and naive models of representation (Foucault 1983), Alfred Korzybski opined:

A map is *not* the territory it represents, but, if correct, it has a *similar structure* to the territory, which accounts for its usefulness. If the map could be ideally correct, it would include, in a reduced scale, the map of the map; the map of the map, of the map; and so on, endlessly, a fact first noticed by Royce. (Korzybski 1933, 58, original emphasis)

Josiah Royce had used the concept of an "ideally perfect" map of England—although he did not actually specify that it had a numerical ratio of 1:1—in a thought experiment to comprehend an infinite series. He was not himself interested in establishing the practical nature of representation (Royce 1899–1901, 502–7; see Cesarz 2012, 69, 78n21). Korzybski now inverted Royce's argument in order to leaven the absurdity of cartographic perfection and to highlight the selectivity of maps and their *structural* equivalency to the world, such that they serve as a model for representation.*

Two decades later, Stephen Toulmin (1953, 105–39) made the same argument in explaining the nature of scientific theories: just as the map captures the structure of a part of the world so that its readers can comprehend and use it, so a scientific theory establishes the structure of a particular aspect of nature to permit understanding and prediction without mere description (also Ziman 1978; also Ortiz-Ospina 2018, regarding economic models as maps). Several map scholars have adopted the concept of map as "theory" (Robinson and Petchenik 1976, 4–14; Turnbull 1993, 1–3; Sismondo and Chrisman 2001), and they have accordingly defined maps variously as, for example, "concentrations" of the "real world . . . in model form" (Board 1967, 672); "abbreviated abstractions" (McDermott 1975, 88–89); "a type of congruent diagram" (Harley 1989, 4); or "depleted homologues" of reality (Fremlin and Robinson 1998, xii, 6). Routinely understood as instances of mimetic representation (see the pictorial preconception), maps have been used to demonstrate the validity of a realist epistemology, such that we believe that true mimesis is in fact possible.

CRITIQUE

Some scholars began in the 1970s to interpret the literary satires as implying that there is something more to mapping than what had been allowed by the ideal. They were not sure what that something was, but their comments were nonetheless the first unambiguous indications of actual dissatisfaction with the normative map. Specifically, Philip Muehrcke and Juliana Muehrcke (1974, 319) recognized the implications of Lewis Carroll's fantasy of a map at one-to-one: maps, they realized, can have significance other than as direct reproductions of the earth's features. However, because the necessary concepts and terminology needed to ex-

---

*It has also been suggested—for example, by Gunnar Olsson (2007, 481n43)—that Royce had influenced Borges's cartographic fantasy. After all, Borges (1981) closely paraphrased Royce's work. But Bruno Bosteels (1996, 121) noted that in his two works, Borges variously "reduces the cartographic ideal of mimetic representation to an alternative between tautology and infinity." Charles Saunders Peirce (1931–58, 8: 93–97, ¶122 and ¶125 [1902]) also logically refuted Royce's map metaphor.

press an unidealized vision of maps and mapping did not yet exist, the Muehrckes could conclude only that maps might also work as "metaphors," without further explanation or consideration of just what they meant by this term (see Rossetto 2014, 515, 522–23).*

Through the 1980s, increasing scholarly frustrations with the normative map induced scholars to engage with Carroll's and Borges's fantasies and other cartographic satires in a more sustained manner. Yet the quality of theorizing about maps remained relatively unsophisticated, and map scholars still treated the satires in an anecdotal and unrigorous manner, as Jeremy Crampton (1990) observed. But as the 1990s proceeded, the satires became increasingly important in the development of new conceptions of maps that effectively challenged the ideal of cartography. As a result, the Muehrckes' realization that *some* maps can have meanings other than as statements of spatial fact was steadily expanded into the argument that *all* maps have such significance (e.g., Turnbull 1993, 3; Montello 1993, 312; King 1996, 1–4; Palsky 1999).

The range of scholars who have subsequently quoted the literary satires indicates the broad disciplinary reach of the sociocultural critique of maps and their history. The critique reaches from geography (Crampton 2001, 240–41; Pickles 2004, 94–95; Fall 2006) to literary studies (Michelson 1995, 9–14; Vivan 2000; Byrd 2009, 29; Cep 2014), art (Harmon 2004; Harzinski 2010), cultural and art history (Reitinger 2008; Weibel 2014), political science (Branch 2014, 39), sociology (Cons 2005, 8), and the history of science (Holtorf 2017, 8). Within its wide range of concepts and theoretical frameworks, the critique has advanced two fundamental and interrelated arguments. On the one hand, scholars from the humanities have addressed maps as human products that demonstrably produce and reproduce fundamental cultural concepts and beliefs; maps are semiotic texts that bear cultural as well as factual significance. On the other, scholars from the social sciences have recognized that maps are made for ineluctably social reasons, to promote and sustain the power of the state and of elites, and that map making is a necessarily social endeavor. These perspectives indicate that the nature and character of maps are defined not, as the ideal maintains, by the parts of the world they represent and by the degree to which they reduce that part of the world, but rather by a variety of nontechnical factors, such as social need, power relations, and cultural conventions

---

*The Muehrckes' insight was not completely original. Just previously, Wilbur Zelinsky (1973, 5) had cited a modern take on the traditional application of maps to political satire—the globe dance in Charlie Chaplin's film *The Great Dictator* (1940)—to exemplify how "the modern map . . . still retains much of its primaeval tactility and choreographic quality." Chaplin, as the dictator Adenoid Hynkel, performs his desire for global domination by dancing gracefully with a globe balloon, to the prelude to act 1 of Wagner's *Lohengrin*, childishly treating the world as his plaything until he accidentally bursts it, whereupon he bursts into tears (see Monsaingnon 2017, 128–29). The second part of the documentary film *The Unknown Chaplin* (1983; dir. Kevin Brownlow and David Gill) includes footage from the 1920s of Chaplin's party trick of dancing with both a globe and a Prussian military helmet.

(Edney 2007b, 118–21). "Sociocultural" map scholars constitute a multifaceted community pursuing several distinct intellectual agendas, and they have generated a large and highly varied corpus of new scholarship.

One measure of the effect of the sociocultural critique has been the significant expansion of the understanding of "map," so that scholars now embrace a wide array of works that had long been excluded from study as trifling irrelevancies or geographical perversions. These include the supposed "monstrosities" of medieval world maps (see Woodward 1987, 288), the "sketch maps" of indigenous peoples (Woodward and Lewis 1998), and the pictorial maps so common in the twentieth century (Griffin 2013; Hornsby 2017). It is now accepted that mapping encompasses a much wider array of imagery and practices than the ideal of cartography had permitted. As a result, the performative mapping practices of indigenous peoples, as well as topologically structured maps produced by industrial societies (see fig. 5.24), are no longer dismissed out of hand as abnormalities or irrelevancies but are accepted as legitimate objects of study.

A PERSISTENT IDEAL (PART 2)

The sociocultural critique of maps has done serious damage to the ideal of cartography. Scholars from across the humanities and social sciences have identified and rejected many of the ideal's distortions and limitations, and their commentaries have gone a long way to expose the flaws of the ideal and to promote ways to approach maps and mapping that are uninfected by the ideal's preconceptions. Yet the ideal has persisted, even as sociocultural studies have proliferated. We can identify three reasons for its persistence. The ideal remains a potent concept within modern culture generally and continues to determine statements about cartography and its history in a variety of print and digital media. Furthermore, many map scholars remain committed to the ideal, especially within academic cartography and professional mapping institutions; some actively reject the sociocultural critique of maps for political or philosophical reasons, but for the majority the critique is simply irrelevant. And, finally, sociocultural critics themselves have been unable to break away from the entire ideal of cartography. I address these points in turn.

Overall, the concepts of "cartography" and "the map" remain thoroughly naturalized within modern culture. In my own experience, public lectures that draw attention to normative conceptions of the nature of maps are often met with antagonistic responses from those audience members who remain committed to the established verities. Audience reactions manifest a certain intellectual and emotional angst. Like existentialism, the sociocultural critique fundamentally rejects the presumptions that meaning is intrinsic to things and that the universe itself has a pattern. Things exist, but it is humans who impose abstractions about the necessity of their existence and who construe the nature of their essence. In par-

20

ticular, maps are devices by which humans have created meaning for the world; they neither reflect nor present some meaning that already exists independent of humanity. This position seems so radical and contrary to modern sensibilities that its rejection is immediate, even visceral (Sartre 1964; see Zynda 2004). Emotional control in the face of such existential angst can only be reestablished by reaffirming the idealized nature of maps in familiar, comfortable, and normative terms.

Cartography offers an angst-free intellectual comfort that can be readily evoked by simply referencing normative maps. Even when faced with indisputable evidence that all maps are sociocultural constructs, people often assert that maps should really be factual instruments of human knowledge. I am reminded of an April 1999 workshop in Atlantic history at Harvard University that addressed the complexity of a sociocultural map history but whose host, the eminent historian of colonial British America, Bernard Bailyn, clung to the idea that any and all maps should be made with the goal of showing geography "correctly." In particular, Bailyn latched onto the hot news of the day that U.S. war planes had bombed the Chinese embassy in Belgrade because U.S. maps had erroneously labeled the building as a Serbian munitions depot. Shouldn't maps be correct? Within the ambits of modern military and aeronautical mapping, accuracy is indeed necessary in both geometry (spatial location) and topography (spatial attributes). The issue, of course, is that such accuracy is neither central nor even germane to *all* mapping, whether now or in the past.

Academic and professional cartographers deal primarily with a relatively small and closely intertwined grouping of formal or official mapping practices. Their institutional interests have led them to argue that these particular sets of mapping practices are precisely coincident with the entirety of cartography; the maps produced are normative. Such ring fencing sustains the ideal's multiple preconceptions within the academic and professional literatures, for example, by means of blanket assertions that political and philosophical criticisms are irrelevant because they do not help to make "better" or even different maps (e.g., Monmonier 2013, 172–73; see Edney 2005, 3–4; Kent 2017, 194).

The basic form of ring fencing has been to deny that there is any need to define "the map" because it is a "banal" concept (Godlewska 1997). On the few occasions when normative map scholars have needed, for disciplinary reasons, to propose definitions, they have also invariably noted how superfluous it is to propose any definition because everyone already knows what maps are. The publisher's blurb on the dustjacket of Arthur Robinson and Barbara Petchenik's *The Nature of Maps* (1976) stated that "'map' . . . is often used, literally and symbolically, without explanation, suggesting that 'map' is so well understood that no definition is needed" (also Zelinsky 1973; Fremlin and Robinson 1998, xi). A later research group noted that "the answer seems so simple and so obvious that it is silly, even, to ask" what a map is (Vasiliev et al. 1990, 119). When forced to propose definitions, normative scholars have adopted either linguistic or lexicographic methodologies that only reiterate their preconceived notions and reproduce the ideal's tenets.

Frustrated by the inability to identify one particular kind of work that epitomized "the map"—would it be a topographical plan or a road map, a world map or a sea chart?—some map scholars adapted a classificatory concept from linguistics. The linguistic argument is that words are grouped into grammatical categories not because they all share precisely the same properties, but because they can be placed in greater or less semantic proximity to a prototype (Lakoff 1987). The prototype does not have to be a thing and, as is the case with "the map," can comprise an assemblage of properties. Having asked groups of people (mostly college students) to identify which of a varied array of images are indeed "maps," the studies found that their prototypical properties include the scaled correspondence of image to world, graphic form, a degree of semiotic abstraction, and spatial functionality. As long as images share those properties, to a greater or lesser degree, they are maps (Vasiliev et al. 1990, 122; MacEachren 1995, 160–62; see also Crampton 2010, 42). Yet the same properties that constitute the prototypical map also constitute cartography: it is the endeavor that creates measured, graphic abstractions whose structural correspondence to the world permits their functional use as tools with which to navigate or modify the world. In other words, these studies only tested the participants' understanding of the ideal.

Other attempts to establish a rigorous definition of "the map" have sought to identify commonalities across multiple dictionary and textbook definitions, and have similarly reified the ideal's preconceptions. One research group analyzed 24 dictionary definitions, finding that what they had in common was that a map is "a representation of the earth's geographic surface" (Vasiliev et al. 1990, 120). Menno-Jan Kraak solicited definitions from 120 participants in the 2013 conference of the International Cartographical Association; analyzing their common terms and concepts, he concluded that "a map is a visual representation of an environment" (Kraak and Fabrikant 2017, 14–15). The most comprehensive of such analyses was undertaken by John Andrews, who collected 321 definitions of "map" from mostly English-language dictionaries, encyclopedias, and geography textbooks from the 1640s through the 1990s (recorded in Andrews 1998). In accordance with the ideal, Andrews did not consider any definitions of *plan* or *chart* and he conflated the specific meaning of *map* as a work of geography, as expressed in pre-1800 definitions, with the normative post-1800 "map." He determined that a universally applicable definition of *map* was as a "representation" (occurring in 64 percent of definitions, 1649–1996), "in a plane" (47 percent, 1649–1996), "of all or part of the earth's surface" (45 percent, 1733–1995). And he further argued that all the complexity and variation within his large data set resulted from modifications of this universally consistent core concept, either by the imposition of lexicographical "motifs" or by the "refinements" made by specialists in line with "changing intellectual fashions," all of which served only to complicate and hide this essential truth (Andrews 1996, 1–2).

Academic and professional cartographers have further deployed particular visions of history to insist that cartography has always been precisely the same as formal and official mapping in the present. One statement by a professional map designer is especially revealing:

> Since man first started mapping our planet cartographers have wrestled with the same challenges: access to the right information, the ability to accurately capture and convey information and the design and effective communication of the knowledge in a format that can be reached and understood by its target audience. Sound familiar? These core challenges repeat through history: the medium, means of delivery, the players and the final output might change but the end game is similar. It is apparent when one delves into the 200-year history of our organisation [i.e., Collins Bartholomew] that the principals, challenges and aims are a constant. (Barclay 2013, 121)

From a historical perspective that looks beyond modernity and the present, these cartographic challenges do not sound at all familiar. They manifest a strictly modern experience, and a heavily circumscribed and idealized experience at that. Unfortunately it is all too common for professionals and lay persons alike to use modern experiences to define our culture's image of past mapping. However, to suppose that the challenges codified by cartographic professionals over the last two centuries constitute the sole concerns of earlier or non-Western mapping cultures—or, indeed, the totality of mapping activities in the present day—is to make a tremendous error.

Most recently, academic cartographers have redirected attention away from the ideal by misapplying the concept of "paradigm." Thomas Kuhn (1970, 176) defined an intellectual paradigm as a composite of three elements: (a) a substantial corpus of concepts, theories, and practices that are held by (b) a defined group of scholars and practitioners, and that are enshrined in (c) a set collection of textbooks and journals. Not as overarching as Michel Foucault's (1970) concept of *épistème*, a Kuhnian paradigm is nonetheless an intellectual formation characteristic of a scholarly discipline as a whole. The ideal of cartography constitutes a paradigm in precisely this sense: it is the paradigm of the academic study of mapping. However, by singling out particular research agendas as constituting paradigms by themselves—whether within the history of cartography (Blakemore and Harley 1980, 14–32; Edney 2015b) or academic cartography more generally (Antle and Klinkenberg 1999; Moellering 2012; Azócar 2012; Azócar and Buchroithner

23

2014, 101–29; McMaster and McMaster 2015; Ormeling 2015; Basaraner 2016, 73–79)—academic cartographers deny that they function within a common intellectual framework. They thus avoid engaging with, and they distract attention from, the ideal.

In addition to determining the conceptions held by scholars and practitioners of the normative nature of their field, the ideal has continued to infect and corrupt the arguments of sociocultural map scholars. The key problem is that the sociocultural critique has emphasized questions about the nature of normative maps rather than about cartography. The resultant discussions have remained bound to banal concepts. Although the critique has succeeded in enlarging the field of images accepted as maps, the old reluctance to engage with foundational matters reasserted itself once the cultural turn in map studies had been achieved. Christian Jacob (2006, xiii) accordingly recorded a range of reactions when he asked map scholars that "so basic, so obvious question," what is a map? "Some raise their eyebrows, others smile (and sometimes laugh), others have an immediate . . . answer, others say: 'Oh no, not again.'"

The ideal's self-effacement and internal complexity are in large part to blame. When taken together, the ideal's preconceptions are herdlike: sociocultural hunters can cull one or two obviously flawed and easily accessible convictions, but the herd as a whole remains intact and resistant. Indeed, the ideal's lack of logical structure means that the active culling of one idealization does not affect the others. Instead, as critics have eliminated one preconception, others have asserted themselves. It is pragmatically difficult to counter the ideal; any critique focused on maps is inevitably partial and incomplete. The inability to appreciate all the issues involved prevents sociocultural scholars from mounting a comprehensive and effective challenge to the ideal, and the ideal's cultural hegemony is sustained.

For example, Denis Wood and John Fels (2008, xv–xvi) forthrightly rejected the conviction under cartography's pictorial preconception that maps are necessarily mimetic:

> We start by replacing the whole idea of the map as a representation with that of the map as a system of propositions. Too long has the eye reigned over cartographic theory. The map is not a picture. It is an argument.

Yet they immediately reaffirmed the ideal's ontological preconception by asserting it to be the source of the uniqueness of "cartographic language":

> The cartographic sign plane differs from other sign planes by virtue of the convention that locations on the cartographic sign plane are themselves signs. Their content is "location x,y in the world," their mark nothing other than their location x,y on the map.

The core to mapping is, therefore, in their opinion, the "indexicality" of the map (Wood and Fels 2008, xvii n3). This indexicality constitutes the "singular logic" that is "not shared by other graphics" (Wood 2012, 137). The degree to which Wood and Fels remain bound to the ideal is clear from the manner in which they deployed Charles Sanders Peirce's concept of "index" with cartographic literality: signs on maps point to, are an index to, the real world.

In another example of an incomplete critique, I once challenged the ideal's presumptions of cartography's innately progressive and uniform character, but in doing so I failed to consider other significant preconceptions about the natures of maps and mapping (Edney 1993). In particular, the ontological preconception led me to conceptualize "mathematical cosmography" as a coherent and distinct mode of mapping (Edney 1994a, 1994b). Further study and reflection have made it clear, however, that far from being a distinct mode, mathematical cosmography was a specific idealization by eighteenth-century geographers (Edney 2011c, 2019b).

The ideal of cartography has been further sustained by sociocultural critics with an overtly political bent, who have turned their attention to cartography as a whole. Sociocultural studies have revealed maps to be tools of powerful states and social elites that have mapped territories in order to regularize and control them, and have presented propagandistic imagery to the masses in order to regularize and control *them*. These findings have prompted some criticism that the entire endeavor of cartography is a crucial element in the formation and persistence of the social and cultural inequalities inherent to modern states. For example, the political scientist Michael Biggs (1999, 377–78) wrote that

> Cartography apprehends space as pure quantity, abstracted from the qualities of meaning and experience. What matters is "the relation of distances" [Ptolemy 1991, 26]. It objectifies the world as a mundane surface, no longer the hub of a sacred cosmos or a succession of tangible places. It differentiates the form of knowledge from its content. A map can represent ocean or land, the entire earth or one parish. Such abstraction, objectification, and differentiation are characteristically modern.

But in criticizing "cartography," such commentaries take the idealized endeavor at face value. They are directed not at the actual ways in which states have engaged in mapping—many of which are indeed deserving of censure—but at the mythic idealization. They have simply lifted cartography *in toto* from atop one plinth, where it has long been lauded and feted as a substantial contributor to Western civilization, to another, where it can be reproached and criticized as innately and entirely immoral, anti-egalitarian, and misogynistic. Like the ideal it castigates, this "maps are bad" critique (Brückner 2008, 30) is remarkable for its ahistoricity. The critics have tended to cherry-pick their evidence, and they have promulgated idealizations in order to support what must be acknowledged as arguments grounded in presupposed political positions (as emphasized by Black 1997, 22–23; Andrews 2001). Even as it might open new and productive avenues of research and investigation, the politically motivated critique of cartography can only be misguided and misleading (Edney 2015a). A true critique of cartography requires, first and foremost, the delineation of the ideal's multiple tenets and all of their flaws.

A further issue is that some of the preconceptions possess an internal structure that seems to encourage conceptual resilience (see chapter 3). Specifically, particular idealizations present each act of mapping as recapitulating the entire process by which cartography as a whole supposedly evolved. Such recursiveness strengthens the ideal's preconceptions and reinforces their apparent logic. The situation is analogous to Ernst Haeckel's nineteenth-century biological formula that ontogeny recapitulates phylogeny—that is, that in its development, the individual repeats the same evolutionary stages through which the species has passed—at least to the extent that any logical inference is necessarily wrong. Just as biologists now utterly reject Haeckel's concept, because ontogeny demonstrably does *not* recapitulate phylogeny, so too is cartography's self-reflexivity quite invalid.

Confusion abounds. The ongoing sociocultural critique of maps has been greatly productive, yet it has failed to achieve a reformulation of our understanding of the multiple practices of mapping. The situation of the field remains largely the same as that decried thirty years ago by Brian Harley (1989, 1): "Despite these symptoms of change, we are still, willingly or unwillingly, the prisoners of our own past." Scholars from across the humanities and social sciences have advanced sophisticated arguments, but their work has been diffuse and partial in scope, so that mapping *as a whole* has remained under-theorized (see Edney 2015a, 11). Even as some map scholars engage with ideas of spatial discourses and semiosis, for example, others still assert that "the map" and "cartographic language" are valid concepts. For every step forward, we take another one backward or sideways.

Breaking Free of the Ideal

In addition to the sociocultural critique of maps, some scholars have sought to emphasize the nature of mapping as process (especially Del Casino and Hanna 2006; Edwards 2006, 5–7; Dodge, Kitchin, and Perkins 2009a). Other map historians who have been influenced by the same conceptual frameworks as myself—notably, the history of the book and actor-network theory—have already made similar arguments (e.g., Barford 2016; Dando 2017; Skurnik 2017). This is a perspective from which we can appreciate that cartography is an idealization of mapping: if what people think they do is cartography, what they actually do is mapping. The perspective applies to contemporary as well as historical map studies. I argue that such a "processual" approach, one whose object of study is mapping rather than maps *per se*, allows us to break free of the ideal and move forward conceptually without any further undue corruption. For me, the approach stems from two empirically grounded realizations that each denies the ideal's image of cartographic universality.

THE MULTIPLICITY OF MAPPING PRACTICES: MODES

The first realization is that there are fundamental dissimilarities between different *modes* of mapping in terms of the conceptualization of spatial knowledge; the technologies used to manipulate that knowledge; and the social institutions that seek, use, and control that knowledge (Edney 1993, 2011c, 2017a). These dissimilarities are well attested within the historical record, and they did not go away during the modern era. In other words, there has never been one endeavor of cartography, even one expressed in a small variety of flavors or dialects; rather, there have always been multiple forms of mapping, whose differences and distinctions have been obscured by the ideal's claims to unity and universality. There are, therefore, multiple kinds of map produced by the various modes, which should always be properly qualified, as geographical maps, marine maps, place maps, property maps, and so forth.

Thus my choice of epigraph for chapter 1, which paraphrased and redirected Steven Shapin's (1996, 1) opening statement: "There was no such thing as the scientific revolution, and this is a book about it." Shapin was reflecting on the historiographical vagaries of the seventeenth-century "scientific revolution." To historians in the mid-twentieth century, the scientific revolution appeared as *the* event that had determined modern culture and the modern mentality, even more so than the Renaissance and the Reformation. Yet, in the hands of later historians, the scientific revolution had become a remarkably elusive phenomenon. Once historians examined in detail the natural philosophy of the early modern period, they discovered "a diverse array of cultural practices aimed at understanding, explaining, and controlling the natural world, each with different characteristics and each experiencing different modes of change." Indeed, Shapin observed, the recent consensus held that it was "dubious" that there was even "anything like 'a scientific method'—a coherent, universal, and efficacious set of procedures for making scientific knowledge" (Shapin 1996, 3–4). Cartography similarly dissolves into a "diverse array of cultural practices aimed at understanding, explaining, and controlling" the special complexity of human existence: modes of mapping.

Modes can be identified by a repeated process of analysis and comparison. Examination of the physical form of maps, the kinds of spaces they depict, and the representational strategies they deploy suggests a coherent arrangement of processes for producing, circulating, and consuming maps, which in turn characterize a mode. Different kinds of map reveal different processual arrangements and, therefore, different modes. Repeated analysis reveals multiple and largely distinct representational strategies that characterize different modes.

27

Consider two maps. Figure 2.3 shows a 1794 map of a small extent of salt marsh, less than three-quarters of a mile (1.3 km) from end to end, in the town of Scarborough, Maine. It was produced as part of a legally mandated procedure to convert a specific parcel of property held in common into discrete lots, their areas to be proportional to the shares held by members of the company. The whole operation was undertaken with plane geometry in a manner that treated this small portion of the world as if it were flat (which such a small area effectively is). The survey of the boundary of the parcel and of the open channel of water running through the marsh required the measurement of the lengths of straight lines and of the angles subtended between them. The surveyor calculated the overall area and then determined how to divide it up in the desired manner; he drew up the final solution neatly to make this plan. Copies of this plan would have circulated among just a few white men: the owners of shares of the common land and their lawyers. There was thus no need to use any other method of producing the map than that commonly used for property maps in the era. The map was drawn by hand, in ink, on paper (two pieces, as it happened, glued together). This specific legal and spatial focus—everyone concerned knew the location of the property in question—meant that it was unnecessary to situate the property within the wider world. No more than a handful of copies of this map would ever have been made, and they would have been preserved alongside other pertinent legal documents in the archives of the surveyor, the property owners, and the lawyers. And, in line with the usual history of private archives in New England, if they left private hands, such property maps generally ended up in local historical societies or record offices.

By contrast, figure 2.4 shows a printed map published in 1793. It is, as it happens, the first printed map to frame the entire province of Maine. Many impressions of the map were printed from a copper plate, and each was tipped into a copy of one of Jedidiah Morse's popular geographical textbooks (Morse 1793; see Brown 1941; Sitwell 1993, 413). The book was sold widely across New England on the open market as an educational resource for the fairly well-off, at a time when printed books were still rare in the region (Hall 2000); the readership for the book and its several maps would have included young women as well as young men. The map encompasses over 50,000 square miles (130,000 km^2), an area that far exceeds what one person could ever hope to observe and measure in person; its author compiled it from several existing maps and surveys. It is surrounded by a graduated frame indicating latitude and longitude, both accommodating the earth's sphericity and situating Maine in its correct location on the earth's surface. From the start, this map and its parent book were relatively expensive and valued by their owners, being stored in small family libraries. Over time, their owners have sought to realize their value by selling them to others. Indeed, impressions of the map have been deemed to have greater value by themselves than when bound into their parent works; many have been extracted from the books for separate storage

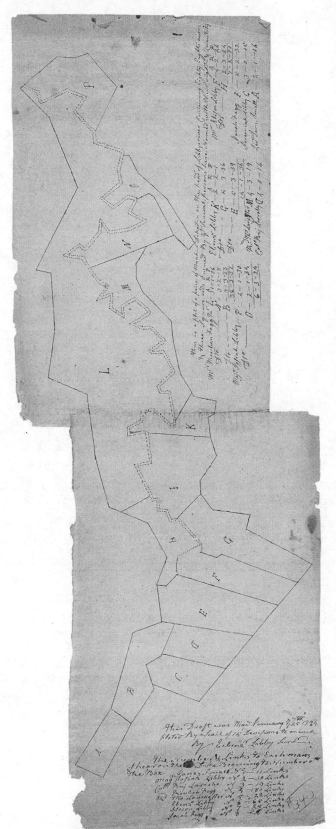

FIGURE 2.3. Ebenezer Libby, "This is a plat of a piece of marsh diked in on the head of Libbys River" (25 January 1794). Manuscript, 23 × 63 cm. Courtesy of the Maine Historical Society (map F 371); www.mainememory.net/artifact/68866.

FIGURE 2.4. Osgood Carleton, *The District of Main*, in Jedidiah Morse (1793, 1: opp. 345). Copper engraving, 27 × 21 cm. Courtesy of the Osher Map Library and Smith Center for Cartographic Education, University of Southern Maine (Smith Collection); www.oshermaps.org/map/2056.0001.

and recirculation. Its value and collectability has made this map a logical candidate for bibliographic description, along with other published geographical maps of the same region (Smith 1902, 33; Wheat and Brun 1978, no. 168; McCorkle 2001, no. Me793.1; Thompson 2010, no. 1).

It is conceivable that one of the salt marsh's shareholders also owned a copy of Morse's book. Even so, there would have been very little reason why this putative landowner would ever have consulted both maps together, at one and the same time, because they related to quite different sets of spatial practices. One was a part of legal and financial affairs, stored with the family archive; the other was a part of intellectual and perhaps political life, housed in a book, kept with the rest of a likely modest library on a shelf or two in a drawing room. Figures 2.3 and 2.4 might indeed show two maps in the idealized, normative sense, but they were maps that were produced and consumed to different ends and within different social settings. Only within the artificial collections of modern libraries, established in the nineteenth and twentieth centuries in line with the ideal of cartography, might such works be brought together and placed in a thoroughly modern dialogue.

What is important in these distinctions is that modes are not defined solely by the manner in which they reproduce the world; that would just perpetuate the ideal's ontological preconception. Each mode comprises a particular kind of knowledge about the world that is mapped for specific reasons and for specific institutions. Each mode comprises, in effect, a particular pattern of processes by which maps are not only produced but also circulated and consumed. Over time, the actual processes change in line with reasons and institutions for mapping. Modes, therefore, change. They intersect as concepts and practices, and personnel are shared between modes. New modes have formed.

There has been an increasing convergence in mapping technologies, especially after 1960 with the application to mapping tasks of digital technologies. But the fact that different organizations generally use the same mapping software— nowadays, ESRI's products dominate the software market for geographic information systems—does not mean that those organizations engage in the same kind of mapping, just as the common use of Microsoft Word does not mean that authors of novels and academic monographs participate in the same form of literacy, with the same style of writing and expression, same audience, same marketing and distribution practices, same business/profit model, and so forth. The history of official mapping policies in the modern era is one of the antagonistic relationship between practical concerns and the desire to establish a single, generic program for mapping. On one hand, the practicalities of making maps for specific governmental mapping needs (civil or military; marine or terrestrial or aeronautical; scientific and analytical or popular and presentational) all deny the ideal's universality through the creation of distinct agencies; on the other, the ideal holds out the potential for an effective, efficient, and cheap single mapping program, grounded in an apparently universal technology, that can meet all those different needs. The ideal of cartography might have promoted a degree of technological unity in mapping, but it has *not* overcome the social conditions that have led to fundamental distinctions in the circulation and consumption of maps.

A mode is a pattern of processes, and that pattern is something to be discerned by scholars. Like the sociological concepts of "class" or "ethnicity" and the linguistic concept of "language," a mode is properly understood as a simplification

31

imposed on complexity to help delineate and make understandable the intellectual contours of a phenomenon. As long as we remember that each mode of mapping is only ever a heuristic—a concept to help us understand the various ways people have mapped their world, for a variety of ends—and as long as we do not get too hung up on fine details, we can use repeated comparisons, such as that between figures 2.3 and 2.4, to distinguish largely discrete groups of mapping practices.

This is not, however, a straightforward process. My first, preliminary cut identified just seven modes (Edney 1993; see Edney 2011c). In the early 2000s, my colleagues and I designed the encyclopedic volumes of *The History of Cartography* (Edney and Pedley 2019; Kain forthcoming; Monmonier 2015) around nine modes, or eleven in the twentieth century (Edney 2015c). I have since continued to refine their classification, in particular through using modes to structure my undergraduate courses. The result is a stable and tested delineation of fourteen modes that have been pursued in the Western world in various eras (table 1).

It is undoubtedly inappropriate to use the same precise demarcation of modes for non-Western societies, but the analytical principle will transfer nonetheless. The analysis of patterns of production, circulation, and consumption of maps will reveal mapping modes. Studies of mapping in Japan, for example, indicate distinct practices of property mapping, urban mapping, regional mapping, and cosmographical mapping (see especially Unno 1994; Wigen et al. 2016).

A stable classification, at least for Western mapping, does not mean that each mode is itself stable. As personnel, maps, instruments, and practices move between modes, they can effect significant changes and they can erode the apparently clear boundaries between modes. Over time, new modes have developed. It can be challenging, looking back from the present, to correctly identify the mode in which a map was produced. A large part of the problem is that two maps that look alike can belong to different modes, so map form cannot be relied upon as a guide. I have long thought that students would appreciate a structured flow chart to identify mapping modes (like Randall Munroe's [2016] guide to dating world maps), but the concept always runs aground on the fact that map form and mode are neither directly nor consistently correlated.

Delimiting mapping modes in a particular society at a particular time requires the understanding of patterns of map circulation and consumption, not just the methods of production. Bearing this principle in mind permits us to achieve clarity, for example, when presented with two apparently flexible boundaries: those between marine mapping and, respectively, geographical and place mapping.

The boundary between marine and geographical mapping was complex throughout the early modern era. Not only did geographers draw on information from marine maps, they also circulated marine maps in manuscript (Fernández-Armesto 2007) and in print among a nonmaritime readership interested in geographical matters (fig. 2.5). The circulation and consumption of these maps, and

TABLE 1. The fourteen modes of mapping

Finer resolution mapping of discrete portions of the world, i.e., places, each potentially observable by one individual:

Place	Physical and cultural landscapes of specific locales in order to create, perpetuate, and reconfigure their distinctive meanings as places; also called "topography" in the sense of "describing place," but clarity requires this term be reserved for one subset of systematic mapping (below).
Urban	Entire urban places in plans or views, recognizing the cultural significance granted to cities as artificial and self-regulating communities.
Property	Landscapes fragmented into discrete parcels of property.
Engineering	In support of planning and building roads, buildings, fortifications, and so forth.
Chorographical	Each region (*choros*) without reference to the global framework of geography, but likely entailing geographical-style compilation of sources.

Coarser resolution mapping of spaces that are beyond the ability of one individual to observe and delineate:

Cosmographical or World	The known world (*mundus* or *oikumene*), depicting the interrelations between humanity, the rest of nature and creation (*cosmos*), and the divine; often astrological or metaphysical.
Geographical	The terraqueous globe of the earth (*ge*) and its regions, including much special-purpose mapping (e.g., road maps).
Marine	Coastlines and features in coastal zones, from oceanic charts to coastal charts to harbor charts, generally made by and for mariners.
Celestial	The heavens and heavenly bodies, from star charts and cosmological diagrams to detailed mapping of the other planets.

Modern, state-driven mappings:

Boundary	Relatively narrow areas over geographical distances along a border or frontier between states.
Geodetic	The earth's size and, after ca. 1700, its shape.
Systematic or Territorial	Mapping based on comprehensive surveys that extend finer resolution mapping across expansive spaces whether landscapes (topography), coasts and oceans (hydrography), properties (cadastral), or for aeronautical purposes (post-1900).
Analytic	Distribution of social or physical phenomena in conjunction with social and natural sciences and governmentality, often called "thematic mapping," but excluding "special-purpose maps" produced for narrowly specific ends within other modes; generally coarse resolution but can be fine.
Overhead Imaging	The earth from above, whether by analog aerial photography or digital remote sensing, not only contributing substantially to most other modes but also engendering distinct spatial discourses.

Modes are defined by the processes of circulation and consumption as much as production. For example, just because a map shows part of a city does not mean that it should be considered as a product of urban mapping; maps of city parts belong variously to place, property, engineering, or marine mapping (see the "Goldilocks model" in Edney 2017c). These modes were previously presented in a somewhat different sequence and hierarchy elsewhere (Edney 2017b, 74–75, table 5.1).

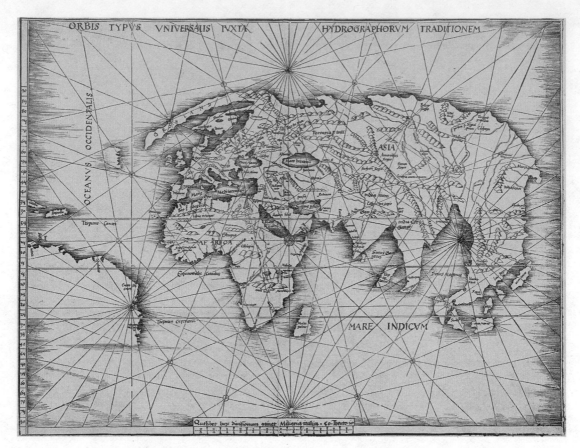

FIGURE 2.5. A world map in the form of a marine chart but distributed as a work of geography. Martin Waldseemüller's *Orbis typus universalis iuxta hydrographorum traditionem* ("Map of the whole world according to the tradition of mariners")—the first map in the supplement of modern maps appended to his edition of Claudius Ptolemy's *Geography* (Waldseemüller et al. 1513, fols. 119v–20r)—replicated the look of a marine chart, with its mesh of rhumb lines representing compass directions, a scale at bottom center for distance, and a latitude scale at left, reflecting the practices of Atlantic sailing. Although this was a world map intended for geographical consumption, the use of marine conventions emphasized how new geographical information had been derived from voyages made by European explorers. Woodcut, 44.5 × 57.5 cm (paper). Courtesy of the Norman B. Leventhal Map Center at the Boston Public Library and Mapping Boston Foundation; collections.leventhalmap.org/search/commonwealth:3f462s23n.

not their subject or look, calls for them to be classified as geographical rather than marine. At the same time, cosmographers and geographers sought to impose their cosmographical framework onto oceanic navigation. Their goal was to supplant the tried and tested techniques that blended the plane geometry of deductive reckoning with the latitude determination of Atlantic sailors, as codified in the era's "plane charts," by integrating the use of longitude into navigation (Chapuis 1999; Ash 2007; Sandman 2007, 2008, 2019; Gaspar 2013). This was the context in which Gerhard Mercator made his 1569 world map on *that* projection, whereby he apparently reconciled the plane chart with the spherical earth (Gaspar 2016). Alas, the projection was effectively useless until about 1800, when longitude could be

34

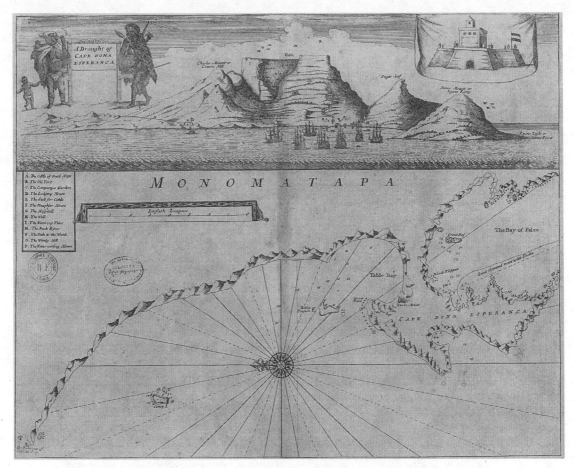

FIGURE 2.6. The landscape view in the upper register of this harbor chart refers to the common practice of including headland profiles in sailing directions and speaks directly to the utilization of place-mapping technologies in the creation of detailed charts of harbors and anchorages. The image was originally created by John Seller, ca. 1675, and was reissued in several versions (Tooley 1969, 104); reproduced here is a late state of John Thornton's version, *A Draught of Cape Bona Esperanca* (with Thornton's name removed from beneath the title at upper left), which was probably extracted from *The English Pilot, The Third Book* (London, [1734–61]). Copper engraving, 48 × 57.5 cm. Courtesy of the Bibliothèque nationale de France (Département des cartes et plans, GE DD-2987 (8286)); gallica.bnf.fr.

readily determined at sea, and it remained the preserve of landlubbers who sought to make chartlike images for their geographical readers (Cook 2006). Map scholars have generally treated early modern maps on the Mercator projection as marine charts, yet in terms of their circulation and consumption, as well as their production, they almost all belong within the mode of geographical mapping (Edney 2017a, 75–77).

Furthermore, marine mapping includes a particular practice of harbor charting that is closely allied with place mapping. This relationship is exemplified by an early English map of Table Bay at Cape Town (fig. 2.6). In its sketching and measurement of relief, place mapping is closely allied to landscape imagery; in fact,

SEEING, AND SEEING PAST, THE IDEAL

they have often been accomplished by the same individuals. Here, the reliance on empowered vision is exemplified in the landscape view in the upper register and, in the harbor plan itself, in both the profiles of the coasts and the soundings of depths in the anchorage. The harbor plan also has the precise and self-contained character of place maps, lacking as it does any reference to where the harbor is located in the earth's oceans beyond its toponym. There is not even a single latitude on the map. The map was prepared by a publisher who specialized in marine mapping and was published in a book of sailing directions to Asian waters for British mariners (Tooley 1969, 104; Maeer and Baynton-Williams 2019); the map must be considered as an element of marine mapping despite its overt similarities to place mapping.

By rejecting the unity of cartography and instead addressing the several modes of mapping, scholars are especially equipped to deal with how, and why, mapping practices have changed over time (Edney 1993). The ideal's progressive teleology—whether presented as a progression from "art" to "science" (e.g., Rees 1980; Thrower 1991) or as a series of "revolutions" (e.g., Robinson 1982, 12–15; Monmonier 1985)—remains the default position for explaining the overall history of maps and mapping. But by examining the histories and intersections of modes, it is possible to write narratives of change that intertwine the social and cultural with the technical (e.g., Edney 2007b, 2011b). For example, the supposedly unitary process whereby cartography became increasingly scientific—whatever is meant by "science"—actually comprised multiple historical trajectories by which each mode accepted instrumental and mathematical techniques (Edney 2017b). The last three volumes of *The History of Cartography*, covering mapping since about 1650, were accordingly structured by modes, so that each could trace the changing institutions and technologies in a meaningful manner without succumbing to presentist preconceptions of inevitable progress (Edney 2015c, 2019e, forthcoming).

THE OPENNESS OF MAPS: DISCOURSE

The second realization underlying my adoption of a processual approach is that because maps are semiotic texts, as is now well established by map scholars (e.g., Wood and Fels 1986; MacEachren 1995), they are also dynamically open. By this I mean that as we start to examine the semiotic character of maps, we are unable to discern any hard-and-fast boundaries between texts the ideal construes as maps and those it does not.

People treat maps as they do other kinds of text: Like works of art, maps are placed on walls; like sculptures, globes are set on plinths and stands. People *read* maps like books and *use* them like instruments. They invest political significance in maps, as they do in slogans and flags; they sell and buy them, or give them away as presents, as they do books and pamphlets and other material goods. People talk around and over maps, discard and preserve them, ritualize and totemize them. In exploring mapping processes, it is impossible to see where maps end and where other kinds of text begin. There are no neat, long-lasting boundaries around "the map," but rather many semiotic strategies that coalesce into maps. This is why I find the concepts of "paramap" and "perimap" unhelpful. Wood and Fels (2008) and others have applied the concept of paratext to maps in order, quite properly, to draw attention to the signs surrounding the map, but in doing so they privilege and close off "the map" as a discrete textual form.

The paper's edge apparently closes off the map: the map is what is on the paper; beyond the edge is everything else. As map scholars have pursued sociocultural approaches, however, it has become impossible to sustain the commitment to the map's self-contained materiality. The words used on maps are the same as those used to write books and to make speeches; the decorative marginalia of maps recapitulate graphic motifs from art and science; the coordinate systems parallel analemmas and other mathematical imagery; the sign systems used for maps are applied to other discourses, as metaphors and satires. As the self-containedness of the individual map frays, so too does the generic category of the normative map and with it the singular concept of cartography.

As an example, consider William Hubbard's map of New England (fig. 2.7). It was printed and published as part of Hubbard's 1677 account of King Philip's War (1675–76). The map was tipped into the middle of the book, to introduce a listing of the war's key events arranged by geographical location. Most towns on the map bore not a toponym but a number. Each number on the map pointed to a paragraph in the book. Each paragraph named the place and listed the events that had occurred at that location, mostly Indian attacks on English settlements; most paragraphs provided further cross-references to pages elsewhere in the volume where more in-depth accounts of the events could be found (Edney and Cimburek 2004, 331–33). Hubbard's indexical chain exemplifies the way in which geographical maps are *never* stand-alone works but are fully integrated into arrays of other written and graphic texts. Indeed, this is why I call geographical maps "geographical," because they are just one of a set of representational strategies that have been deployed to collect, organize, and communicate knowledge of the wider world. Geographical maps are not read in isolation. Hubbard only made explicit what is implicit in other geographical writing, that readers are expected to move from map to narrative and back again so that maps blur semiotically with the written word. Where does the map end?

The same uncertainty applies to modes other than geographical mapping. Early modern marine maps were integrated within systems of written sailing directions (pilot books) and headland profiles (as fig. 2.6); after 1800, they were augmented by new technological systems that featured, among other things, the installation of lighthouses and buoys that effectively become, for the user, physical extensions of the graphic images. Bill Rankin (2014; also 2016, 205–51) has demonstrated

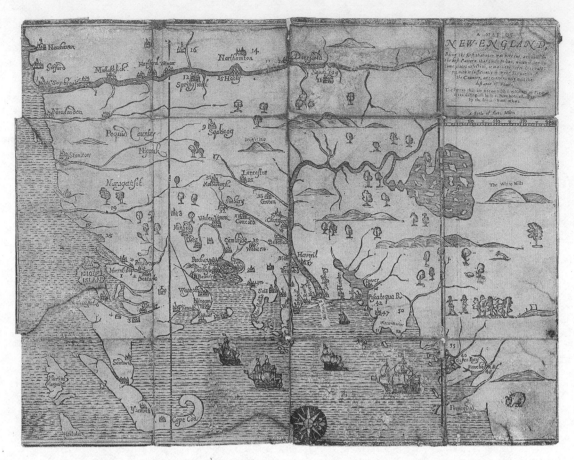

FIGURE 2.7. [William Hubbard], *Map of New-England*, cut and printed by John Foster, extracted from William Hubbard (1677, opp. sig. T1r). Woodblock, 30 × 38 cm. Courtesy of the Osher Map Library and Smith Center for Cartographic Education, University of Southern Maine (Osher Collection); www.oshermaps.org/map/492.0001.

that twentieth-century systems of radionavigation embedded the navigator in a mesh of tangible and intangible objects. In property mapping, the graphic plan has always supplemented verbal metes-and-bounds descriptions of boundaries running between marks and monuments physically installed in the landscape; those monuments are the real focus of the maintenance of property rights and are also mapped through a variety of rituals and performances (e.g., "beating the bounds"). Place mapping is tied not only to landscape art, and to practices of looking out at landscapes, but also to certain forms of poetry. And so on.

When we study the ways in which users and readers have variously consumed maps, we find that what have been taken as uniquely and specifically cartographic images intersect constantly and blend into other inscriptions, whether in graphics, words, or numbers, and into gestures and performance. Maps do not even need to be graphic in form. We need look no further than Hubbard's book to find, introducing the last, separately paginated portion detailing the Indian wars in northern New England, a geographical map of the river estuaries along the coast; using words alone, the map gives the distances and some bearings between the rivers, identifies and names other geographical features, and describes the English settlements and some of the local Indian tribes (Edney and Cimburek 2004, 334–36). Hubbard's verbal map is little different from Joseph Nicolar's mapping, some two centuries later, of the Penobscot River in Maine by reciting the indigenous Penobscot toponyms and economic activities along the river (Nicolar 2004 [ca. 1887]). In a completely different vein, Michelangelo's painting of the Last Judgment on the altar wall of the Sistine Chapel, begun in 1536, has been persuasively read as presenting a "sacred topography" of Rome: it was a place map (Burroughs 1995). Modern authors work spatio-temporal layers into novels and poems that function as literary maps (e.g., Sorum 2009; generally, Bulson 2007). And are tactile maps for the blind "graphic" (Thomas 2017)?

Moreover, anthropological studies of mapping practices by indigenous peoples have concluded that inscriptions, if present, were not the primary means of conveying spatial information. The key signifying element in indigenous mapping was generally performative. If an inscription was used, then meaning was incorporated through the inscription's deployment within an exchange or ritual. This was the case, for example, in the eighteenth-century "Pawnee [i.e., Skiri] star chart" that was used only within certain rituals and in markedly different ways in each—as a flag in one, as a baton for a race in another—to represent different aspects of the Skiris' multidimensional cosmos (Gartner 2011). When Western explorers requested information from local informants, the answer might have featured lines drawn in sand or on paper, but the weight of spatial explanation was oral and gestural. Indigenous societies have sustained distinct reasons for mapping, so that it is possible to differentiate various modes of mapping—between, say, the mapping of property (Pearce 1998) and the mapping of political relationships (Waselkov 1998)—but they commonly rely on the incorporation of spatial meaning within oral and bodily performance rather than on the inscription of spatial meaning in some medium (Rundstrom 1991; compare Bernstein 2007).

The existence of nongraphic, noninscriptive, and incorporative strategies for communicating spatial information poses a serious challenge to the insistence that maps must be graphic artifacts and that mapping is solely about directly recovering meaning from those graphic artifacts. In the first volume of *The History of Cartography*, founding editors Brian Harley and David Woodward (1987, xvi) proposed a new definition of maps as "graphic representations that facilitate a spatial understanding of things, concepts, conditions, processes, or events in the human world." They intended this definition to be culturally all-inclusive, and it has certainly been widely adopted, but it nonetheless proved inadequate to characterize indigenous mapping strategies that are neither material nor graphic (Woodward and Lewis 1998, 3–5).

Performative mapping practices pervade modern life as well; they are a function of humanity, not indigeneity. The performativity of everyday, modern life has prompted a concern for "nonrepresentational" theory among academic geographers (Thrift 2007). This term exaggerates for effect both in reaction to human geography's "linguistic turn," with its emphasis on representation, and in calling for a return to the study of lived experience (e.g., Glennie and Thrift 2009). Map scholars, who deal with spatial representation, have adopted the same approach under the less dramatic label of "postrepresentational theory" (Kitchin and Dodge 2007; Kitchin, Perkins, and Dodge 2009; Rossetto 2015). From this perspective, we can finally dispose of the modern fixation on maps necessarily being graphic objects. The consumption of maps, the multitude of moments when their meaning is fixed, is not a passive absorption of information but an incorporation of the map into actions, and those actions further carry meaning and shape the meanings accorded to the map. So, maps are themselves of multiple forms and map consumption is a performance. There is no room here for a strictly and solely material map.

Sociocultural analyses have disproven maps' supposed stability and self-containedness. Maps are not just graphic images or things but are variously integrations of words, graphics, numbers, gestures; installations of multiple objects; and even intangible artifacts. Maps are not simple, self-contained "objects" but multicomponent "things" (Rankin 2014, especially 626–27, 662–64). Consideration of the myriad ways in which maps are consumed indicates that map reading is as intertextual a practice as any other kind of reading: people read and perform maps according to other works they have previously consumed. If maps were indeed normative and defined solely by their relationship to the world, then they would be just like scientific laws, which are equally intelligible to anyone, regardless of linguistic and cultural background. Yet as Mark Twain suggested (see fig. 2.1), maps are neither equally intelligible nor self-evident. Readers require an understanding of the cultural context to interpret any map; no map can be understood except by reference to other texts and cultural forms (Turnbull 1993, 19–27). Conversely, one map, regardless of its form, does not have a singular meaning determined by its correspondence to the world. Every map exists within a web of texts that provide the map with different shades of meaning. Like any other cultural product, one map can sustain multiple interpretations at the hands of its consumers, and those interpretations change with the circumstances of consumption. Maps are best understood as works in progress (Kitchin and Dodge 2007; Dodge, Kitchin, and Perkins 2009a; Kitchin, Gleeson, and Dodge 2013).

Maps are not, however, texts that float free in a sea of signs, open to any idiosyncratic interpretation a reader might wish to impose. Their interpretation is limited by the circumstances in which they are consumed, circumstances that are

40

intertwined with those in which maps are produced and circulated. The form of each map, the semiotic strategies it deploys, and the manner in which it integrates with other texts together constitute a regulated network of communication, which is to say a "discourse" in its most restricted sense. This is the narrowest definition of "discourse" offered by Michel Foucault: "a regulated practice that accounts for a certain number of statements." I endeavor to limit my use of "discourse" to this precise heuristic, although I undoubtedly lapse on occasion and use the term in Foucault's less narrow sense of "an individualizable group of statements" (Foucault 1972, 80; see Mills 2004, especially 55–56).

Each precise spatial discourse comprises a network of people who produce, circulate, and consume a suite of texts, including maps; who effectively regulate the forms and particular semiotic strategies of those texts; and who regulate membership of the network. This narrow definition is, as ever, a heuristic, but one driven by the empirical study of how the circulation of specific texts determines the bounds of specific networks of people interested in mapping.

Thus, "mapping" is a function of spatial discourses, "map" its product. In the most general sense it is possible to define them as follows:

"mapping" is the representation of spatial complexity;
"map" is a text representing spatial complexity.*

My use of "representation" and "text" will perhaps be contested. To be clear: I use "representation" in a constructivist sense. It is the process by which meanings are constituted and communicated. It is, moreover, an ongoing process, as texts are continually interpreted and reinterpreted by their consumers. "Representation" should not properly be used for the final product, so that it is quite inappropriate to refer to "*a* representation." The product of representation is a "text" in the most generic sense, which is to say, a complex of signs that has been assembled through a process of semiosis; those signs can variously include words (oral and written), graphics, physical installations, and performances. That is to say, my interpretation of "representation" is neither as limited nor as restricted as that perhaps adopted by those who advocate for a nonrepresentational or postrepresentational theory.

These definitions are underspecified. They do not rely on formal or functional criteria and so cannot be used to judge whether an artifact or a ritual is, or is not, a

---

*I will fully explain "spatial complexity" in *Mapping as Process*. For now, it serves as an indicator of perceived need or function: if a spatial situation is not complex in some way, if it does not require explanation, then there is no occasion for mapping. Denis Wood (1992a, 67) suggested that if a painting is "something produced through the process or art of painting," or a "writing" is "something written," then a map is simply "'something produced by the process of mapping' or 'something mapped.'" He offered this succinct statement as a rhetorical flourish to a larger argument about the supposed neutrality and "virgin birth" of "the map," but it nonetheless has a powerful simplicity and directness that I find very useful.

map. Scholars, therefore, need to identify maps in precisely appropriate discourses and, conversely, to explore how specific discourses produce particular forms of maps. This is a recursive process; as the circulation of maps between producers and consumers is traced, and as mapping processes are delineated, new maps and processes are identified. Ultimately, what mapping entails—the potentially multiple semiotic strategies and intertextual practices—depends not on a universal "cartographic language" but on the specific semiotic formations unique to each spatial discourse.

Spatial discourses change over time. Consider a communal discourse of urban mapping pursued in Portland, Maine, before the U.S. Civil War. It was communal in that the maps were consumed almost entirely within the wealthier elements of the urban community and did not circulate among U.S. society in general; the maps were accordingly part and parcel of that community's self-construction of their city as a moral and energetic commercial and political center. In the 1820s, when the maps were produced locally, this discourse quickly developed a very specific set of conventions for delineating the city. Most of these maps were integral to the city's published directories of residents, businesses, and services. On each map, the peninsula ran horizontally across the page, so the simple north arrow pointed to the upper right corner; the title block was placed in the Back Cove, above the peninsula; and the Fore River was shown in full, to emphasize the whole harbor and to be wide enough to accommodate the north arrow and the list of references to churches and civic buildings (fig. 2.8). These local conventions for representing the city would be abandoned after 1850, as both civil engineers and representatives of the lithographic publishers in Philadelphia took over the production of maps of the city, reconfiguring the discourse. Even though the maps continued to be directed to local consumers, and the new producers tried to imbue them with local flavor, the discourse nonetheless shifted to impose national standards, expectations, and conceptions: north at the top of each map; all notable buildings labeled and shown by their footprint; elimination of the reference key; and so on (Edney 2017c).

If there is sufficient evidence, one can delineate very precise discourses, each concerned in some way with spatial variation and organization, that have sustained the production and consumption of certain maps. For example, we can identify a tight circuit of politicians and lawyers in London who commissioned, circulated, and used printed maps of the provinces of eighteenth-century New England in the adjudication of legal disputes over colonial boundaries. This particular discourse developed its own conventions. In particular, although the maps were not published, in that they were not sold in the marketplace, they were nonetheless printed as part of the entire legal proceedings for each intercolonial dispute. Several were then completed in manuscript to distinguish disputed features and to echo the manuscript plans previously attached to Privy Council orders about

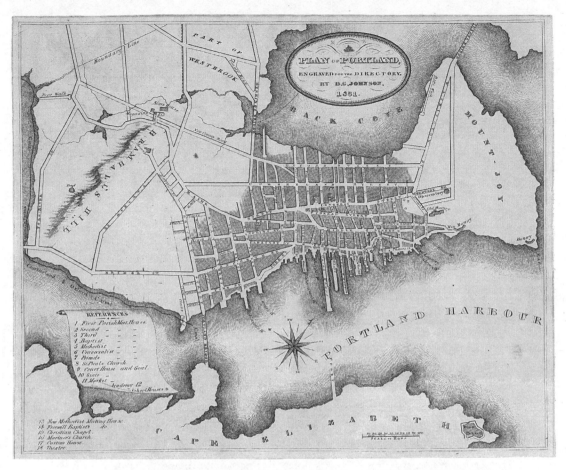

FIGURE 2.8. One of the locally produced maps of early nineteenth-century Portland, Maine, that adhered to a locally developed set of conventions for depicting the city. David G. Johnson, *Plan of Portland . . . 1831*, extracted from Anonymous (1831, opp. title page). Copper engraving, 23 × 28 cm. Courtesy of the Osher Map Library and Smith Center for Cartographic Education, University of Southern Maine (Osher Collection); www.oshermaps.org/map/12023.0001.

colonial boundaries. The network of people who commissioned and consumed these maps was so small that we can even discern one individual at its center, the solicitor Ferdinando John Paris, who acted on behalf of several colonies against the territorially aggressive province of Massachusetts Bay (Edney 2007c).

It is only within each precise discourse, within each regulated network of communication, that texts might achieve semiotic stability. As maps are relocated to new discourses—for example, when indigenous performances are translated onto the pages of an explorer's notebook, or when landlubbers consume marine maps, or when archived maps are served up to the historian—their new discursive contexts require new readings. Such relocations can be subtle, involving a shift from one precise discourse to another within the same mode. One example occurred when London publishers in 1677 reprinted William Hubbard's history of King Philip's

43

War for an English audience; they also reprinted Hubbard's map (see fig. 2.7), but in line with the general lack of geographical knowledge of New England among the English, they bound the map at the very front of the book as a reference image (Edney and Cimburek 2004, 338). This relocation of the map from a local to an imperial discursive context transformed the map. In the Boston book it served as a geographical index integral to the overall work. In the London book it functioned in accordance with other geographical maps of the American colonies, as an image whose primary function was to establish the geographical stage on which historical action took place. More generally, maps can be relocated from a discourse within one mode into an entirely different mode, as when marine charts were absorbed into geographical and cosmographical discourses (see fig. 2.5).

The meanings of maps undergo substantial alteration when maps are repackaged and redistributed many centuries later as historical artifacts. Hubbard's map was long of interest to historians, not because of how it functioned within his book but because of the manner in which its rough-hewn carving by John Foster captures both the hesitant transfer of Western civilization to the wilds of the New World and an understanding of the crude and unrefined nature of material life in early New England that contrasted markedly with the Puritan's refined life of the mind (Edney and Cimburek 2004, 319–20). Neither significance would have been recognized by the map's original readers in New England. Antiquarian dealers' catalogs and sales pitches, and the recent online proliferation of commentary about early and contemporary maps, reveal many reasons to appreciate and desire early maps, all of which have little to do with why and how they were originally produced and consumed. The openness of maps must therefore become a cause of self-criticism by map scholars as well as the cause of careful historical analysis.

A PROCESSUAL APPROACH

The realizations that there are a multiplicity of mapping practices and that maps themselves are semiotically stable only within particular spatial discourses indicate that the proper subject of analysis is not maps, in whatever forms they might take, but the mapping practices that produce them. After all, the most effective way to understand a phenomenon is to identify and explicate the processes that give rise to it. An understanding of underlying processes and the time periods over which they operate permits explanations of the phenomenon, of how the phenomenon contributes to other processes that generate still further phenomena, and of how the processes reconfigure over time to engender changes in the phenomenon.

Philosophically, a processual approach provides a conceptual framework within which scholars can develop research agendas without falling into the many traps laid by the ideal of cartography. It insists that scholars explicitly and equally consider the processes of producing, circulating, and consuming maps as they develop models and explanations in general or particular studies of maps and mapping. It provides a new intellectual ontology of mapping that replaces the inadequate ontology of the normative map offered by the ideal of cartography.

Methodologically, a processual approach requires scholars to discern coherent sets of mapping practices. With those established, scholars may bring to bear any number of topical or theoretical concerns—iconology, feminism, military-fiscal state analysis, textual analysis, national identity, book history, to name just a few—in order to interpret maps or to investigate changes in mapping practices. A processual approach is not methodologically prescriptive and does not preclude other approaches to interpreting and contextualizing maps. It sustains both synchronic and diachronic analyses. What it requires is that scholars take nothing for granted but rather explicate the precise social, cultural, and technical contexts within which people have sought to represent spatial complexity. It ensures that maps and mapping practices are related and compared only when they are indeed comparable and relatable in terms of their underlying processes.

Analysis of mapping processes takes place at three levels, each with pros and cons. The first, macro-level of inquiry recognizes the discrete modes of mapping as coherent sets of mapping practices and processes whereby maps are produced, circulated, and consumed. Modes themselves are actually a rather crude heuristic for thinking about the "diverse array of cultural practices" (in Shapin's terms) by which humans have sought to understand, organize, and communicate the spatial complexities of their existence. It is helpful to understand that Carleton's 1793 image of the state of Maine (see fig. 2.4) is a geographical map, because it means we can avoid the mistake of attributing to it the characteristics and functions of, say, marine charts or boundary maps. As a geographical map, it was not a work that would have helped mariners negotiate the area's rocky coasts, nor can it serve today in a dispute over the state's boundaries. As such, Carleton's map should not be grouped or mentioned together with those other kinds of map. Yet the identification of this work as a geographical map does not help historians figure out just who might have had access to it, how they might likely have understood it, and how they acted accordingly; such an identification does not indicate how the map differed from or adhered to the conventions of contemporary geographical mapping.

We should pursue precise, micro-level analyses that sustain meaningful historical and cultural interpretations and explanations. In other words, we need to outline and study each particular spatial discourse that sustains mapping practices. Yet it is pragmatically difficult and time-consuming to delineate specific spatial discourses from the bottom up. For example, each of the studies mentioned above, of the regional mapping of intercolonial disputes in eighteenth-century New England and of the antebellum mapping of Portland, Maine, required a great deal of archival sleuthing and careful data collection that did not promise much obvious intellectual return. And, most important, we cannot hope to analyze all of the innumerable spatial discourses that need to be studied in this precise manner.

Fortunately, a useful, meso-level heuristic lies between overly crude modes and overly precise discourses. Specifically, the individual filaments of discourses tend to intertwine to create what we might consider to be discursive threads; such threads interweave to constitute modes. Threads of spatial discourses are readily visible to the historian and are still sufficiently coherent to permit interpretation and explanation. We can work now on the threads and worry later, as appropriate, about picking them apart to get at the individual filaments of specific spatial discourses. Of course, if a precise discourse should happen to become visible in the course of a research project, then the scholar should not miss the opportunity of pursuing it.

Martin Brückner (2017) provides a thorough delineation of some of the patterns of production and consumption in geographical mapping, in the process distinguishing three main threads. In order to narrate the history of the commercial market for geographical maps in the United States before 1860, Brückner balanced the technological narrative of the increasing industrialization of map printing against the three main arenas—threads—of the contemporary consumption of geographical maps: first, as wall decorations that served as theatrical imagery in businesses and homes, a means of consumption often overlooked by historians because of the high mortality of wall maps but which was nonetheless the dominant form of geographical map consumption in the early republic; second, as works in atlases and travel guides, the latter again being downplayed in traditional histories because of the high mortality of the guides; and, third, as integral elements of educational textbooks and classrooms.

Brückner's third category permits an alternative means of analyzing the weave of some of the threads that constitute geographical mapping. Different pedagogic philosophies have promoted different ways of using maps for educational ends, and we can accordingly discern a certain thread of interrelated geographical discourses in which different institutions and philosophies have promoted the educational use of geographical maps. Detailed studies have explored some of the precise discourses. For example, Judith Tyner (2015b) examined the discourses within which American girls embroidered regional and world maps, and even globes, in the eighteenth and nineteenth centuries; Susan Schulten (2017) traced the origins of the pedagogic practice in early America of having boys and girls draw manuscript maps (see fig. 4.14); Sumathi Ramaswamy (2017) explored the pedagogic role of globes in the adoption and reconfiguration of Western science in British and independent India; and Jordana Dym (2015) investigated the contributions to later twentieth-century Guatemalan education by the high school teachers Julio and Oralia Piedra Santa, who parlayed their cheap *mapitas* for schoolchildren into a larger pedagogic publishing company.

We might further trace threads of discourse via analyses of the physical form of pedagogic materials. Much educational mapping has emphasized the reading, copying, and creation of regional and world maps within the study of geography, and all of the pertinent discourses are part and parcel of the weave of geographical mapping. Educational discourses that relied on general, stand-alone atlases sold through the marketplace are also part of the thick weave forming the huge arena of public discourse. Alternatively, educational discourses that emphasized the detailed measurement of classrooms and villages are part of the largely distinct weave of place mapping. And there are also the discourses of formal training of surveyors, engineers, navigators, and geographers that intersect and connect to wider considerations in quite different ways.

A careful attention to spatial discourses, or at least to threads of such discourses, places map studies in line with established scholarly practices in other fields, especially book history, actor-network theory, discourse theory, science and technology studies, language history, material culture studies, ethnography, the geography of science, and in general histories of knowledge creation (see the review by Skurnik 2017, 15–23). As with book history, a processual approach offers the prospect of what D. F. McKenzie (1999) called the "sociology of texts" and David Hall (1996, 1), "the social history of culture." Such approaches have long since transcended book history's narrow focus on the hand-printed codices of the early modern era to encompass even the incorporative practices of orality and performativity that circumscribe and blend with the inscriptive practices of writing. A processual study of mapping should do the same.

A processual approach brings map history into line with Paul Carter's (1987) "spatial history," which is to say, the study of the creation of spatial concepts through past acts and activities, as opposed to "imperial history" in which those spatial concepts are taken as *a priori* categories used to frame historical narratives (see also Fraser 2008). It is not enough, as Latour (2005) argued with respect to actor-network theory, to demonstrate that mapping is a "social practice"; rather, the production, circulation, and consumption of maps are constituents of social relations, and they need to be studied accordingly. Sociocultural map studies have shown how most maps before the later nineteenth century were consumed by the wealthier and more educated elements of society, but instead of simply explaining map consumption as predominantly a function of the middle and upper classes, scholars need to show how map consumption contributed, probably in various ways, to the formation of class identities.

Furthermore, the processual approach gives significant insight into historical events. Sociocultural studies have been largely concerned with the institutional and epistemic structures of knowledge. They have established the historical significance of mapping at the *longue durée* of social and economic change. But what about the *courte durée* of events? A processual approach presents mapping practices as part of a complex and far-reaching network or mesh of actors and actions, broadly construed, and permits a greater precision in determining which specific maps and related texts people had access to, how they would have read and used

47

those works, and to what effect. Jeffers Lennox (2017), for example, has studied the circulation of spatial knowledge in maps and memoirs, oral and written, among Native Americans (Mi'kmaq and Abenaki confederation), the English (both colonists and in London), and the French (local Acadians, those in New France, and those in Paris) to reconfigure their interactions and decisions within the contested regions of Mi'kma'ki, Nova Scotia, and l'Acadie, without relying on the modern conception of "Nova Scotia" as the well-delimited province within the confederation of Canada. A processual history brings maps of all sorts into the center of social, environmental, cultural, and political history (also Mapp 2011).

A processual approach construes mapping to be necessarily active and dynamic. By contrast, the ideal of cartography insists that maps are innately stable and static. Likening maps to photographs, commentators have described them as snapshots of the world, distilling one moment in time that is already past before the work is complete; maps thus appear immediately out of date. Their apparent stability is such that Latour (1987, 227) thought them the prototypical "immutable mobile" in which knowledge is recorded and then carried—unchanged—from the field to centers of calculation. But people are always undertaking mappy acts: making maps, circulating them, using them, ignoring them. As maps continue to circulate within their discourses, or cross between discourses, and are found to still be meaningful, they remain valid and up to date. Even the storage and destruction of maps are dynamic processes, requiring decisions to be made and actions to be taken; archives and libraries are not just places of storage but are sites of further knowledge production (Skurnik 2017, especially 10–11).

The very foundation of mapping is thus inherently dynamic and fluid, as spatial discourses constantly form, reconfigure, and dissipate. Those discourses promote new techniques and technologies, new conventions and functions. The processes of mapping constantly change. While the history of cartography has been written as long periods of stability and golden ages, punctuated by periods of rapid change, even revolution, the history of mapping is actually a story of continual flux and fluidity, of many small alterations and occasional gross rearrangements. A processual approach is therefore synonymous with a historical perspective.

By extension, there is no one "cartographic culture." In positing that an entire culture might profess a particular "understanding of and attitudes towards maps as representations of spatial knowledge," I once again fell afoul of the ideal by making the otherwise unwarranted assumption that the idealization of spatial knowledge promoted by eighteenth-century geographers was applicable across all mapping modes (Edney 1994a, 384; Edney 1997, 36; also Edney 1994b). But each thread of spatial discourse features its own particular set of attitudes to specific map genres, and those attitudes can vary among the participants according to the conditions of their participation. It makes sense, therefore, to expect that people generally limited in their access to and consumption of maps—such as women (Richards 2004;

Dando 2017), ethnic minorities (Hanna 2012), colonial and postcolonial peoples (Ramaswamy 2004, 2010, 2017), and members of the lower classes—developed approaches to maps that differed from those of more socially privileged map consumers, to the point where map work by the excluded and disadvantaged might form distinct spatial discourses. Overall, to talk about *one* cartographic culture is to run afoul of the ideal.

Fundamentally, a processual approach pays *equal* attention to the three elements of mapping: production, circulation, and consumption. The ideal of cartography has long induced scholars to study map production above all else, and to see map production as an innately technical process to be assessed by the quality and quantity of information produced. This has applied as much to evaluations of early maps as to the development of new and more effective map-making techniques. The sociocultural critique of maps has tended to emphasize map consumption. In between, circulation has been barely considered, although historians of science have pursued the matter in depth to bring out connections otherwise hidden by Eurocentric narratives (e.g., Cañizares-Esguerra 2017). A processual approach integrates all three processes and does not overemphasize one to the detriment of the others. Particular studies might, of course, focus on just one process, but the general intellectual framework within which research agendas are developed and evaluated must recognize the importance of all three. A processual approach is theoretically informed in terms of how research problems are construed, but each study must be empirically driven. Its goal is not to perpetuate the political critique of "maps are bad," but to explore and demonstrate how mapping processes have contributed, and can contribute, to human cultures and societies in myriad ways.

49

Cartography's Idealized Preconceptions

Cartography is myth

The ideal of cartography, as it grew increasingly elaborate over the course of the nineteenth and twentieth centuries, came to comprise a series of preconceptions that construe the diverse practices of mapping to form the singular and coherent endeavor of cartography. While pervasive in both academic and popular discussions of maps and their history, these idealized preconceptions produce unwarranted presuppositions and unhelpful assumptions. Sociocultural map scholars, in particular, have worked hard in recent decades to expose the inadequate and distorted conceptions of maps that these preconceptions have engendered, yet their scholarship continues to be infected by other lingering, persistent preconceptions.

I began to appreciate how the ideal's preconceptions intertwine to form a resilient web of convictions when I explored the literature surrounding John Smith's relatively brief voyage in 1614 from Monhegan, an island off the Maine coast, to Cape Cod, and the map that resulted (Edney 2010). Smith's map is a unique map-portrait of himself and New England together, the first printed work to show Cape Cod and many other details in an apparently accurate manner, and hence an icon of early colonial exploration, encounter, and mapping (fig. 3.1). Since the mid-nineteenth century, I found, historians have told substantially the same stories about this map, stories that consistently and often grossly misinterpret the well-known and limited empirical record. Historians have twisted and altered that record to make the map fit preconceived but nonetheless sincerely held notions about the nature and history of maps. Specifically, I found that historians have presumed that

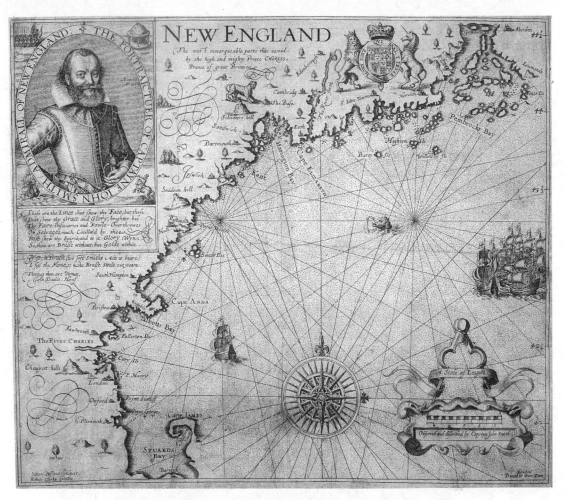

FIGURE 3.1. John Smith, *New England*, state 1 (London, 1616/7). Copper engraving, 30 × 35.5 cm. Courtesy of the William L. Clements Library, University of Michigan, Ann Arbor (atlas E1b).

Smith was a visually and culturally empowered observer ["OBSERVATION"] who, finding existing maps to be inadequate ["DISCIPLINE-A"], directly and faithfully re-presented ["PICTORIALNESS"] his own ["INDIVIDUALITY-A"] improved observations ["DISCIPLINE-B"] of a pre-existing region ["ONTOLOGY"] in a map ["MATERIALITY"]; he then published this map to disseminate it widely ["PUBLICITY"], in the process keeping it up to date ["DISCIPLINE-C"], in order to improve mankind's communal archive of spatial data ["MORALITY"], to facilitate future navigation by other mariners ["EFFICACY"], and to enable future colonists ["INDIVIDUALITY-B"] to use New England in an effective manner. (Edney 2011a, 18)

Yet, each of these preconceptions is belied by the archival record. This is more than just a problem of imposing modern conceptions of cartography onto the early modern era where they prove to be inadequate. No: the modern conceptions are themselves wrong.

This chapter exposes and delineates the ideal's preconceptions and gives some examples of their implications so that map scholars can notice and avoid them. My intent is less to correct these flawed habits of thought and more to demonstrate their continuing power over map studies. Isolating the preconceptions effectively counters the persuasive power of the whole web of idealizations. The flaws in each can be readily exposed and countered. Isolating the preconceptions is especially useful in exposing how several are apparently stabilized by the recapitulationist argument that the production of an individual map seems to recapitulate the history of cartography as a whole; such recapitulationism has long been discarded by biologists, and should have been discarded by map scholars, too (see chapter 2). I therefore present the preconceptions in a sequence designed to show how they support each other to create that potent and resistant web.

This chapter is the first attempt to comprehensively elucidate the ideal's complex web of beliefs (although Biggs 1999, 377–78, and Delano Smith 2001, 286–96, usefully identified some of the preconceptions). There is no one way to present the preconceptions, and my ideas for condensing a host of interrelated misconceptions into an effective account of the preconceptions have inevitably changed as I have developed my understanding. (The quotation just above contains my current labels for the preconceptions, to prevent confusion.) Other map scholars might well argue for alternate delineations and definitions. Corrections, additions, or even wholesale reconfigurations are welcome.

My examples come from a wide range of literature, from popular commentaries as well as from scholarly statements by both normative and sociocultural critics of maps. Some are anecdotal, stemming from my various discussions about maps and mapping with colleagues, students, and members of the public. I encourage readers to identify their own examples of how scholars have variously fallen afoul of the ideal. To that end, I offer the following checklist of the ideal's preconceptions and their more particular convictions. They are frequently self-contradictory. *Every one of them is wrong*.

CHECKLIST OF WRONG CONVICTIONS SUSTAINED BY THE IDEAL, GROUPED BY PRECONCEPTION

"Ontology": the map is a reduction of the world/archive.
- The map is a database, a scaled subset of the world/archive.
- The archive itself is scaleless and infinitely precise.
- The archive has one geometry (projected coordinates $[x(\phi, \lambda), y(\phi, \lambda)]$) that combines properties of both plane and cosmographical geometries.
- The plane of the map is the same as the infinitely extensible plane of Euclidean and Cartesian geometries.

- Cartography's core process is generalization, the degree of which is map scale (expressed as 1:x).
- Maps at similar map scales are directly comparable and equivalent.

"Pictorialness": the map is an unmediated, graphic mimetic presentation of the world.
- Maps are graphic images.
- Geographical features look the same at all map scales.
- Cartography has its own, distinctive, special "language."
- Map reading is algorithmic.
- Maps picture the world in a direct and self-explanatory manner.
- Maps are all concerned with "visualizing" distributions; maps make the invisible visible.

"Individuality": making maps and using maps are acts solely of individual cognition.
- "Cognitive maps" are directly expressed as "mental/sketch maps."
- "Cognitive mapping" is the *ur*-process of all map making: all individuals construct cognitive maps, therefore it is a fundamental human urge to make maps.
- Map making is a transcultural practice pursued by all cognitively developed humans.
- A map's meaning is determined by its (singular!) author.
- The cognitive capacity, and degree of rationality, of individuals and entire cultures can be judged by their maps.
- Women have less well-developed cognitive maps and mapping abilities compared to men (who constitute the default standard for spatial ability).
- Non-Westerners, past and present, have markedly less well-developed cognitive maps and mapping abilities compared to Westerners, and are irrational and primitive.

"Materiality": maps are things made at a specific moment.
- Maps are things.
- Maps are stand-alone, self-contained documents.
- Maps are "sovereign."
- Maps sharply divide mapping into two processes: map making and map using.
- Maps are stable and "immutable."
- Maps possess chronological fixity.

"Observation": all maps are grounded in observation and measurement.
- Map making and map reading share the same visual regime of perspectivism.
- Modern maps are made by observation from above.
- The default map is the fine-resolution map of the environment, as experienced by the individual.

- The map is a mirror of nature, setting the map reader outside nature.
- The map is "sterile" and denies human experience; cartography is inherently "totalizing," that is, it reduces all human experience with a limited straitjacket.

"Efficacy": maps are made to be used, especially for guiding movement.
- Maps are instrumental.
- Even coarse-resolution world and regional maps are intended to aid personal mobility.
- People used maps in the past in the same ways, and to the same ends, that people in the present use them.
- Maps are the best devices for showing spatial relationships and for guiding movement.

"Discipline": all maps and mapping practices are disciplined by being tested against the world and are corrected as necessary.
- Continual disciplining has given cartography a "metaphysics of presence."
- Maps are routinely kept up to date.
- New surveys are prompted when errors are found in maps.
- The cultural importance of early maps is defined by how much previously unmapped information they provide.
- New maps are made when existing maps become "out of date" or when new data become available.
- Maps are properly dated to the last new feature they contain.
- Cartography is inherently progressive, both in quantity and quality of information in the world/archive, and inexorably improves over time.
- The more correct of two maps is the later.
- Maps that look modern must have been based on modern techniques.
- The cultural importance of early maps is defined by their "firsts," whether of content or of technique.

"Publicity": maps are made to be disseminated widely.
- Printed maps are more efficacious than manuscript maps.
- Maps surviving in multiple versions must have been influential.
- Manuscript is the precursor of print, in terms both of map making and of the whole history of cartography.
- The history of cartography is properly told through printed maps.
- Early modern printed maps are integral to the formation of European rationality and "Western culture."
- Cartography can be meaningfully divided into "surveying" and "mapping": manuscript surveys made in the field (periphery), printed maps made in the office (core; center of calculation).

"Morality": the making of maps is an innately moral act.
- Cartographers' moral duty is to strive to make the best maps possible.
- "Bad" maps are at once inaccurate and immoral.
- Cartography disciplines and regulates its practitioners.
- Society disciplines and regulates cartography, leading to phases of national leadership.

"Singularity and universality": cartography is a singular endeavor pursued by all map makers—one world/archive, one technology of observation, one perfectible goal.
- Cartography is innately a science.
- Maps adhere to cartographic norms.
- Cartography is an ahistorical universal.
- There exists a coherent "cartographic reason."

Ontology

The ideal of cartography claims to depend on several axiomatic principles: that each map is necessarily and directly tied to the territory it depicts; that the map is a store of spatial data about that territory; and, most important, that the character of the map is determined by that territory. That is to say, maps present preexistent categories of reality (regions, types of spatial features, and so on). Under this "ontological preconception," *all* characteristics of a map stem directly from the territory, as mediated by the cartographers' technologies: the map's spatial frame is determined by existing territorial limits or lines of latitude and longitude; its content is determined by the features on the ground. Maps are the "theater" of the world, a reduced version of the world on which history plays out. This metaphor has early modern origins, as in the title to Abraham Ortelius's 1570 atlas, *Theatrum orbis terrarum* ("theater of the sphere/orb of the lands [i.e., earth, globe]"). However, the concept has taken on a more intense and pervasive aspect under the ideal, contributing to what Paul Carter (1987) called "imperial history."

As distillations of reality, normative maps supposedly preserve, store, and present that reality; they are understood to be comprehensive repositories of geographical facts, both literally, as the sum of geographical information known about a region, and also metaphorically, as the sum of all information known about a topic or phenomenon. This is a different metaphorical deployment of "map" than that common in the early modern era, in which the coarser resolution geographical map was a metaphor for systematically constructed knowledge (Withers 2019b). Within the ideal, "the map" is a metaphor for comprehensive knowledge, so that if something is "off the map," it lies beyond our ken. Cartography thus appears to be cognate with the "logical empiricist" philosophy of science (Rouse 1987, 3).

The ontological preconception has been expressed in two extreme positions taken by academic cartographers since the 1950s. First, studies of how people perceive and understand particular kinds of map signs have been based in part on the assumption that maps should reveal precise information, that the map is simply a table or spreadsheet of data in graphic form. Thus, studies of the perception of analytic maps with graduated circles sought to determine a design strategy that would permit map readers to correctly perceive the precise value represented by each circle without reference to other signs on the map; each circle was to be read as a number, not as a symbol of a value (Chang 1980). Second, the development of digital computers has promoted the argument that the computerized database of geographical coordinates, with locational and attribute data, is itself a map. The database comprises a selection of the world/archive that can be queried and used, so it must be a map (Tobler 1959; Aumen 1970; see Woodward 1992; Board 2015; Guptill 2015).

Yet, although the ontological preconception appears to be universal, it is contradicted or subverted by some of the ideal's other preconceptions. As such, the ontological preconception cannot be axiomatic and does not actually serve as a foundational logical justification for cartography.

In a naive way, commentators know that conceptions of the modern archive as an idealized, immaterial, Platonic form are inadequate; after all, the archive is made up of actual maps. So, by default, the modern archive is treated as the world itself. The all-too-common formula of "the map" necessarily conflates the world with the map image via the abstracted archive. This conflated world/archive is properly scaleless and capable of indefinite precision. This is why it is meaningful to conceive of a map at 1:1, and why readers do not just reject out of hand the idea of such a map: it already exists as the world itself! The effectiveness of Jorge Luis Borges's and Lewis Carroll's satires of a map at 1:1—"So we now use the country itself, as its own map, and I assure you it does nearly as well" (Carroll 1893, 169)—thus relied not only on the obvious, functional impracticalities of such a map, but also on the logical recognition that the territory already comprises a perfect map of itself.

Each and every normative map is conceptually tied directly to the world as a subset of the spatial archive. This assertion is at once expansive, in that it is readily applied to other planets in the solar system, and yet restrictive, in that it construes maps of nonphysical places as being merely maplike objects. The map is essentially a statement of fact, a store of geographical data, but one that distills, condenses, simplifies, reduces, or otherwise concentrates reality. As one author recently stated most succinctly, "a Map is a condensed history of God's creation and man's accomplishments" (Williams 2015, 1).

The core process of cartography is, therefore, the act of generalization—the reduction of the complexity of data derived from the world/archive in order to create a map/archive, as explicated by Joel Morrison (1976) and Waldo Tobler (1979). Map scale—the metric of the relationship of any map to the world/archive—is handily expressed in shorthand by the universal and unitless numerical ratio, such as 1:63,360 or even 1:1. The numerical ratio is a metric for the degree to which a map has generalized the world/archive and so establishes its fundamental character. All maps can be meaningfully compared according to the degree of their generalization of the world/archive, regardless of their actual nature or date and culture of origin: map scale has no history because *all* true maps are necessarily derived from the world/archive. This accounts for the insistence by map librarians and modern cartobibliographers on calculating and recording in map catalogs the numerical ratio for all early maps, regardless of their origin and nature. At the same time, images that are not structured with the same geometry as the world itself cannot be actual maps (as fig. 5.24) and are excluded from consideration.

More generally, we can identify three particular ways in which the ontological preconception has prompted inappropriate historical interpretation. First, it has encouraged the acceptance of regions as preexistent and natural concepts, which are then mapped, rather than as human concepts that are created through being mapped (Carter 1987; Lewis and Wigen 1997). For example, a 1972 study of the European exploration and mapping of the "New England coast" before permanent English settlement, which began in 1620, was framed in such ontological terms: the era "began with the region's existence barely proved," while "its extent, configuration, relationship to adjacent lands, and physical and cultural geography were total unknowns" (McManis 1972, 1, 112; see Edney 2011a, 13). Traditional map historians have generally structured their works according to modern regions, notably nation-states and their provinces, and have imposed those modern spatial categories onto periods when such regions either did not exist or existed in quite different forms. When the history of the mapping of each region is then told, the result is a double teleology of the development of both cartography and the region being mapped: "This simultaneous narrative of national/cartographic home-coming, where home is the here-and-now of modernity, permeates [the] account of American cartography, just as it does many other older histories of the map" (Edwards 2003, ¶7, referring to Schwartz and Ehrenberg 1980).

Second, map historians have overemphasized, to the point of exaggeration, historical attempts to create and curate comprehensive geographical or marine archives, as representing major advances in the push toward the modern triumph of the cartographic archive: Claudius Ptolemy's *Geographia* in the second century CE; al-Sharīf al-Idrīsī's "Book of Roger" in the twelfth; the *padrón real* of Spain's Casa de la Contratación in the sixteenth; or Giovanni Domenico Cassini's *mappemonde* drawn on the floor of the Paris Observatory at the end of the seventeenth. Each pre-1800 archive could be attempted—before they failed because of their creators' inability to incorporate new data and, as necessary, restructure the archives

(Thrower 1972, 74; Turnbull 1996; Jacob 2006, 94; Sandman 2007)—because each featured solely low-resolution information within a consistent set of mapping practices; they combined like data with like data. The ideal, however, necessarily holds out the potential for creating detailed, comprehensive knowledge of the world.

Third, when combined with the temporal flexibility of European languages, such that a term like "*historical map*" refers equally to a map made *in* the past and a map made *of* the past (Liberman 2013), the ontological preconception has led historians to conflate early maps of a region made in one era with the maps they themselves have made of the same region in the past. For example, three recent histories have reproduced the same map of the site of the modern city of Portland, Maine (fig. 3.2). All three gave the impression that the map was made in 1690. As the caption to the map in the first of the three histories stated:

> In 1690 tiny Falmouth Town was a cluster of some 40 houses, mostly at the foot of Fore and Broad (now India) streets. From the lonely ridges along Queen (now Congress) Street, one could watch the waves break against the narrow peninsula, north and south. (Holtwijk and Shettleworth 1999, 25; see also Conforti 2005, xiv; Levinsky 2007, 26)

Yet this map was actually made in the 1880s as a historical reconstruction of early settlement on the Machigone peninsula and illustrated an account of the 1690 destruction of the English fort at the site by the local Abenaki (Hull 1885). Moreover, it adhered to specific conventions for framing, orienting, and organizing maps of the modern city that had been developed in Portland only in the early nineteenth century (Edney 2017c; see fig. 2.8). In such instances, the ontological preconception—the insistence that a map's character is determined solely by the area mapped—ignores chronological separation and permits scholars to presume that a modern map of a place in 1690 is essentially the same as a map of the same place created in 1690.

Pictorialness

The conviction that maps "stand [in] for" or "represent" the world has a long history in the European tradition. According to this perspective, and in line with the ontological preconception, a map represents the world in a mimetic manner, which is to say in a manner determined solely by the world. Maps, at whatever resolution, appear to reveal the world directly and in an unmediated fashion. Maps are thus deemed to be "pictures" of the world. As one Soviet propaganda tract exclaimed,

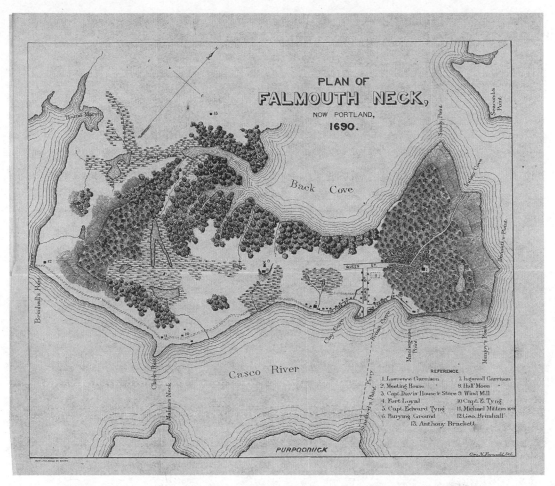

FIGURE 3.2. George N. Fernald, *Plan of Falmouth Neck, Now Portland, 1690*, in John Hull (1885, opp. 9). Lithograph, 32 × 37 cm. Courtesy of the Osher Map Library and Smith Center for Cartographic Education, University of Southern Maine (Osher Collection); www.oshermaps.org/map/7940.0001.

A map is hanging on the wall. Let us go up to it as we would to an open window. . . . The whole of the boundless Soviet Union lies before our eyes! (Mikhailov 1949, 9)

Or, as an early Anglophone textbook in cartographic design introduced its subject matter, the map is "a conventionalized picture of the earth's pattern as seen from above, to which lettering is added for identification" (Raisz 1938, 1; see "Observation," below).

As a mimetic picture of the world, the map becomes the site of debate about the world. When there are discrepancies between the map and the world, perhaps the world itself needs to be corrected. Certainly, fiction authors have played with this point, reifying the ideal even as they satirized it. Garrison Keillor (1985, 90–92) developed a territorial satire to sustain his fictional Minnesotan town of Lake Wobegon: an error in an 1866 survey led to the fifty square miles of Mist County, and the town, being omitted from the state map; the debate in the statehouse between the "accurates" and the "moderates" was eventually won by the latter, who did not want to change the state map, as that would require the renegotiation of all property and social relations (see King 1996, 1). Susanna Clarke (2004, 361n7) suggested, in her fictional account of the rebirth of English magic during the Napoleonic Wars, that it was "more convenient" for a wizard assisting the British army to move a town "rather than change all the maps" (see Edney 2007b, 157). And a recent vision of the political fragmentation of Europe in the near future imagined—with the assistance of a tongue-in-cheek historical account and carto-bibliography, complete with a fake secondary literature—how a private territorial survey undertaken in tandem with the official Ordnance Survey since the late eighteenth century had modified the map of England so as to create an entirely new territory topologically parallel to Great Britain (Hutchinson 2004).

The pictorial preconception differs from the ontological in that the mechanism of the map's replication of the world is through the specifically *graphic* modeling of the world's geometrical structure, rather than through the reduction of its complexity as a scaled derivative by generalization of the world/archive. The pictorial preconception enshrines the conviction that maps are, necessarily, strictly visual devices:

> The map is a written language expressed in a system of shorthand. In its earlier forms it was usually pictographic, and this one trait has clung to it throughout the ages. (Goode 1927, 1)

There is no scope within the pictorial preconception to permit consideration of nongraphic maps, such as verbal (oral or written) and performative maps. Nor does the preconception leave much room for pondering or defining just what is entailed by "graphic."

The pictorial preconception promotes the uniqueness and universality of the "language of maps." It is unique in that the mimetic system of *carto*-graphic rules is quite unlike verbal languages or other semiotic systems (Robinson and Petchenik 1976, 43–67). It is universal because it is direct and unmediated. Since the early nineteenth century (Edney 1997, 96–97), but only since then, commentators have held that maps "have their special language" (Dallet 1893, 11) and that "maps speak a universal language" (Anon. 2015c). This cartographic language is distinct from other kinds of texts, is specially and uniquely qualified to depict spatial relationships, functions independent of resolution or mode, and is independent of culture. In this supposedly universal linguistic system, map reading is a practice not of interpretation but rather of the "algorithmic" re-creation of the "literal meaning" of each map (see Olson 1994, 168–69).

The pictorial preconception has found frequent expression in the writing of map history. It has, to begin with, led to the common presumption that the techniques of relief representation are scaleless, such that the hachuring of mountains on low-resolution regional maps is the same as hachuring on high-resolution topographical plans, so their histories can be told together (e.g., Imhof 1982 [1965], 1–14; Cavelti Hammer, Feldmann, and Oehrli 1997). This conviction drives the insistence of map librarians that they have to catalog how early maps graphically depict relief. Inevitably, the scholarly result is a historiographical mess (see especially Wood 1977). A similar conflation has been made of the graphic representation of urban places, in which the images of towns and cities on low-resolution maps are discussed in the same historical sequence as high-resolution urban views and engineering plans (e.g., Hodgkiss 1981, 133–34; Delano Smith and Kain 1999, 181–86).

The pictorial preconception has long underpinned the reaction of laypersons to the shapes of features shown on early maps. Apparent similarities of outlines between early and modern maps have, especially when combined with a failure to appreciate early mapping technologies (see "Observation," below), fueled many arguments along the lines of "it looks like X, so it is X." These arguments are a hallmark of controversial, pseudoscientific works. I am not saying, of course, that map historians who adhere to problematic concepts have sunk to the same intellectual level as Erich von Däniken and his ilk! Rather, I argue that the work of both groups has been shaped, to differing degrees, by the precepts of the ideal of cartography, and those precepts are most readily apparent in the works of marginal figures.

In particular, Charles Hapgood (1966) argued that the outline on Oronce Fine's 1534 heart-shaped world map of the suppositional *terra australis*—a mythical continent created from the contemporary belief that a large, southern landmass had to exist to counter the weight, as it were, of the northern continents—was so like the "real" coastline of Antarctica as delineated on some modern maps, with the ice sheets removed, that that Renaissance map had to have been a completely factual statement based on actual survey and measurement. Furthermore, a portion of South America on the surviving fragment of Piri Reis's world map of 1513 precisely matched the modern coastline of Queen Maud Land. Therefore, Hapgood argued, because both sixteenth-century maps were manifestly accurate representations of the world made before any European had yet surveyed Antarctica, they had to have been based on some ancient tradition of knowledge that had somehow mapped Antarctica without ice. Hapgood did recognize that Fine's *terra australis* was not in the same location as modern Antarctica, but he explained away this inconvenient fact by positing that the southern continent had at some point shifted laterally, and rapidly, across the earth's surface. Hapgood's book has enjoyed continued

61

readership, being most recently reprinted in English (2001) and translated into German (2002) and Italian (2004). Criticism is much less pronounced (Jolly 1986; Anon. n.d). As a result, and despite their spatial, historical, and geological ineptitude, Hapgood's arguments continue to inspire arcane and foolish arguments about ancient global mapping, further credited by some to Plato's mythical utopia of Atlantis (Flem-Ath and Flem-Ath 2012; Anon. 2014).

The mode of argument of all such works, and also of lay attempts to dismiss them, is simply to discern or dispute visual resemblance. Indeed, there have been so many attempts to identify things as maps that are not, especially in the context of prehistoric petroglyphs and rock art (see fig. 3.4), that John Krygier (2008) coined a new term, "cartocacoethes," for the "mania, uncontrollable urge, compulsion or itch to see maps everywhere" (see also Krygier 2013). This practice is perhaps a form of pareidolia, the psychological condition that causes the mind to perceive meaning where it does not exist.

Hapgood's arguments were exactly of the sort that the prominent map librarian R. A. Skelton had criticized just a few years earlier. However, when Skelton objected about "the ease with which a heavy structure of theory can be built on foundations which are too narrow to support it," he was actually complaining about trained map historians who allowed themselves to be misled by facile visual inspection. Scholars have tended to be misled, not so much by visual comparisons of early with modern maps, as by readily apparent similarities among early maps. But, as Skelton also warned, "visual impressions suggesting affinity or development of the outline in two maps may be misleading" if actual techniques of compilation and drawing are not taken into account, especially when examining coarser resolution maps. He adduced several examples to demonstrate that visual resemblance, without further support from independent evidence, is unreliable (Skelton 1965, especially 4 and 15).

A more pernicious manifestation of the pictorial preconception is the assumption that the same feature should be shown in the same way on different maps, regardless of map scale or discursive differences. A cartometric analysis of maps of the South Carolina coastline studied several maps from the eighteenth and nineteenth centuries, which the researchers argued were "reasonably similar in scale," although they actually varied from a scale of one inch to five miles to a scale of one inch to about eleven miles—a factor of two! More important, they used both sea charts and geographical maps without considering the different levels of care, and therefore geometrical accuracy, with which modern hydrographers and geographers treat coastlines. So when they found that "one map was . . . less accurate than expected, given its date," it should not be a surprise that the errant map was not only the smallest scale map in the sample but also a map of North and South Carolina together from a geographical atlas (Lloyd and Gilmartin 1987, especially 1; see Edney 1993, 56).

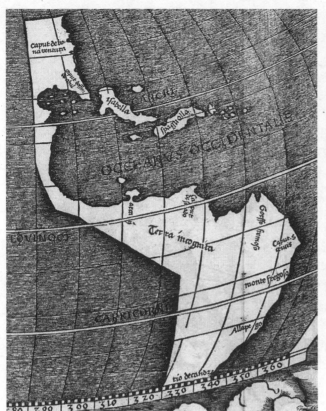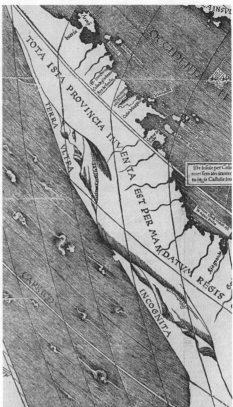

FIGURE 3.3. Details of the Americas from Martin Waldseemüller, *Universalis cosmographia secundum Ptholomæi traditionem et Americi Vespucii alioru[m]que lustrationes* [St. Dié, 1507]. Left: From the small, marginal map of the western hemisphere; on the original, the western coastline of the Americas is about 8 cm long. Right: From the map itself, showing the putative, but unknown, lands beyond the mountains; the length of the Pacific coast on the main map is about 69 cm long, the portion shown in this detail being about 45 cm. Woodcut on 12 sheets, each 46 × 63 cm or smaller; assembled, 128 × 233 cm. Courtesy of the Geography and Map Division, Library of Congress (G3200 1507 .W3); www.loc.gov/item/2003626426.

Again, much has been made in the last couple of decades about the two delineations of the western coast of South America on Martin Waldseemüller's great world map of 1507 (Hessler and Van Duzer 2012). In a small world map set in the upper margin that depicted two hemispheres—the one as known to the ancients, the other then being revealed to European voyagers—the western coast of South America was delineated with two straight lines that connect at an angle of a magnitude and location that seems to emulate the actual bend in the coastline at the modern border of Peru and Chile (fig. 3.3). It has therefore been argued that some European sailors must have sailed the Pacific coast well before either Balboa crossed the isthmus of Darién in 1513 or Magellan negotiated Cape Horn in 1520 (Dickson 2007). Alas, this argument fails to take into consideration how, in the much larger representation of South America on the main map, the western coast appears without the shading applied to all the coastlines known to European

63

voyagers and bears the legend *terra ultra incognita*, or "furthest unknown land," beyond a range of mountains depicted as if seen from the east. That is to say, the argument that the small, sketchy map of South America reveals unknown European knowledge, and therefore unrecorded voyages, makes sense only in light of the ideal's expectation that maps properly depict the structure of geographical features, regardless of their nature and degree of generalization.

The ongoing sociocultural critique of maps has not dispelled the pictorial preconception, other than to demonstrate that maps are not in fact "mimetic." Even so, map scholars have retained the general conviction that maps depict the world graphically, and they have extended it to argue that maps also reveal phenomena and patterns that are otherwise hidden or imperceptible. "The map is a mechanism that shows what no eye could ever see even when the map represents the most familiar territory—the space marked out by daily experience" (Jacob 2006, 2). In this sense, maps appear to parallel scientific instruments that make visible natural phenomena so that they can be measured, as a thermometer measures temperature or a barometer, air pressure. Academic cartographers have accordingly come to emphasize mapping as a process of visualization, as a process of revealing the invisible (e.g., Antle and Klinkenberg 1999; Hallisey 2005), and some map historians have similarly turned to visualization as the core function of mapping (e.g., Koller and Jucker-Kupper 2009; Hansen 2015). Reconfiguring representation as visualization is a welcome refinement in freeing ourselves from the ideal of cartography, but we cannot permit the pictorial preconception's insistence that maps are necessarily graphical in nature to restrict our understanding of mapping to its being solely a process of visualization.

Individuality

The ideal construes the making and using of maps to be strictly individual, cognitive work. Moreover, it holds that maps themselves, the things that have existence external to their makers' minds (see "Materiality," below), are unmediated replications of their makers' own, internal "cognitive maps." In other words, the individualist preconception refuses to accept that maps are semiotic texts that are determined by social needs and cultural conventions, but insists that they are the product strictly of intellect-driven observation. If maps are the product of intellect, then they can be used to diagnose intellectual capacity. As a result, simplistic accounts of how groups of people make maps have served as the foundation for unwarranted racist and sexist characterizations about the intellectual capacity of entire groups of people.

The individualist preconception emphasizes the work of the map maker's own intellect, which selects the data, abstracts the world, and encodes the map's symbols. The map maker's intellect authors the map and determines its meaning. Map historians have accordingly sought to identify the precise authorship of each early map. If no one individual can be clearly identified as the work's creator or designer, then historians will settle for anyone else whose name appears in concert with the map, with little regard for consistency or appropriateness (see Lane 1986).

The map's reader is also an individual. The reader's intellect decodes the map's symbols in order to gain access to the abstracted image of the world, with the aim of deriving knowledge from the map that the reader can add to his or her own cognitive store of knowledge and spatial comprehension. However, in accordance with the pictorial preconception, map reading appears to be an algorithmic and almost mechanical process that is much less intricate than the task of actually making the map.

This preconception is central to the arguments of normative critics who oppose the sociocultural critique of maps; they insist that social factors come to bear only in terms of the organization of mapping institutions and are irrelevant to how individuals read and use maps. John Andrews explicated this objection when he sought to rebuff the sociocultural critique of maps offered by Brian Harley and others. It is the map maker's solitary mind, Andrews (2001, 6) wrote, that "registers impressions from the external world and translates them into graphic form." He then pursued a counterfactual scenario positing that the quintessentially solitary individual—Robinson Crusoe—could have mapped his (fictitious) island and concluded that

> cartographic philosophers should certainly remember [Crusoe] when they read that the map is a "socially constructed image" or "inherently political." They might then become aware that map making, in Harley's eyes an essentially interactive exercise, is easier for a one-man community than say starting a family, voting in an election, serving on a jury, or betting in a game of chance. (Andrews 2001, 7)

Andrews's comments demonstrate the deep-seated conviction that map making and map using are both necessarily and solely *individual* and *intellectual* acts. They take place within a closed system that leaves little room for social or cultural factors. The only external factors allowed are those that do not impinge on the individual/intellectual processes of map making and map reading, such as the interactions of individual cartographers with each other and with their clients or patrons (see Edney 2005, 8–12).*

More generally, in the 1960s and 1970s, academic cartographers relied on the individualist preconception to define and model the process of "cartographic

---

*These beliefs permeate the repeated accounts of map making in Kiran Millwood Hargrave's young adult novel, *The Cartographer's Daughter* (2016).

communication," which they thought to use as the foundation for an autonomous discipline. Their various communication models all posited that cartography is a process by which the contents of the cartographer's mind—the cartographer's understanding of reality—is transferred to the mind of the map reader (Edney 2005, 36–41). Their models all reveal the conviction, building on the pictorial preconception's implication that cartography has its own language, that the look of the map itself is a cognitive construct. They hold that each map maker undertakes an "intellectual transformation" of the "contents of [their] mind" into "cartographic language" *before* that language is "objectified" as an actual map; map use follows the reverse process, first de-objectification and then, within the map reader's mind, comprehension (Koláčný 1969, 48, steps 3–6). In other words, the graphic map is a direct and unmediated externalization of the map maker's own, internalized spatial knowledge. Maps seem to exist not only as immaterial images, but also as direct and unmediated extensions of their maker's mind (see Blakemore 1981 for an example of the logical complexities that this stance engenders).

The conflation of internalized and externalized spatial knowledge is apparent in the common usage of "mental map" to refer to what are properly discrete phenomena (Curtis 2016, especially 341–42). "Mental map" is widely used both for an individual's internal and intangible neurological construct and for an individual's externalized expression of such knowledge, whether through speech, gesture, or graphics. We should instead distinguish the two phenomena: the internal neurological schema is a "cognitive map," with the understanding that "map" can only ever be a metaphor without explanatory significance (Kitchin 1994; see Poeppel 2012); the individual's externalization is a "sketch map." The term "mental map" itself should be restricted to its third common usage, as the graphic expression of spatial conceptions held in common by a group of people, generally constructed by academic researchers from social surveys or compiled from multiple individual sketch maps (Gould and White 1974).

But the assumption that these three phenomena—cognitive map, sketch map, mental map—might all be referred to by the same term reveals the easy conflation of an individual's internal cognitive map with an individual's sketch map with more formal spatial structures construed by societies or cultures. Externalized representations, whether sketch maps or researchers' compilations, are mediated by their discursive conditions and are in no way direct expressions of internalized knowledge schemas. Comments that "the urge to map is a basic, enduring human instinct" (Brotton 2012, 4, citing Blaut et al. 2003) or that "I map therefore I am" (Harmon 2004, 10) fundamentally confuse personal, cognitive acts with social, semiotic acts. Within this confusion, "the map" becomes a symptom of both individual and social conceptions. (To be clear: most scholars who focus on individual spatial cognition keep precise control of the scope of their work and do not fall foul of this profound misconception.)

The failure to recognize this confusion accounts for an apparent dichotomy that has long bothered academic cartographers. If the content of a map is determined by the map's maker, if each map's meaning is inherent to its image and lies latent until a map user retrieves it, then the map user should not be able to extract more, or different, information out of the map than was put into the map by its maker. Yet map users manifestly create new knowledge. Barbara Petchenik (1977) was perhaps the first map scholar to argue that the person who gives meaning to a map is actually the reader, a position that largely parallels basic semiotic principles, which in turn have generated the current argument among some critical cartographers that the map reader is the map author (Dodge, Kitchin, and Perkins 2009a). Yet the map maker continues to be privileged. The central question posed in one culturally minded study—"Why, in the process of communication," are maps "able to convey interpretations that go beyond not only the original intentions of the cartographer but also beyond his actual knowledge?" (Casti 2000, 10)—reveals the persistence of the presumption that the map maker is the sole determinant of a map's meaning. Attempts to resolve this conundrum while adhering to the individualist preconception have produced only tortured logic (e.g., Neumann 1994).

ORIGINS OF CARTOGRAPHY

The individualistic idealization that map making is founded in acts of cognition rather than semiotics has substantially determined discussions of the origins of cartography, as scholars of various stripes have sought to ascertain the nature of the putative "first map" without wondering how some isolated, ancient act could have inaugurated the endeavor of cartography. Consider the succinct history provided in the 1969 edition of the *Encyclopaedia Britannica*:

> Human existence would be impossible without knowledge of the sort that maps convey. Primitive folk carry such knowledge in their minds—mental "maps" of the areas where they live or hunt, fish or fight. As culture develops, such mental maps no longer suffice, and real maps are made to meet countless practical needs, to satisfy scientific curiosity and to give aesthetic pleasure. (Wright, Kish, and Skelton 1969, 827)

The thing about accounts of the supposed origins of cartography is that, without evidence, they are necessarily assertions that, as a form of counterfactual, are highly revealing of their authors' intellectual concerns and preconceptions (Evans 2013). In this respect, arguments advanced since at least the later nineteenth century for what "the first map" would have been have consistently manifested their authors' own interests. Scholars interested in world or regional mapping have imagined that the first maps were of expansive spaces (e.g., Ruge 1883, 515); those more committed to mapping as a practice of observation and measurement have construed

67

the first maps to have been of local topographies or routes (e.g., Reeves 1910, 1–2). Such arguments have been further influenced by older, eighteenth-century arguments that the origins of cartography lie in state mapping programs of early civilizations, whether Egypt or Rome.

The disparate arguments are nonetheless readily compatible because the individualist preconception permits a seamless transition from talking about the individual to talking about the social. The shift is implicit in the above quotation from 1969 and explicit in a longer discussion from the 1890s:

> For prehistoric man must have possessed a local geographical knowledge, just as in our own day the savage is intimately acquainted with the neighbourhood of his home, whose hill-sides and forests, rivers and lakes supply him with food and protect him from his enemies. This 'local geography' is the German *Heimatskunde*, which has been defined by Dr. Rüge as the 'sum-total of the individual's imaginative perceptions of the surrounding phenomena of Nature.' *Heimatskunde* is the seed out of which has grown the science of geography.
>
> Prehistoric man stood centred in a narrow circle, whose circumference was the limits of his individual experience, and it was only when men began to unite together in settled communities, sharing the same language and customs, that the separate geographical experiences of individuals were gradually fused into the larger geography of the state; and the Fatherland, in the place of the individual, became the centre of the universe.
>
> The belief that one's own country is the centre of the world—the hub of the universe—has existed at all times. It has been held alike by the Babylonian, the Greek, the Chinaman, the Indian, the Arab. . . . (Philip 1895–96, 313–14)

Some much more recent work has emphasized that the first mapping was cosmological and conceptual in nature, on the argument that "prehistoric maps are most likely to have been made in the process of communicating esoteric knowledge, through the use of cosmological symbolism" rather than being "in aid of the same purposes as modern maps, that is, as way finding aids or objects for information storage" (Meece 2006, 10, who also cited Delano Smith 1987, 59). Most recent opinion, though, holds that the first maps must have been maps of place (from Wood 1973 and Zelinsky 1973, to Clarke 2013). This insistence has been strongly influenced by another element of the ideal, specifically, that the topographical map is the default or prototypical map: what else could prehistoric peoples map except what they knew from firsthand experience and had already organized in their cognitive schema? (See "Observation," below.)

The pictorial preconception is relevant to this discussion. The special "cartographic language" of maps stems in part from their graphic nature, and they accordingly provide a "more direct" way to express an individual's thought than writ-

FIGURE 3.4. This is not a map. Many scholars have taken this repeating pattern, with leopard skin motif, on the wall of room 14 on Level VII of the Neolithic settlement of Çatalhöyük, Turkey, to be the oldest known map constructed to scale. Length of the original, about 3 m. Redrawn from Mellaart (1964, pl. VIa).

ing. Accounts of the origin of map making have, therefore, generally rested on the assumption that the graphic map necessarily predates writing. Malcolm Lewis (1987, 50, 52), for example, argued for the innate graphicacy of spatial thinking, communicable also by gesture and dance, by contrast to the linearity and ephemerality of speech and musical performance. However, a recent inversion of this chronology, such that graphic maps are held to follow "linguistic acts of spatial description," still does not challenge the fundamental preconception that maps are an unmediated expression of individual cognition rather than the product of social semiotics (Rochberg 2012, 44).

The image often held up as the "first map" was indeed of a place, being supposedly also the first "town plan." The image is a wall decoration, approximately eight to nine thousand years old and three meters long, from the Neolithic town at Çatalhöyük in Turkey (fig. 3.4), whose repeating pattern is similar to the archaeologists' own plans of their excavations at the site. The archaeologists who first found the design in 1963 admitted that this interpretation was perhaps controversial and that, if the image was indeed a map, it did not match up with any part of the town as yet excavated. They nonetheless held that "it seems likely that the eighty or more squares drawn . . . in rows or terraces represent a view of a town," together with a view of the source of the obsidian in which the townspeople traded, the two-peaked volcano, Hasan Dağ, seemingly imaged during an eruption (Mellaart 1964, 55). This interpretation has been widely accepted by many map scholars (e.g., Delano Smith 1987, 73–74; Casey 2002, 132, 225–26; Rochberg 2012, 9–11).

Stephanie Meece (2006) argued that this cartographic interpretation is actually unwarranted and presumptuous. First, why would people eight thousand years ago make a map of their small town? It was indeed small, no more than ten thousand inhabitants at its peak, and lacked the large public spaces and public structures that would be mapped in the third and second millennia BCE. It is possible, however, that there might have been a ritual and religious reason to map the town, as the archaeologists suggested. Meece's second point is more telling: why, if they did make a map of their own town, did inhabitants of Çatalhöyük do so in a manner akin to

69

that of present-day archaeologists, which is to say in a manner that has nothing in common even with the style of urban maps made by present-day town planners and others? Given all the problems with the image's correspondence to the town and the archaeologists' presumptions, and also given that chemical analyses had long since revealed that the two-peaked volcano Hasan Dağ was not actually the town's obsidian source and that the volcano had last erupted fifteen hundred years or so before the town flourished (summarized by Barber 2010), Meece argued that archaeologists should set cartographic fixations aside, cease to impose thoroughly modern conceptions, and instead follow good scholarly practice by seeking to interpret the image in light of all the other remains found in the same archaeological level. In fact, the image adheres closely to the art in the rest of the town. It is just one more figure formed from the repetition of geometrical shapes, a very common form of decoration throughout the site; at the same time, the volcano is more properly taken as an instance of another common motif, of what look like leopard skins to the modern eye. Meece concluded that the image was just another work of decoration with indeterminate original meaning, no different in style from any other mural found at the site.

Nonetheless, the individualistic preconception has led academic cartographers to insist that the design is indeed a map. Keith Clarke (2013) and Dan Dorling (2013) both conflated cognition with semiosis and asserted that the Neolithic mural was self-evidently grounded, like all maps, in observation and measurement (see "Observation," below). It looks like the town—even if it looks only like the present-day, excavated remains of the town—so it must be a normative map (see Krygier 2008). Shortly thereafter, the discovery by geologists that Hasan Dağ had likely last erupted during the period when the design was made reinforced the argument that the image's upper register does indeed realistically depict the volcano; that implied that the rest of the image must also be realistic and must therefore be a map of the town (Schmitt et al. 2014; Boyce 2014). Unfortunately, this further argument still presumes that maps are necessarily "realistic" and "natural" representations of the world ("Pictorialness") whose nature is defined not by the culture of their makers and by their functions but solely by the place or space being mapped ("Ontology").

The flaws in this position are easily exposed. The assumption by more recent scholars that the earliest maps were undoubtedly fine-resolution images of local topography has failed to address the fundamental question of *why* early peoples would have thought it necessary to make such maps. The assumption has been that because maps express individuals' cognitive maps, then maps are just made (see "Efficacy," below). Indeed, Clarke (2013) criticized Meece for evidently thinking that the Neolothic townspeople were not cognitively competent humans because they did not make maps; such misplaced criticism is only possible if human cognition and map making are presumed to be one and the same.

The sociocultural critique of maps has clearly demonstrated, by contrast, that maps are made for specific reasons within particular discursive contexts. In the case of a society with only a small economy close to subsistence, whether in the distant or recent past, it is reasonable to expect that everyone in a community knew the same terrain and so could refer to it through oral and gestural maps, and would not need to make graphic maps. And there remains absolutely no expectation that ancient peoples would have followed modern cartographic strategies, when the historical record is full of sophisticated, noncartographic traditions. Meece was correct.

RACIST AND SEXIST ATTITUDES

The individualistic preconception further manifests in a series of essentialist arguments about how certain categories of people think about space that are undeniably racist and sexist in nature. These arguments deserve special attention.

Studies of early mapping have been shaped by the long established practice of equating the "primitive" cultural forms of ancient, preliterate peoples to the cultural forms of contemporary indigenous peoples, which appear equally primitive to Western eyes. Leo Bagrow, for example, discussed prehistoric and indigenous mapping in the same chapter, using the latter to intuit the character of the former (Bagrow 1951, 14–17; Bagrow 1985, 25–28). The same argument is made by the use of the present tense in the quotation, above, from the 1969 *Encyclopaedia Britannica* (Wright, Kish, and Skelton 1969, 827). It remains tempting to place indigenous mapping first within summary histories, in order to suggest the prehistoric origins of mapping (e.g., Thrower 1972, 4–8; Thrower 1996, 1–11).

The equivalency of indigenous and preliterate mapping practices has led sociocultural critics to suggest that the basic act of mapping—both the *Ursprung* of the cartographic endeavor and the impetus of each act of mapping, in which respect the individualist preconception is recapitulationist—is the individual's expression of their internal, cognitive map as an external, sketch map. Thus, the cover design of the paperback edition of Franco Farinelli's (2009) study of "cartographic reason" featured a detail of a prehistoric petroglyph from Bedolina, Valcamonica (see Delano Smith 1987, fig. 4.28).

The usual occasion for such expressions is the communication of travel directions and routes (see "Efficacy," below) or some vague impulse to record the features of the earth's surface that is supposedly innate to humanity. Norman Thrower thus imposed thoroughly modern and idealized reasons for map making on both ancient and present-day indigenous cultures:

> The maps of early man were attempts to depict earth distributions graphically in order to better visualize them; *like those of primitive peoples*, these maps served specific needs. (Thrower 1972, 1; emphasis added)

In such statements, no consideration is given to the ineluctably social character of making maps, of mapping, beyond communication between two individuals, and the individualist preconception has been sustained.

The individualist preconception has engendered racist and sexist positions. If the maps made by individuals or entire societies are unmediated expressions of cognitive schemas, then, in a significant contribution to the power dynamics of the modern world, white, male, Western scholars have inevitably used maps to assess and rank cognitive development not only of individuals but also of entire cultures or other groups of people.

Jean Piaget stands as the exemplar of racist interpretations based on the false equivalency of cognitive schemas with maps. He assessed indigenous peoples as being "childlike" because their sketch maps were akin to those made by children in Western, industrial societies. He then enshrined these sentiments in his model of individual cognitive development, a model that served to classify individuals according to their cognitive maps as revealed through external expressions, whether verbal or graphic. Sufficiently aware to recognize that scholars should pay attention to prehistoric peoples on their own terms, but still nonetheless bound to the individualist preconception, Malcolm Lewis (1987, 50) incautiously used Piaget's model of cognitive development to argue that

> mapmaking appears to have remained undifferentiated in those cultures in which cognitive development, even in adults, terminated at the preoperational stage, which is distinguished by the topological structuring of space.

This would mean that fully adult humans had the cognitive skills of children between the ages of two and seven, a ludicrous characterization against which Denis Wood (1993, 2–3) rightly railed.

In a similar manner, in line with the persistent sexist argument that women are "intuitive" while men are "rational"—and in line with the surprisingly common act, bordering on the pornographic in its objectification, of mapping women as if their bodies were landscapes (Edney 2007a)—Western scholars have asserted that apparent differences in how men and women navigate in space and talk about spatial relationships necessarily indicates different cognitive capacities and capabilities. A 1947 publication by the National Geographic Society, highlighting how its maps had contributed to the war effort, contained several pictures of women who looked on as men made and did things with maps. But two pictures did show women actually doing things: one, a row of women slipping "hundreds of acres" of map supplements into huge, mountainous stacks of *National Geographic* magazines; the other, a white woman relating Inuit maps, formed from "bits of driftwood tied to sealskin," to modern coast charts, being able to use her supposedly liminal position to mediate between the indigenous other and Western science (Chamberlin 1947, 24, 40).

A more recent instance of the sexism engendered by the ideal occurred during the "Symposium on Cartographic Design and Research" that the Canadian Institute of Geomatics organized at the University of Ottawa in August 1994. One presenter used sketch maps made by his students as *prima facie* evidence that men and women think about space differently—women more topological and oriented to local landmarks; men employing topography and gridded, north-oriented space, and so on—but without controlling the conditions in which his students had made the sketch maps and without considering the possibility that the differences in their maps reflected gendered differences in education, responses to authority, and expectations. The subsequent discussion, featuring both male and female participants at the conference, did not question or challenge the underlying presupposition of cognitive difference between men and women (although the published paper took a significantly different tack [Kumler and Buttenfield 1996]). In this situation, and indeed within all of cartography, the default for study and discussion is always the cognitive schema of men; women are seen as diverging from men. In this respect, cartography is as much patriarchal and misogynistic as it is racist.

The individualist preconception has also been at the center of a misunderstanding of "cognitive maps" from the standpoint of historical materialism. The argument is that the manner in which individuals have perceived and understood space and the manner in which they have experienced place and space have together undergone profound alteration in the era of late capitalism. Frederic Jameson (1991) argued that postmodern urban life has fragmented and alienated individuals' "cognitive maps"; new cartographic strategies are necessary to reassemble those cognitive maps. David Harvey argued that those new strategies had already developed after 1848, as impressionism and then cubism overturned the older verities of realist perspectivism. Jameson's and Harvey's arguments both relied for their power on the assumption that maps made in earlier economic eras were directly relatable to, and indeed manifested, how individuals *thought about* and experienced space. In this respect, Harvey also drew on the observational preconception (see below): in early and late capitalism, he declared, "the connection between individualism and perspectivism is important" (Harvey 1989, especially 245).

In general, any argument that an entire culture or group "thinks of space" in just one way should serve as a logical and ethical red flag. To think that cognitive schemas are the same as semiotically constructed maps, and to equate personal experience with discursively structured spatial knowledge, is to adhere blindly to a flawed idealization.

73

Materiality

Maps are images, but they are also objects. They are stable and fixed. Of course, they might be torn in two, wear out, or be consumed by fire, but until their loss or destruction, they carry their maker's vision to their users. They are thus the stereotypical "immutable mobile" (Latour 1987, 227). The materiality of maps is so naturalized that academic cartographers, when first faced with the ephemeral maps displayed on computer screens, initially insisted that such maps constituted an entirely new and different class of "virtual" maps (e.g., Moellering 1984). Christian Jacob (2006) referred to the presumptions of the map's necessarily self-contained materiality as the map's apparent "sovereignty."

The normative map's material essence is enshrined in modern terminology. Renaissance Europe might have deployed a variety of terms for what we today call maps in the normative, cartographic sense, such as *descriptio* (description) or *typus* (image, form, model), but the few labels used in the modern era relate specifically to distinctive physical aspects (Van der Krogt 2015, 125–26). The Latin *carta* or *charta*, from which derives the term common to many European languages (e.g., French *carte*, German *Karte*), originally meant a sheet of paper or parchment and eventually came to mean any kind of formal or official document. Indeed, *carte* encompasses a wide array of paper documents with varying degrees of authoritativeness, from blank sheets (*cartes blanches*) to playing cards to restaurant menus to royal charters, as well as maps. And the English word *map* derives from the medieval Latin *mappa*, "large cloth," that is, a large and mobile thing that can be laid flat, although this original meaning has not survived in the form of other usages.

The material preconception holds that maps are objects made at fixed points in time. The apparent physicality and self-containedness of maps forms a barrier between the two sets of individuals construed under the individualistic preconception: the map maker and the map reader. The materiality of the map has continued to appear to interrupt and break up mapping processes even as sociocultural map scholars have increased their attention to map use, or at least to the wider social situations in which maps have been used. For example, having reviewed the field of map history and its traditional concerns, Denis Cosgrove proposed two new "sets of questions" that should

> bear heavily on any history of mapping. The first is the complex accretion of cultural engagements with the world that surround and underpin the authoring of a map, that is, treating the map as a determined cultural outcome. The second is the insertion of the map, once produced, into various circuits of use, exchange and meaning: that is, the map as an element of material culture.

Cosgrove outlined two sets of processes: one to produce the map and the other, once the map has been made and has material existence, to govern its use. He thus bifurcated the entire process of "mapping" at the map, "around which pivot whole systems of meaning, both prior and subsequent to its technical and mechanical production" (Cosgrove 1999, 9).

As creations of a specific, liminal point between misconstrued mapping processes, maps seem to possess an innate chronological fixity. Each map's maker makes it at a certain time, it thereafter remains fixed and immutable, and it remains so even as it is used. As such, "maps are usually outdated the very moment they go into print" (Schlögel 2016, 56). This moment of completion cleanly distinguishes the epoch of the making of the map from the epoch of the using of the map.

Intersections with other preconceptions have, however, led to some confusion over how the moment of a map's completion is to be determined. Publicity (see below) suggests that the map's true origin occurs when its potential is fulfilled with its first reproduction. Printed maps, therefore, are commonly dated to their first publication, regardless of when a particular impression actually came off the press. Variants of John Smith's map of New England (see fig. 3.1) are accordingly dated to the first impression, in 1616 (Old Style), rather than to the dates when its variants actually appeared, sometime between 1617 (New Style) and 1639 (see Edney 2010, 202–8). Alternatively, Observation (see below) suggests that a map should be dated to the time of its survey. This has led some scholars to ignore the fact that Smith's map appeared in conjunction with his *Description of New England* (1616) and to insist on dating it to 1614, the year when Smith sailed down and observed the coast of northern Virginia from Penobscot Bay to Cape Cod (Edney 2011a, 10–13).

These historiographical practices are informed by, and reinforce, the expectation that every map was created by an individual at a set time. But they are wrong, or at least misleading and inadequate. Mapping processes are ongoing and dynamic, and are not divided by the physical map. For example, the common act of annotating maps (e.g., Akerman 2000) manifests the manner in which readers determine meaning, and the manner in which all participants within a discourse—whether producers or consumers of maps—use the same semiotic system. It is not just that multiple people in multiple roles contribute to the production of a map, such that one really must refer not to authors but to "authorities" (the pairing of a name found on a map with a role), but that in chronological terms there is no divide in processes applied to the material map before and after its moment of creation. As one explores the multiplicity of people who have input into the creation of the map and its meaning, one finds oneself on the same ground as literary scholars and cultural theorists who have proclaimed the "death of the author." Ultimately, each spatial discourse is a unified network rather than one repeatedly split in half by the maps it produces.

75

Observation

The complexity of the ideal of cartography is readily apparent when we turn from the individualist and material preconceptions, and their inappropriate promotion of discrete arenas of activity between the map producer and the map consumer, to the observational preconception, which suggests, among other things, that producers and consumers are subject to the *same* visual regime of perspectivism. Rehearsing the individualistic conflation of internal cognitive schema with externalized map, some have argued that a map "is a representation of what was in the retinal representation of the man who made the map" (Bateson 1972, 460, quoted by Brotton 2012, 7).

The crux of the observational preconception is the conviction that all maps are, ultimately, grounded in the observation of the world. How else is a map to be made? The default or prototype map is thus taken to be the general-purpose, topographical map of landscape produced from precise observation and survey (MacEachren 1995, 220). But this perspectivism is then adopted as the way in which people produce and consume coarser resolution maps as well. Thus, a general article on map history in the Smithsonian Institution's popular science and history magazine could conclude by asserting that ancient people looked at the highly schematic "Babylonian world map" of ca. 600 BCE (see Rochberg 2012, 32–34) with "the same perspective" as someone today who used Google Earth's presentation of aerial imagery to zoom in on their own home (Thompson 2017, 22).

The result is significant confusion, both in the chronology of the origins of cartography and in the mechanisms of perspectivism. In terms of chronology, recent sociocultural critics have taken the apparent conjunction of the two resolutions of perspectivism in the Renaissance to be *the* origin of the modern endeavor of cartography—the moment when marine mapping, property mapping, and geographical mapping all collapsed into a single process that "treated" space "as the dead, the fixed, the undialectical, the immobile" (Harvey 1989, 204; see chapter 1, above). This argument runs counter to the empirical evidence adduced in this chapter and the next, which points to cartography as the idealized creation specifically of the nineteenth and twentieth centuries. In this respect, sociocultural critics of the map have been thoroughly misled by the ideal.

The ideal of cartography promotes the sense that maps are created from the "view from above"—the planimetric perspective was taught as the world seen from above—which, of course, presents a certain problem when dealing with early maps. As people from all educational backgrounds have plaintively asked me after public lectures, "How could people map the world so well before they could get up into the air and see it?" Setting aside for now the problems with impressionistic evaluation of map accuracy (see "Discipline," below), the commitment to the view from above has broadly manifested in modern commentaries on the "God's-eye view," on what Donna Haraway (1988, 581–82) called the "god trick," and on the critique of modern science's claims to omniscience. Implicit here is one of the fundamental flaws of the ideal: the presumption that the act of observing a map, so as to think one sees the world from above, is the same as the act of observing the world in order to map it. This conflation was clearly made by David Harvey (1989, 247) and Geoff King (1996, 14). In terms of the distinct discourses of mapping, however, looking at coarser resolution maps constructed through projections that supposedly accord with the rules of visual perspective is a quite different process from the fine-resolution observation of landscape. The fixation on the view from above is strictly a post-1800 product of the ideal and otherwise is historically invalid.

Much of the false perspectivism accorded to map readers stems from the interaction of the observational, pictorial, and material preconceptions. As a mimetic picture, the map presents the world to the reader's sight, permitting the reader to view and observe not just the map but, through the map, the world. Readers can, in most instances, take hold of the map as an artifact, turn it whichever way they desire, and physically dominate it. And, through the map, they can imagine they dominate the world depicted. This is an empowering relationship: "There is nothing you can *dominate* as easily as a flat surface of a few square meters," argued Bruno Latour (1990, 45); "there is nothing hidden or convoluted, no shadows, no 'double entendre.'" Even when individual maps are too big to be held, modern maps are generally positioned or mounted to permit readers to examine them in detail, even to crawl over them, in order to visually dominate them and so examine them and the world at the readers' pleasure. The physical domination of the map promotes the intellectual conviction that one dominates the depicted world. The map permits its readers to explore the world as they wish, without being hemmed in by guides or restricted by the demands of actual travel (Edney 2003a, 2007a, 2009).

Modern maps give readers a panoptical *sense* not only of vision but also of the ability to discipline, control, and regulate territory (Edney 2003a; 2009, 24–26). In a like manner, the strategies of detailed, topographical mapping have been extended to the pornographic imaging of female bodies by configuring those bodies and their landscapelike curves as objects upon which male readers can impose their desires (Edney 2007a). To be precise, this sense of perspectival vision sets the viewer not so much above space as *outside* it, so that maps seem to function as "mirrors" of nature (Biggs 1999, 378, drawing on Gombrich 1975). This metaphor has an early modern origin with, for example, the title of Gerard de Jode's 1578 atlas, *Speculum orbis terrarum* ("Mirror of the world").

Conversely, assumptions of empowered vision on the part of readers have been transferred to map makers, to the surveyors who observed and measured the world. Sociocultural critics have commented on the manner in which surveying (from the Anglo-Norman French *surveier*, "to overlook") produces a commanding and domineering perspective. In imperial situations, the surveyor assumes still further authority as the agent of the controlling power, standing on high places to look down on foreign lands and peoples in order to map and control them. Mary Louise Pratt (1992, 201–8) drew on the facetious opening line of William Cowper's 1782 poem, "The Solitude of Alexander Selkirk," to call this illusion the "monarch of all I survey" syndrome. Surveyors appear innately empowered, able to see and measure all manner of things; they seem to be equipped with special vision and access to the world, although in practice this is rarely the case. William Wordsworth revisited Cowper's conceit and in the process highlighted some of the limits to vision in his 1811 poem, "Written with a Slate-Pencil, on a Stone, on the Side of the Mountain of Black-Comb" (Carlson 2010a, 2010b).

The assumption of empowered vision has gained new currency in the present age of remote sensing and mass surveillance. Just watch any police procedural or spy thriller, whether movie or TV show, made since the early 1990s. Problems with vision—e.g., a crime was not caught on tape because the CCTV surveillance cameras were not working—are the exception and are admitted only as a device to complicate and propel a plot. In many respects, Andrew Davis's *The Fugitive* (1993) established the modern trope, with its lingering pans across the cityscape of downtown Chicago, shot vertically from a low-altitude aircraft, and its depiction of federal marshals who, spiderlike, occupy the center of a web of surveillance, tensed and ready to spring into action as soon as the fugitive's presence sets the web vibrating.

More generally, the ideal posits that all mapping is grounded in unmediated observation and measurement. Combined with the ontological and individualist preconceptions, the observational preconception supports assumptions about the nature of the earliest maps. It also has generated a popular image of the progression of mapping techniques and the ordered generation of the spatial archive. In one example, the career of the medieval Arab geographer, al-Sharīf al-Idrīsī, is completely and ludicrously reenvisioned in accordance with the ideal of cartography. This fictional history of al-Idrīsī's cartographic work recapitulates the idealized sequence of cartography as a whole, from detailed surveys to global observation and mapping: al-Idrīsī supposedly began his career by mapping "every stitch, thread, and embellishment on the queen's royal robes," then moved on to map "every plant, herb, fruit, root, tree, and grove in [the royal] garden," a variety of architectural works, and the whole city, before finally setting out on a grand expedition "to draw a map of the known world" (Fasman 2005, 15–16). Here is another instance of cartography's apparent recapitulationism.*

A more subtle and mainstream expression of the observational preconception—

---

*Al-Idrīsī actually lived in Sicily, at a crossroads of Mediterranean trade routes, where he could gather and compile coarse-resolution geographical information with no reference to finer resolution surveys (Ahmad 1992; see Jacob 1999).

the assumption that the only way to map the world is through coherent and comprehensive surveys—can be found in Jerry Brotton's *History of the World in Twelve Maps*. Brotton presented eleven analyses, in chronological order, of maps or atlases of the world from Ptolemy's *Geography* in the second century CE to *Google Earth* in the twenty-first. Yet he interrupted his global narrative by inserting a twelfth account—between his chapters on Joan Blaeu's *Atlas maior* (1662) and Halford Mackinder's 1904 map of the "geographical pivot of history"—of the tremendous but plainly nonglobal eighteenth-century survey of France that produced the *Carte générale et particulière de la France* (Brotton 2012, 294–336; see figs. 4.2 and 5.16–18).[*] Pragmatically, the insertion of the French surveys allowed Brotton to discuss the formation of modern states, a phenomenon of central importance to modern history but one that is not readily accessed through world maps. Yet, as the first statewide systematic territorial map grounded in a comprehensive triangulation, the *Carte générale et particulière* initiated what would become, after 1800, a new form of territorial mapping that would be applied across the world, region by region, and that would become the benchmark for the new ideal of cartography. Although Brotton properly avoided making the argument that world mapping after 1800 has been directly grounded in such careful observation of the landscape, his ability to discuss the finer resolution *Carte générale et particulière* within a book otherwise dedicated to coarse-resolution imagery of the cosmos and the world reveals the widespread conviction that all modern mapping, even modern global mapping, is properly and solely grounded in observation.

In recent decades, the observational and ontological preconceptions have been combined within a visual trope of sequences of images from above of the same spot, but at different distances and therefore at different map scales. The trope seems to have been created as a pedagogic tool by Kees Boeke in his 1957 book, *Cosmic View* (fig. 3.5). Each of forty black-and-white drawings depicted the earth at different resolutions, starting with a girl holding a large cat, first "outward" via progressively distant vantage points to show the whole universe, then "inward" via ever more microscopic ones down to a single atom; the intervals were each one order of magnitude, creating a "scale of nature" (Doiron 1972, 11). Boeke effectively implemented the series of factors for graphic reduction and enlargement proposed in 1802–3 by Pierre-Alexandre-Joseph Allent, factors that underpin the modern concept of map scale (see chapter 5). Charles Eames and Ray Eames (1968, 1977) later developed Boeke's work into short color films, the second of which was then turned into a famous book, *Powers of Ten* (Morrison et al. 1982).

Many similar animations have been prepared, especially in the present age of digital animation, and can be found online. The modern stoic philosopher Massimo

---

[*] For reasons that will become clear in chapter 5, I prefer to use César-François Cassini de Thury's own, original name for the great eighteenth-century map of France, rather than the name it later acquired, of *Carte de France*.

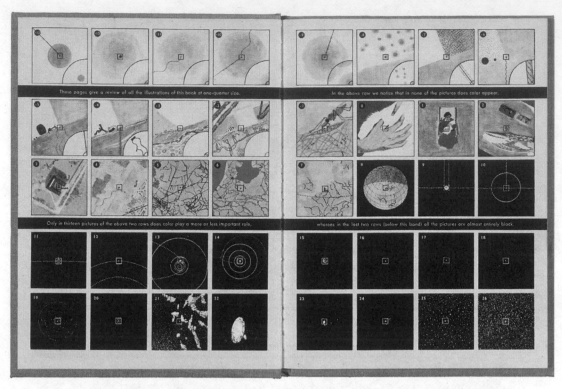

FIGURE 3.5. Kees Boeke's views of the cosmos in "forty jumps" (Boeke 1957, endpaper), diagramming the full sequence from an atom (−13, top left), through the 1:1 picture of a girl's hand grasping a cat (0), to the universe of galaxies (26, bottom right). Within Boeke's book, the images are all monochrome, but the endpapers colored the views visible to the eye—whether directly or indirectly through maps—in an attempt at realism (from −3 to 8). Colored lithograph, 21.5 × 33.5 cm. Image courtesy of the Osher Map Library and Smith Center for Cartographic Education, University of Southern Maine.

Pigliucci has indicated that such sequences enable, if not an actual sense of control, then at least an understanding of the sheer size of the world, such that we can focus our anxieties on more immediately personal concerns (reported by Burkeman 2017). More particularly, science fiction films and TV shows often feature the view of the descending spacecraft, starting with a picture of the whole earth and then proceeding rapidly to ground level and a (usually hapless) protagonist. (The sequence is sometimes reversed.) In many respects, these animations mimic the "satellite view" available in Google Maps, with which the user can zoom in and out. These books and animations provide powerful ammunition to the cultural conviction that maps are all views from above, while the seamlessness of the animations reinforces the conviction that there is just one process—cartography—of turning these views from above into maps.

Yet it is salutary, when watching such animations, to pay attention to the depiction of clouds. Distant views of the whole earth depict clouds, as they are omnipresent and it is impossible to photograph the whole world from very high orbits without including them. As the animation closes in, but still from the high altitudes of satellites, the clouds suddenly disappear, as the constituent satellite views are selected for cloudlessness; this is the upper photopause, as I like to call it, which is to say, the abrupt transition from high-orbital to low-orbital imagery. Soon the viewer/descending spacecraft passes through a high-altitude, wispy, but previously invisible layer of clouds, marking the abrupt and unavoidable transition, the middle photopause that must be obscured, between overhead imagery taken from satellites and the imagery taken from atmospheric airplanes. Finally, with the lower photopause, the viewer reaches ground level and clouds once again return to the sky to bring the geographical action to a close and to set the stage for the dramatic action. The presence and absence of clouds remind us that these visualizations come from multiple sources that were created for different reasons and have different geometries, and have all been subject to human manipulation. They are most emphatically not unmediated.

Finally, the observational and individualist preconceptions combined in the postwar era as the basis for a wide-ranging criticism of cartography that once again takes the ideal at face value. The social movements of the late 1960s thoroughly implicated cartography within the network of capitalism, militarism, and perversions of technology that had produced the waste and horror of the Vietnam War and imminent environmental collapse; anticartographic sentiments reached deep into the contemporary antiscience/antiwar movement (e.g., Roszak 1972, 407–12). This particular critique of cartography was then absorbed by a new movement within academic geography that was concerned with human beings' lived experiences, in reaction both to geography's traditional regionalism and its positivistic turn in the 1960s. Humanistic geographers contrasted the multifaceted experience of place, of what it means to be alive in the world, with the apparent sterility of mapped space. The one is grounded in humanity, the other in a distancing perspectivism. Within this framework, personal experience, cognitive maps, and works of art were held to be far more authentic than the formal maps made by government agencies and commercial companies. It was permissible to compare personal cognition to social production because of the individualist preconception's insistence that cognitive mapping is equivalent to map making, and the observational preconception that perspectival vision is the key to all mapping (Woodward 1992, 53–54; Edney 2005, 34 and 104–5).

In leveling their criticisms, humanistic geographers did not question the ideal of cartography. They accepted it as a valid statement of the nature of maps and mapping, to which they took exception as a perversion of individual vision and cognition. Thus, Yi-Fu Tuan (1979), in a survey of the role of the visual in geographical studies and in the individual's appreciation of place, made no mention of maps at

81

all (see Muehrcke 1981, 39n10). Otherwise, the only maps that Tuan treated positively were those, such as medieval *mappaemundi*, that obviously eschewed cartography's rigid geometries (Tuan 1977, 34–50, 85–100, 118–23).*

The humanistic critique has extended into postmodern studies. The contrast of medieval with modern mapping, to the detriment of the latter, has become common especially within literary studies (e.g., Avery 1995; Mitchell 2008, 27–76; Doherty 2017). As a geometricized abstraction, "the map" has appeared as a hallmark of modernity, capitalism, and imperialism that needs to be abandoned in favor of postmodern strategies that represent the experience of place in more authentic ways. For example, Michel de Certeau (1984, 120) stated:

> If one takes the "map" in its current geographical form, we can see that in the course of the period marked by the birth of modern scientific discourse (i.e., from the fifteenth to the seventeenth century) the map has slowly disengaged itself from the itineraries that were the condition of its possibility.

And, he argued, scholars should address the experience of passing through places, and especially of walking through urban places, as the only authentic form of mapping. Moreover, "actual maps" appear as "totalizing devices" that constrain individual experience within a straitjacket of grid lines created by modern states for their own nefarious, statist ends. Thus, Pierre Bourdieu succinctly defined "cartography" as "the *unitary* representation of space *from above*" (1994, 7; emphasis added). In his view, cartography is just one more strategy by which "the state concentrates, treats, and redistributes information and, most of all, effects a *theoretical unification*. Taking the vantage point of the Whole, of society in its totality, the state claims responsibility for all operations of *totalization* ... and of *objectivation* [*sic*]" (ibid., original emphasis). Yet, when we leave the distortions of the ideal behind, we can see that the newly geometricized images of space did not, in fact, replace itineraries and that the European mentality (whatever that is) did not

---

*When in graduate school, during the 1980s, I repeatedly witnessed geography faculty criticize cartography as being a "sterile" and "unintellectual" "technique." The underlying mindset manifested in other ways: the comment, "not bad for a cartographer," from a senior professor in reference to my final paper in the required methods and theory course; the continual denigration of maps and mapping in the seminar on geographic thought required of all doctoral students, led by a different senior professor; the absolute refusal—until it became a matter of raising external funds—to deal with remote sensing and geographic information systems, which were otherwise dismissed as merely applied technology with no intellectual merit; and the frankly shabby treatment routinely meted out to my advisor. I admit that my jaundiced attitude toward geographers' treatment of cartography was shaped by these formative experiences, and might not be properly representative of the larger discipline of geography. Cosgrove (2007a, 203; 2008) maintained that geographers in general have remained positive about maps and mapping, but several other scholars have independently confirmed my own, negative perspective (Muehrcke 1981, 4–5; Rundstrom 1989, 185–86; Openshaw 1991; Wheeler 1998). Since 2010, however, the growth of web-based mapping, big data, and informational graphics has brought geographers back to mapping in exciting ways.

CHAPTER 3

become geometrically sterile in the Renaissance (see, e.g., Woodward 2007a, 12, *contra* de Certeau). Once again, the ideal's essentialist tendencies must be resisted.

Efficacy

None of the preconceptions discussed so far has suggested any reasons for why maps should be made, other than as inherently necessary records or presentations of the world. At best, the individualist preconception holds that if maps are direct expressions of cognitive maps, then they are inevitably made once a culture reaches a certain degree of complexity. The common expectation would therefore seem to be, to paraphrase Douglas Adams, "It's a map; you have to make maps!"[*]

To be sure, academic cartographers have pragmatically differentiated between "general purpose" and "special purpose" maps. Special-purpose maps are those intended for specific uses by particular communities of users; relatively rare before 1800, thereafter they became increasingly common with the growth of professional specialization within industrialized societies. By contrast, all maps that are not specialized—from coarser resolution atlas maps and wall maps to finer resolution systematic territorial surveys—fall into the catch-all and frankly meaningless category of general-purpose maps. This distinction has been expressed as one between a functionless "base mapping" and an overtly focused and perhaps even persuasive or propagandistic "thematic mapping." In this respect, academic cartographers have generally and improperly assumed that the *selective* character of special-purpose maps was necessarily the intellectual (even cognitive) precursor of the *abstracted* character of analytic maps (see chapter 4).

Yet the instrumental use of maps by the general public in modern, industrial societies has engendered the preconception that maps are in fact functional and that, more particularly, their default function is to be navigational instruments. This presupposition underpins recent as well as traditional accounts:

> Space represented on geographical maps is ordered and navigable, allowing one to "know" features and layouts of places before visiting, and finding landmarks, pathways and spaces of importance that connect you from one point to another, reducing possibilities for disorientation. Maps "work" because they represent a reality already ordered and structured. (Cowell and Biesta 2016, 431)

Maps do indeed order space, in accordance with their parent discourses, most of which are not concerned with navigation. However, since the early nineteenth century, most of the kinds of maps the general public has encountered and used seem to be about aiding mobility: sea and air charts and, of course, special-purpose

---

[*] Adams (1978, "fit the first"): "It is a by-pass. You have to build by-passes!"

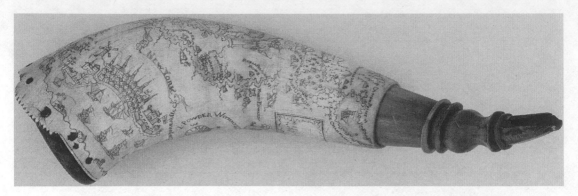

FIGURE 3.6. A map of the Hudson River corridor inscribed on a powder horn, 1763. From left, the map shows New York at its harbor, the Hudson River to Saratoga north of Albany, the portage to Lake George, and at far right Lake Champlain with the fort at Crown Point. Not visible are a second map of the Mohawk River and the arms of Great Britain. Cow horn and pigment, with wood fittings, 31.5 cm (length) × 8.5 cm (diameter). Courtesy of the Metropolitan Museum of Art, New York (Collection of J. H. Grenville Gilbert, of Ware, Massachusetts; Gift of Mrs. Gilbert, 1938; accession 38.57.2); www.metmuseum.org/art /collection/search/29501.

maps designed to aid mobility, from road maps to plans of hospitals and shopping malls, but also ostensibly general-purpose topographical maps, which have served to guide not only military personnel but also hikers and orienteers (Akerman 2007; Akerman and Nekola 2016). One could also argue that this practice points to the original purpose of systematic topographical surveys, to aid generals in directing the movement of military units, although from the mid-twentieth century on, in the West at least, the practice was taught to individual soldiers, so that topographical map reading and navigation was popularized through mass enlistment.

A study of maps inscribed by hand on powder horns in eighteenth-century North America reveals the deep-rootedness of this efficacious preconception. Such maps were generally crude, even by contemporary standards (fig. 3.6). The study's author accordingly suggested that it was "questionable" whether the horns were actually used, as previous scholars had assumed, as a guide to "the location and direction of forts and landmarks in the area in which [the militiaman] was serving." Yet, even as the author admitted that it was "likely" that "the more elaborate [horns] were professionally engraved as a typical American souvenir to be carried home by British officers," he nonetheless maintained, as a matter of course, that other maps on powder horns "were unquestionably of some use" in way finding (Du Mont 1978, 5). Even crude maps must be of some instrumental value.

The idea that maps are inherently efficacious as guides has been further sustained by the innumerable popular and metaphorical maps for "finding one's way," both culturally and metaphysically, as well as by the mapping of imaginary places (Akerman 2007; Padrón 2007; Reitinger 2008). The efficacious preconception has found critical expression in the argument that "the map" constitutes a technological extension of human cognition and experience, so that humans who use maps are, in effect, cyborgs or integrated human-machine organisms (Piper 2002, 1–5). The argument that map users are cyborgs is also grounded in other preconceptions of the ideal that hold that maps are material objects that are read only in an algorithmic manner.

Map historians, as well as the general public, have routinely assumed that the default function of early maps is navigational. Beyond this general misconception, two particular errors are common. The first is the unwarranted conviction that maps not intended for navigation were in fact used for navigation. In an egregious example, at the very end of the film *The Da Vinci Code* (2006), the professor of symbols realizes that a meridian, otherwise a strictly geometrical line of longitude, also functions as a pointer toward a sacred site. In a more complex manner, we can find many instances when map historians have taken geographical maps to be "charts." Historians have routinely presumed that John Smith's map of New England (see fig. 3.1) was actually used to guide later colonial settlement (Edney 2011a, 16); less academically, Smith's map appeared in the title credits to *Catwoman* (2004) to evoke the supposed historic migration of cat-worship from Egypt to China. Map historians have accepted Gerard Mercator's great wall map of 1569, prepared on his famous projection, as a chart even as they adduce evidence that clearly places the map within landlubberly practices of geographical mapping (most recently Seed 2015; also Crane 2003; Taylor 2004). Willingly misled by the opening words of the title of Olaus Magnus's *Carta marina* (1539), a geographical map of Scandinavia with many icons of northern life, map historians have thought of that work as a sea chart whose many sea monsters are readily and naively interpreted as warnings to mariners of the dangers of the northern Atlantic waters (Nigg 2013); it is more appropriate to use Olaus's own title for the map, *Carta gothica* (Ehrensvärd 2006, 58–76).

Second, there is the misconception that maps produced before 1800 in support of navigation were used in the same manner as those from the modern era of complex technological systems. We readily find assumptions that early navigators had "chart tables" on which they spread their charts to plan and to record the actual paths taken. Conversely, the privilege accorded to maps as apparently stand-alone documents makes it easy to forget that modern charts are intended to be used in conjunction with modern navigational aids, such as lighthouses and buoys.

There is a further persistent corollary to the efficacious preconception, specifically that premodern peoples must have used maps in the same kinds of ways, and for the same kinds of tasks, that modern people use maps. One event—the division of the Holy Land among the tribes of Israel after the Exodus (*Numbers* 26:53–56, 33:54, 34:1–29)—was the subject of much unwarranted presumption in the nineteenth century, as in this example:

85

It is to the *Hebrews*, however, that we owe the three oldest and most interesting of early geographical records. . . . The description of the march through the wilderness is the first itinerary we possess, and we can safely assume some kind of topographical map to have been the basis for the division of the Promised Land among the twelve tribes. (Philip 1895–96, 314)

Alas, we cannot "safely assume" anything of the sort. More recently, the historically inept ideas of Charles Hapgood and others (see "Pictorialness," above) that early maps of an apparently precocious nature must have been based on detailed surveys undertaken by ancient peoples depend on the expectation that those ancient peoples would have made maps of the same nature as those we make today. See also the debate (see "Individuality," above) over the interpretation of a mural from Neolithic Çatalhöyük as a map made somehow to modern standards (see fig. 3.4).

Discipline

The corollary of the observational preconception, at least as mediated by preconceptions of efficacy, is that of discipline. Any use of a map, especially for the assumed default operation of navigation, necessarily requires the user to test the map and thereby prove its accuracy and worth. Map evaluation is a routine and repeated exercise undertaken across modern society whenever people use paper or digital maps to direct their driving, hiking, traveling, orienteering, geocaching, and touristic wandering. In the twentieth century, popular culture featured innumerable cartoons and complaints about the accuracy of road maps; in the twenty-first, popular ire is directed at the inaccuracies of online mapping services (Mapquest, Google Maps, or Apple Maps) and of GPS navigational systems. Errors are to be corrected and new information added to keep the world/archive current, so that the spatial archive ineluctably progresses.

Just as the observational preconception is concerned with the disciplining of the landscape, the disciplinary preconception entails the constant evaluation, correction, and regulation of maps as unmediated incarnations of the world/archive. The user, guided by a map, constantly checks it against the world, discovers errors or mistakes, and otherwise affirms its correctness; any errors that are identified are to be corrected, assuming that the user is in a position to make corrections. In effect, the map user is as empowered as the map maker. Map use constitutes the core ritual of cartographic examination that lies at the heart of disciplining the world/archive (see Foucault 1977, 184–92).

Map evaluation sustains the validity of the ideal of cartography. Every time a map is tested against the world—when military pilots use maps to guide them to their targets, or when a driver consults a road map to navigate through a strange city—the instrumentality and factuality of maps is affirmed. Indeed, scholars who have reacted negatively to the sociocultural critique have done so in part because such repeated acts of evaluation and the continued effectiveness of maps apparently refute the argument that maps are culturally and socially mediated representations.

Discipline, of course, requires the identification, through examination, and then the extirpation of erroneous behavior. So, too, for cartography: the identification of errors in a map requires the map to be corrected and updated. Before the sociocultural critique began to alter presumptions, map historians commonly held that the identification of errors in maps was sufficient to motivate new surveys. They have not extended such assumptions to low-resolution maps that loudly boasted to have incorporated the "latest" discoveries or "most accurate" surveys; such claims are easily revealed as fraudulent and intended to sell maps to gullible purchasers. But for finer resolution maps that seem to be grounded in observation and measurement, map historians have long asserted that surveyors and mariners, like John Smith—who had explicitly complained that the six or seven charts of northern Virginia that he had seen before 1614 had all proven to be so unlike the actual coast as to be mere waste paper (Edney 2011a, 6)—were motivated not by institutional or commercial reasons, but rather by their desire to discipline existing maps and thereby improve spatial knowledge. Such purely cartographic motivations are attributed to map producers regardless of the quality of the supporting evidence or of their appropriateness in the historical circumstances. After all, cartography is an expression of the Platonic ideal of measurement.

Similarly, the key disciplinary ritual for normative map historians has been the analysis of the geometrical accuracy of early maps. For the most part, the assessment has been visual and has been either relative (against other early maps) or absolute (against present-day maps). Some graphic and statistical methods have been used (Andrews 1975; Ravenhill 1976; Murphy 1979; Blakemore and Harley 1980, 60–68; generally, Kishimoto 1968; Maling 1989). The result has been the classification and arrangement of maps according to their apparent historical importance. Maps that for their time seem to show an appropriate degree of accuracy are taken to be "important"—a term never explicitly defined—and, taken together, constitute the map historical canon; they are arranged in chronological order by area shown. Maps that appear not to show an appropriate degree of accuracy are simply disregarded.

If maps are defined by the world/archive and intended for instrumental use, changes in the world must necessarily generate new or updated maps. Cartography must properly maintain maps' currency. Map historians have thus presumed that new or updated maps are required whenever geographical change—whether natural or anthropogenic—makes maps outdated, whenever new surveys are undertaken, or whenever changing practices or circumstances reveal a current map to be incomplete or inaccurate. This presumption of inevitable currency combines with other preconceptions (see "Observation," above; "Publicity" and "Morality," below) to establish the conviction that coarser resolution mapping of the world and its regions is driven by the undertaking of higher resolution surveys, further contributing to the conviction that cartography is indeed a singular endeavor.

Consider, as one example among many, a study that asserted that Joshua Fry and Peter Jefferson undertook their map of the colony of Virginia in the mid-eighteenth century because "there was an urgent need for a modern map of the colony" as "no comprehensive map based on new surveys had been produced in the eight decades since Augustine Herrman's map of 1673" of Virginia and Mary-land. It was to "resolve" this "crisis," the author averred, that the Board of Trade and Plantations in London directed the colonial authorities to commission a new map of the colony, a task duly accomplished by Fry and Jefferson in 1751. Further-more, the author saw no contradiction when he noted immediately thereafter that "only a few" impressions of Fry and Jefferson's map were actually printed in 1753 (Cresswell 2000, 83). A true crisis in the representation of the colony and an out-standing demand for maps would surely have required a larger printing in order to disseminate the updated geographical knowledge. The limited print run instead suggests that the colonial officials in London had motives other than a rather vague and idealized desire to do good by updating the stock of human knowledge; indeed, the order for new geographical information was a blanket request sent to all the British colonies in North America. Fry and Jefferson's map would not be printed in large numbers until the simmering Anglo-French competition for empire once more erupted into open war, in the colonies, in 1755 (Edney 2008a, 71–72; Taliaferro 2013).

Conversely, map historians have generally presumed that the existence of a new or updated, but undated, map must have been necessitated by some geographical change. As just one example, the fourth variant of John Green's *A Map of the most Inhabited Part of New England*, which still bore the copyright date of the first variant, published in London in 1755, is distinguished solely by the application of hand-coloring to towns west of the Connecticut River and a printed note that these colored towns fell within the territory (now the state of Vermont) granted to the province of New York by the privy council in 1764. Since the late nineteenth century, cartobibliographers have accordingly assumed that this particular variant had to have been created in 1764 in response to this territorial change (Stevens and Tree 1985 [1951]). However, impressions of the third and fourth variants were both included in Thomas Jefferys's postbankruptcy atlas, *General Topography of North America* (1768); the additions that distinguish the fourth variant were added not to record geographical change but to give the impression of updatedness (Edney 2003b).

This is not to say that some map makers did not regularly update their maps and keep them current. For example, contemporary evidence suggests that, starting in the sixteenth century, some European map sellers updated their maps with the locations of battles as soon as they received word about them. But such acts need to be carefully documented and cannot be taken as the historical or even present-day norm.

The constant disciplining of maps, the presumptions of currency and of the innate morality of cartography (see below) combine to create the grand conviction that cartography is inherently progressive. Morally, map makers strive to make the best maps possible; the constant evaluation of maps exposes their flaws and promotes corrections; new surveys are incorporated to keep the maps current. Over time, maps inexorably get better. This progressiveness is evident whenever one compares two maps of the same area, at the same level of resolution, but from different eras. This is the kind of comparison commonly found in the historical introductions to modern texts and atlases (Edney 1993, 56). The older map is invariably less geometrically accurate and less detailed than the more recent map. Given the cumulative and seemingly undeniable increase over time in both the quantity and the quality of geographical data, cartography's history is necessarily progressive.

Faced with two maps of the same area, or even from the same printing surface, one with mistakes not found in the other, map historians have assumed that the map with mistakes must be the earlier. For example, two maps of New England were printed from woodblocks and tipped into different editions of William Hubbard's account of King Philip's War (1675–76) that were published in the same year (1677), the first in Boston and the second in London. The profligate practices of nineteenth-century antiquarian dealers so obscured the original relationship of the variant maps to the two books that by the 1880s historians could no longer determine which of the two maps was the first to have been made in British North America (Deane 1887, 14; Adams 1939, 26–27). But one variant has multiple errors, all of which are correct on the other. The two are therefore commonly known by their respective labels for the hills shown in northern New England: "Wine Hills" (incorrect) versus "White Hills" (correct) (see fig. 2.7).

Initially, historians took it for granted that the Wine Hills variant was the original, cut in Boston, and that its mistakes were fixed in the White Hills variant, cut in London (Deane 1887; Green 1905, 18–19; Fite and Freeman 1926, 164–66). But after a careful bibliographical analysis, Randolph Adams (1939, 30) concluded that the White Hills variant had in fact accompanied the Boston edition; he supposed that the errors were introduced into the Wine Hills variant by a careless London craftsman with no knowledge of the geography of New England. Adams did not presume to know the nature of past mapping, but used carefully acquired bibliographic facts to reconstruct the particular method by which the White Hills map was reconfigured as the Wine Hills map. But his analysis flew in the face of the common-sense understanding that correct maps must replace incorrect ones. To resolve this paradox, Richard Holman (1960; 1970, 42–43) proposed a contorted sequence of events. Claiming quite erroneously that no one in London still printed maps from wood, Holman supposed that the erroneous Wine Hills variant was cut first, in Boston, to be shipped to London for printing there, although he had no evidence that the book's creators planned or implemented such an undertaking; while the first woodblock was still in transit, the second, corrected block was also cut in Boston and then printed before the first could be printed in London. Holman thus argued that the Wine Hills map was the first cut in North America, the White Hills map the first printed. Finally, David Woodward (1967) undertook a careful analysis of contemporary practices for making woodblock copies of maps and of the use of printer's type for the title block of each, to prove that the Wine Hills map had indeed been cut and printed in London as a corrupted derivative of the White Hills map (Edney and Cimburek 2004, 320–22).

Map historians have also presumed, because more spatial information is apparently acquired over time, that of any two maps of the same region, the one showing more information and place names will necessarily be the later, more up-to-date one. But, as R. A. Skelton (1965, 12) observed with respect to toponyms,

> the density of nomenclature is an unreliable guide to the relationship—chronological or otherwise—between two maps. Of two undated maps, it cannot be assumed that the map with the greater number of names is the later; and if the later maps in a dated series show more names, we are not justified in postulating new sources. Variations, both in nomenclature and in design, as between one map and another may—and often do—reflect only differences in rendering a common source or in selection from a common stock, and not differences of content due to the accretion of new information.

Overall, the altruistic and automatic extirpation of error is neither a cartographic necessity nor a cartographic universal.

A persistent implication of these convictions of progress and currency is that maps that seem to make a significant advance on early images of the same region are accorded an unwarranted level of technical sophistication. One particular, recurring problem has been the claim that explorers or surveyors used triangulation to provide a rigid geometrical framework for their maps, when they did not in fact do so. In part, this habit stems from a common misunderstanding that any surveying process involving the construction of triangles, including graphic construction on a plane table or the trigonometrical solution of a simple system of intersecting lines, is "triangulation." This misunderstanding seems to explain the comment by Kim Sloan (2007, 104) that Thomas Harriot had used "the triangulation method, carried out on board ship" along the coast of Virginia and the Carolinas in the 1580s, when he had undertaken a traverse along the coast and perhaps multiple angles to the same locations onshore to fix their location by intersection. But at times scholars have attributed actual triangulation—the observation of one or more baselines and then a series of triangles, to permit the trigonometrical calculation of the position of each vertex (see fig. 5.16)—to works when it is quite unwarranted, from Christopher Saxton's surveys of the English counties in the 1570s (see fig. 5.6), to James Cook's charting of the coasts of Newfoundland and New Zealand (Heiser 2003, 41), to eighteenth-century military surveys (Kaplan 2018, 35).

Finally, the convictions of currency and progress stemming from the disciplinary preconception have influenced map history by directing historians to seek out and privilege the first inclusion of specific features on maps. This persistent concern has often been characterized as an aspect of antiquarian puffery, a means used by dealers to promote and increase the sale price of their stock. But it is also an effective marker of progress and reaffirms all the presumptions the ideal sustains about the circulation of manuscript and print materials.

Publicity

The normative abstraction of the concept of "the map" within the ideal extends to its physical materiality. The materialist preconception holds that maps exist as things. And when maps are *made*, or at least when the grand catch-all category of general-purpose maps are made, the presumption is that they are made to be distributed, and distributed widely. This is the preconception of publicity. It is telling that when the academic cartographer J. S. Keates (1982, 117–20) proposed his own model of cartographic communication, he began from the position that the distribution surface for maps is frictionless, except for the pragmatic, institutional barriers that he enumerated.

The preconception of publicity rests on a thoroughly modern understanding of publication as broadcast and indiscriminate, in which the printed work goes out to inform the world. Coolie Verner (1985, 136) succinctly expressed the basic conviction:

> The early printed map was the principal instrument for the spread of new geographical information. . . . If the original explorer's map remained only in manuscript the diffusion process was much slower than if the original was printed and distributed.

These sentiments and various corollaries are readily apparent throughout the literature on map history. For example, a map that was unprinted could not have been "effective" in that it could not have reached a wide audience (Fite and Freeman 1926, 127), while a map that went through many states "must have had a large circulation" and contemporary impact (Lukens 1931, 435; echoed by Cumming 1980, 80). A map that was demonstrably accurate and useful but remained in manuscript poses a quandary: why was it not printed? Other fields of study, such as the history of science, have similarly privileged print, with the same problematic results. In just one example, Thomas Hankins identified several early forms of "diagrams" that he otherwise excluded from consideration as having little impact on later scientific practice because they were "more like maps" or had "remained in manuscript" (Hankins 1999, 52n3).

Fundamentally, manuscript maps are presumed to be merely the precursors of printed maps (Akerman 2000, 27). Only through print are maps capable of attaining their proper status. This is more than evident in comments that require maps to be printed. For example:

> One early figure, a Scotsman named John Ogilby, created the first large-scale [*sic*] map of Carolina in 1673, based on a manuscript map provided to him by the wealthy proprietors of the colony. (Blanding 2014, 85)

What distinguished the "first" map by Ogilby from the map(s) on which it was based was that Ogilby was a London-based publisher who had the map printed for commercial sale (see Cumming 1998, 17–18, nos. 65 and 70, regarding the map). Without the innate privileging of print, this statement makes no sense. Furthermore, assumptions of the natural expression of manuscript maps in print also foster the conviction that the printed map is a direct and unmediated copy of the original manuscript drawing; time and again, map historians have suggested that the act of engraving (in monochrome) simply replicates the manuscript original (often in color).

One might argue that the historical bias in favor of printed maps stems from the style of map history promoted by dealers and collectors of antiquarian maps and books—what Michael Blakemore and Brian Harley (1980, 23–26) called the "old is beautiful" bias of map history. Printed maps make up the vast majority of maps in the antiquarian marketplace, and they appear to have constrained the disciplinary vision of map historians. Such constraint has occurred through the preparation of cartobibliographies of printed maps, whether by region or by map seller. Not only is the inclusion of manuscript maps in cartobibliographies uncommon, but the preparation of cartobibliographies seems to involve a limited vision that emphasizes printed maps. I think in particular of Barbara McCorkle's (2001) bibliography of printed maps of New England. The original planning conference had decided to reserve manuscript maps for a separate study, which made practical sense, given the effort that would be involved in locating them. Yet in executing the bibliography, McCorkle set up a divide between archives (supposedly manuscript only) and libraries (print only), and so failed to locate several maps in the former that were indeed printed. McCorkle's working assumption that printed maps are to be found today only in libraries manifests the presumption that printing necessarily entails widespread distribution, although the printed maps she missed were actually intended for very limited circulation (Edney 2007c).

The development of map printing in Europe presents a fairly sharp divide. Of course, map historians have paid attention to the production of manuscript maps well after the first printed maps of the 1470s, but only as they pertain to particular areas of research. It is accepted, for example, that sea charting continued as a manuscript endeavor through the sixteenth century, or in England for most of the seventeenth century. But once a particular genre of mapping gives way to print, the assumption is that maps are henceforth always properly printed. The falsity of this position is soon evident when we consider that, although the Dutch began printing marine maps in the sixteenth century (Schilder and Van Egmond 2007; Schilder 2017), they continued also to produce marine maps in manuscript well into the eighteenth century, as evidenced by one Dutch firm otherwise remembered for its printed charts (Guiso and Muratore 1992; De Vries et al. 2005; see Bom HGz 1962 [1885]). Recapitulationism rears its ugly head, again, as the great transition from manuscript to print is apparently recapitulated in the practice of drafting the individual map in order to print it.

The historiographic fixation on the schematic, tripartite *mappamundi* printed in Günther Zainer's 1472 edition of the *Etymologiae*, a seventh-century CE encyclopedia by Isidore of Seville, as constituting "the first printed map" is a function of Western exceptionalism (fig. 3.7). After all, maps were reproduced as rubbings taken from stone steles in Tang dynasty China (618–900 CE) and by printing in ink from woodblocks in the twelfth century (Woodward 2007c, 591–92, citing especially Yee 1994, 46–50). Moreover, both the printing of maps in Europe, as emblematized by the 1472 Isidorean map, and the spread of map printing to Europe's colonies play into the misguided arguments propounded after 1950 that modern Western culture and rationality are a function of the development of printing with moveable type (see Warner 1990, 5–9; Edney and Cimburek 2004, 317–20).

The argument has thus developed that the necessary element in the formation of the endeavor of cartography was neither the appropriation of Ptolemaic ideas of space and projection nor Leon Battista Alberti's system of linear perspective—as commonly asserted—but the development of printing:

> These ventures are significant for what they reveal about the mapping impulse of the quattrocentro: Because they languished in obscure manuscripts, their subsequent influence was slight. (Biggs 1999, 379; see also Skelton 1972, 12–13)

What is unquestioned in such arguments, and by the publicity preconception as a whole, is the size of intellectual networks: printing's assumed efficacy is to reach out to and to shape Western culture in the broadest way, which in turn assumes that there is such a meaningful category of analysis as "Western culture."

93

FIGURE 3.7. The schematic T-O map from Isidore of Seville (1472, fol. 177v). Woodcut, 6.5 cm diameter. Courtesy of the Newberry Library, Chicago (folio Inc. 1532).

The supposed widespread, outward distribution of printed maps is countered within this preconception by an inward flow of manuscript maps to centers of map production and publication. Just as water must flow downhill, so manuscript maps seem to move naturally and inevitably from the cartographic periphery—colonies, provinces, and "the field" generally—to the metropolitan, cartographic core, to be printed and then returned *in multiplied numbers* to the periphery. The inward flow is enshrined in the common phrase of "surveying and mapping," which posits separate and individualistic acts of observation and measurement in the field, recorded in manuscript journals and maps, and of cartographic production in metropolitan offices, which process the field work so that it can be added to Western culture's great printed storehouse of human knowledge. The inward flow also underpins Bruno Latour's (1987, 215–57) concept of the gathering of stabilized documents from the field—stabilized in the sense of being inscribed and immutable as they change hands—within "centers of calculation." Still, by foregrounding the issue of the circulation of manuscripts, Latour did direct scholarly attention to the processes and patterns of the circulation of knowledge, even as historians of the book were beginning to problematize the circulation of printed works (Darnton 1982).

Morality

The abstraction of the map and its resemblance to the world—a world that has been disciplined by the manner of its production from observation and measurement—together impart to maps a certainty and truthfulness lacking in other texts. If maps are to be accurate and are to be used for navigation, then it is a moral duty for cartographers to ensure that they are indeed correct, comprehensive, and trustworthy. Cartography is thus inherently moral and cartographers are innately altruistic. As a university newspaper recently stated, although with the goal of denying this preconception:

> We all look at the world subjectively, from our own point of view. Mapmakers, we assume, do the opposite: They portray the world objectively, as it actually is. (Singer 2016)

One commentator adopted a more figurative understanding of morality and mapping: "The map is 'moral,' for it implies a denigration and valorisation of the locations within its frame; the map puts forward an argument for what should be desired and what should be devalued, guiding behaviours and beliefs" (Smith 2014, 801).

The aviator Beryl Markham recognized the innate morality of cartography in her 1942 memoir, *West with the Night*, when she compared maps with the written word in a passage that deserves to be quoted at length:

> A map in the hands of a pilot is a testimony of a man's faith in other men; it is a symbol of confidence and trust. It is not like a printed page that bears mere words, ambiguous and artful, and whose most believing reader—even whose author, perhaps—must allow in his mind a recess for doubt.
>
> A map says to you, "Read me carefully, follow me closely, doubt me not." It says, "I am the earth in the palm of your hand. Without me, you are alone and lost."
>
> And indeed you are. Were all the maps in this world destroyed and vanished under the direction of some malevolent hand, each man would be blind again, each city be made a stranger to the next, each landmark become a meaningless signpost pointing to nothing. . . .
>
> Here is your map. Unfold it, follow it, then throw it away, if you will. It is only paper. It is only paper and ink, but if you think a little, if you pause a moment, you will see that these two things have seldom joined to make a document so modest and yet so full with histories of hope or sagas of conquest. (Markham 1942, 245–46; see Hansen 2013)

That these comments celebrate the idealized map—or at least its instantiation in the early aeronautical maps of modern Europe—is made clear by Markham's complaints elsewhere in her memoir about the poor quality of the only maps available for sub-Saharan Africa, all of coarse resolution:

> Moreover, it seemed that the printers of African maps had a slightly malicious habit of including, in large letters, the names of towns, junctions, and villages which, while most of them did exist in fact, as a group of thatched huts may exist or a water hole, they were usually so inconsequential as completely to escape discovery from the cockpit. Beyond this, it was even more disconcerting to examine your charts before a proposed flight only to find that in many cases the bulk of the terrain over which you had to fly was bluntly marked: "UNSURVEYED." It was as if the mapmakers had said, "We are aware that between this spot and that one, there are several hundred thousands of acres, but until you make a forced landing there, we won't know whether it's mud, desert, or jungle—and the chances are we won't know then!" (Markham 1942, 35)

Another early aviator, Antoine de Saint-Exupéry (1931), similarly disparaged the coarser resolution maps he used for their inability to depict the landscape in sufficiently fine resolution.

Pragmatic issues aside, the idealized map seems implicitly truthful because the idealized cartographer is, at root, an altruist. To be used properly, without sending ships onto rocks or travelers astray, maps must be correct. Cartographers' moral duty is to strive to make the best maps possible. Cartography disciplines spatial data, improving them so that they fit a norm and uplifting them until they are correct and truthful. Any other process denies the propriety of the scientific search for truth. A "bad" map is not only inaccurate, it is also immoral.

A variety of cartographic *bêtes noires* have been held up as being bad and immoral because of the ways in which they willfully distort the image of the world and actively misdirect their users. There is, of course, the perversion of cartography to purely commercial ends, to make shoddy or even unnecessary maps and fob them off on an unsuspecting public, simply to make a profit. For example, a later nineteenth-century exposé of the "county history, atlas and map scheme" revealed that it had "taken millions of dollars from farming communities" for no useful return (Harrington 1879, unpaginated preface). An integral element of commercial perversions is the unethical stealing of another cartographer's work and infringement of their copyright. Harley (1991, 9–10) objected to what he saw as the naive equivalency drawn by academic cartographers, when they finally began to discuss issues of ethics, between "ethics" and "respect for copyright" (see McHaffie, Andrews, and Dobson 1990). I remember, at that particular moment, a certain concern for the practice of makers of road maps to insert "copyright

96

CHAPTER 3

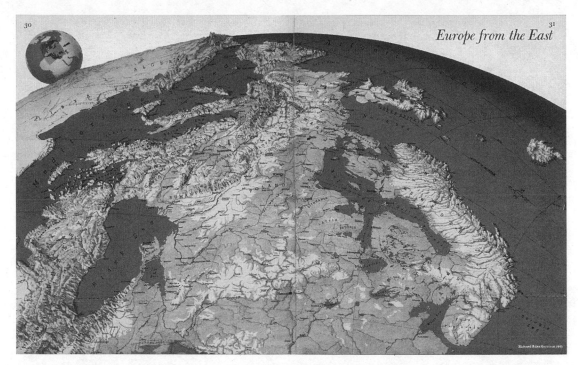

FIGURE 3.8. One of Richard Edes Harrison's global views of war: *Europe from the East* (Harrison 1944, 30-31). Color lithograph, 35 × 55.5 cm (paper).

hooks" into their maps as a guard against the theft of their intellectual property (see Rice 2015), but in retrospect I don't know whether the concern was that cartographers admitted that they stole others' work or that falsehoods, however inconsequential, had been consciously introduced into maps. The exemplary "bad map" is, of course, the political propaganda map, whose geographical content has been manipulated by its makers in accordance with a political agenda; the pursuit of geographical truth is unavoidably corrupted (Tyner 2015a). Much of academic cartography has been undertaken in rejection of propaganda mapping and especially that perpetrated by the Nazis before and during World War II (Pickles 1992; see fig. 4.13).

Opprobrium has also been heaped on maps not made to certain standards, regardless of their effectiveness. For example, the maps created by Richard Edes Harrison to show the different theaters of World War II, each showing the curved earth as if seen from high altitude (fig. 3.8), ran directly counter to formal cartographic practices and were criticized by academics even though they proved highly popular with the U.S. public (Schulten 1998; see also Cosgrove and Della Dora 2005; Barney 2015, 36–58; Rankin 2016, 70–80). In the 1980s, when many U.N. agencies and nongovernmental organizations interested in global affairs heeded Arno Peters's (1983) claims that his world map corrected the overt biases and Eurocentrism of standard world maps made on the Mercator projection, academics

97

once again criticized the map as much for Peters's overt political agenda as for the map's technical deficiencies (Crampton 1994; Kretschmer 2015b).

The active rejection of bad maps has generally been accompanied by calls for propriety and correct practices by map makers. In this respect, the ideal of cartography demands the disciplining not only of maps and spatial data but also of its own practitioners and its own techniques. The disciplining of maps is as much an assessment of and passing of judgment upon the morality of the maps' makers as it is on the quality of the maps themselves. Scholars have often, and explicitly, scolded past map makers for the failings evident in their maps.

Only properly disciplined practitioners can be called "cartographers." This sentiment runs deep in map history. For example, when, in the 1930s, Harvard University's Erwin Raisz (1938, 45–46) reconfigured the basic, progressive narrative of the history of cartography, he did so through a telling comparison between the seventeenth-century "Dutch school," otherwise unlabeled, who he argued were strictly commercial and motivated solely by profit, and the scientifically minded geographers of the eighteenth-century "French school of cartography," who he thought were motivated solely by an altruistic concern for accuracy and rectitude. David Bosse (1995, 163) expressed this aspect of the moral preconception when he observed that Osgood Carleton, a mathematical practitioner active in Boston during the 1790s and 1800s (see fig. 2.4), was "a commercial mapmaker not associated with the printing or engraving trades" so that he "stands apart as one of the first professional American cartographers." Apparently by avoiding an overt taint of commerciality, so much so that he went bankrupt, Carleton appears in hindsight as a properly scientific map maker.

More generally, the disciplining of cartographers has been the *raison d'être* of academic cartography. Max Eckert (1907, 1908) revealed that the early drive to codify and regularize mapping practices was driven by a need to impart the "logic" of cartography to map makers, to ensure that they adhered to the correct standards. After World War I and the division of Prussia, and working in concert with his colleagues, Eckert developed his major text, *Die Kartenwissenschaft* (1921–25), in large part to assert the need to apply cartographic logic in a proper manner so as to show the German state and *Volk* in their proper, truthful extent (Herb 1997, 34–48). Two decades later, J. K. Wright (1942) reiterated Eckert's (1908) arguments in the face of Nazi propaganda mapping to urge U.S. map makers to adhere to a moral and scientific code, to discipline themselves to avoid subjective idiosyncrasy and political bias, and so generate truth. Even so, the apparent need to discipline cartographers existed before academic cartography developed, and academic cartography has not cornered the market, as it were.

There was, and remains, a diffuse expectation that some social institution holds authority over cartography. And that institution can only be cartography itself. The supposedly innate morality of cartography—innate because, if not moral, then mapping cannot constitute cartography—transforms the field from a technology of observation and measurement that fixes the spatial location of terrestrial and celestial features into an ineluctably social phenomenon. Only within such *disciplinary* terms is it meaningful to construe the history of cartography as comprising a series of phases in which "cartographic leadership" shifted from one European country to another, or to suggest that the eighteenth century was the era "when France was king of cartography." One might ascribe such unfortunately pervasive and ahistorical narratives to the nationalistic tendencies of historians of cartography, as they have inevitably focused on their own nation's past cartographic triumphs (see Blakemore and Harley 1980, 26–32), but the internal logic of these narratives requires the more fundamental conviction that cartography has a social structure that binds all cartographers within a self-regulatory whole. The moral preconception is thus a major element in the overall commitment to the singularity of cartography.

A Singular and Universal Endeavor

The manner in which the ideal of cartography took root and intensified over the course of the nineteenth and early twentieth centuries has given rise to persistent and mutually reinforcing, though often contradictory, preconceptions about the nature of maps and cartography. Satirists might have highlighted some of the tensions between these preconceptions, in particular between ontology and pictorialness, but their humor prevented any serious criticisms from developing, at least until the 1980s and 1990s. For most of the last two centuries, however, the ideal's complex web of interlocking and intersupporting convictions about the nature of maps and map making has been so thoroughly naturalized that it has been barely visible. The overall belief is that cartography is truly a singular endeavor pursued by *all* map makers: one world/archive, one technology of observation, one perfectible goal, one moral purpose, one institution. Attention shifts ineluctably to "the map."

Cartography's underpinning morality and its quest for true images of the world's features appear to be a historical, cross-cultural constant. The *why* of cartography thus appears constant: to correctly express cognitive maps in material form so as to enhance spatial action by individuals. Variability is understood to exist in the *what* of cartography, but only to exclude the apparently nonscalar ontologies of indigenous cultures that give rise to maplike oddities. In this respect, real maps uniformly manifest a scalar ontology. The only variability actually permitted within the ideal lies in the *how* of cartography: how technologies are deployed in order to picture the world in an ever less depleted and more homologous manner; how cartography disciplines the world, maps, and map makers. And that methodology is subject to a teleology of technological and informational progress, as the quality and quantity of spatial information inexorably improved over the centuries.

99

This teleology overcomes the paradox between the preconceptions of universality and individuality. If there is a difference in terms of how individuals "think about space" between, on the one hand, Westerners and more particularly Western men and, on the other, non-Westerners and Western women, then cartography cannot be universal. To counter this, the teleology has been further construed to be cognitive as well as technical. Individual mental capacity for understanding the spatial extent of the world has itself supposedly changed and improved in lockstep with cartographic technology, but only within those portions of humanity that have actively pursued a scientific cartography. It is thus possible to exclude the "prehistoric mind," the "primitive mind," and perhaps even the "female mind" from the ambit of real cartography. The core essence of cartography is, therefore, its "scientific" quest to observe, to measure, and to know the world.

Preconceptions of the singularity of cartography underpin the long-standing manner in which all kinds of surveying are lumped together, whether for place, property, boundaries, chorography, or territory, and more especially in the equally long-standing confusion of geographical and marine mapping. Consider a statement in a recent essay by a historian of mathematics on the possible origins in the classical world of sea charts otherwise known only from medieval archetypes:

> Although there are still a few proponents of an ancient origin, the charts themselves do not contain even a hint of such an origin. They show no similarity to the maps of Claudius Ptolemy. (Nicolai 2015, 521)

This statement presumes an equivalency between marine mapping and geographical mapping that is valid only within the presumed singularity of cartography. Throughout this particular essay, the author's focus on the map image, and more especially on the image of a particular, isolated feature (coastlines), permits the conflation of otherwise distinct mapping processes. In another example, Patrick Chura (2015) studied two instances of literary place mapping from two works published in 1854, but he began by improperly and anachronistically comparing one of the two—Henry David Thoreau's own surveyed plan of Walden Pond—with a 1979 map of the whole world. The belief that such comparisons are permissible stems from the conviction that cartography possesses basic practices—of observing, measuring, and recording physical reality—that are all part of the same, singular historical process. That is, there is one and only one endeavor of cartography.

Some exception is made within the ideal for what Anne Godlewska (1997, 31) called the "dialects" of "cartographic language." An institutional divide has often been drawn between fine-resolution surveying by engineers, on the one hand, and coarse-resolution map design and data visualization, on the other. (One of the determinants of these distinct dialects is the variability of map scale: consistent on fine-resolution maps and variable on coarse-resolution maps, but nonetheless always subject to the idealized conception of map scale itself; see chapter 5.) Yet the existence of such dialects has never been allowed to undermine the overall coherence of cartography and the apparent consistency of cartographic language. Even as Arthur Robinson (1982, 12, 24), for example, maintained that these two cartographic dialects had diverged in the nineteenth century, he also maintained that it should be possible to draw a single graph of all "cartographic activity" over time.

The presumption of the singularity of cartography undermined my own study of the British mapping of India before 1843, as I unwittingly conflated the geographical mapping of South Asia ("how India came to be painted red on the map") with the detailed territorial and geodetic mapping activities whose history I traced (Edney 1997). One reviewer correctly pointed out that I had misapplied the idea of "geographical construction" across multiple scales (Ludden 1998); although this flaw has no effect on my empirical study of how the British mapped India in detail, and what they thought they were doing as they did so (see Edney 2009), it does cast doubt on how I dealt with the coarser resolution mapping of India. More recently, Daniel Foliard (2017, especially 19) adopted a similar and improperly inclusive understanding of mapping—"mapmaking and mapping are two interdependent elements of *a single process of arranging the world*" (emphasis added)—in his otherwise excellent detailed analysis of British colonial mapping in the Middle East.

The singularity of cartography finds expression in the normative map, the "singular, universal record of geographic fact that includes everything worthy of attention, and nothing more" (Rankin 2016, 3). All maps must, therefore, have the same essence regardless of any variation in their scale, content, or artifactual form. Whether they are survey plans, sea charts, geographical maps, triangulation diagrams, or analytic maps, all cartographic images are taken to be fundamentally alike and to share a common cartographic language. This accounts for the emphasis on creating comprehensive and appropriate definitions of "the map," especially as the first stage in academic attempts to ring fence the field. Inevitably, definitions construe the map either in terms of form or presumed function, but are always in accord with the ideal's preconceptions (especially ontological, pictorial, and functional).

The ultimate conviction that cartography is a single endeavor has long been sustained by the histories that have been told of its origins and development. These histories developed in the nineteenth century, with their origins in the 1830s soon after the neologism "cartography" took hold in the 1820s. The "history of cartography" flourished as a field of study after 1860, by which time the ideal of cartography already had a strong hold in popular and academic culture. From the start, the history of cartography was told as a story of the triumph of Western rationality over non-Western irrationality, of the technological and scientific achievements specifically of the West. The continual repetition of the ideal's preconceptions within both major narratives and specific interpretations upheld and validated them.

But the preconceptions are *all* wrong. Some of the preconceptions have stemmed from particular practices and so might be valid for specific mapping processes within certain threads of spatial discourse; if so, they need to be carefully qualified. But the preconceptions are all historically and conceptually inadequate when applied across the board. Each and every one turns the intellectual gaze away from the myriad ways in which people produce and consume spatial knowledge, whether in the past or present, and fixes it instead on a shiny, multifaceted, impossible gemstone of an ideal. It is tempting to make some comment that what seems to be diamond is only synthetic cubic zirconia, but to do so would imply that there is still something *there*, no matter how cheap and nasty the form. Rather, the preconceptions are all particular aspects of an illusion—a simulacrum—about the nature and history of maps that does not exist outside of our collective discourses.

4

The Ideal of Cartography Emerges

Cartography is a *modern* myth

The fundamental proof that cartography is not a universal and transcultural endeavor is that the ideal has a history. It developed only after about 1800. Before then, there was no conception of a universal endeavor of map making; the words "cartography" and "cartographer" did not exist. Idealizations about the nature of maps and mapping crystallized over the course of the nineteenth and early twentieth centuries. The ideal of cartography grew ever more elaborate and resilient as it added further layers of belief and increasingly permeated modern culture.

The propagation of idealizations through modern society can be traced in the definitions of "map" adopted by dictionaries and encyclopedias over the course of the nineteenth century. All addressed the normative map, but they did so with several styles and formulas. Each style manifested new social developments that increased and extended popular map consumption and thereby propagated the ideal of cartography through modern culture. Beginning in the first half of the nineteenth century, the rise of compulsory mass primary education produced "popular" definitions; the metaphorical use of the normative map in mathematics and philosophy, starting in the late nineteenth century, produced "philosophical" definitions; finally, the twentieth-century institutionalization of academic cartography gave rise to "professional" definitions (Andrews 1996, 4–7).

This chapter explores these and several other factors that contributed to the construction by modern Western society of cartography as an apparently coherent, moral, and universal science of observation and measurement. The discussion of the contributing factors is necessarily brief; each probably deserves its own monograph. I have grouped the factors within several categories arranged in approximately chronological order. The result is something of a narrative of the formation, intensification, and elaboration of the ideal's preconceptions over the course of the nineteenth century.

Several of the ideal's contributing factors were present in some form before 1800. For example, Gilles Palsky (1999, ¶5) identified an early conceptualization of an enlarged and all-encompassing cartographic archive, albeit not at 1:1, in Jean François's (1652, 349–50) vision of the "largest and most analogous" map of France.* Even so, beyond the general principle of perfecting the map, there is no direct connection between this idiosyncratic comment and the formation of the ideal some two hundred years later. Rather, François stands as the immediate precursor of the core argument of early modern mathematical geography: that the geographical map's cosmographical network of meridians and parallels provided a coherent framework for constructing the geographical archive at coarse to moderate resolutions. Attempts shortly after 1800 to incorporate moderate- to fine-resolution surveys into the geographical framework would expose the framework's limited capacity, with the result that the concept of the archive would be significantly reconfigured and idealized over the nineteenth century, in the process shifting from an aspiration specifically for geographers to an aspiration for all cartographers.

The difference is indicated in the shifting understanding of map language. In the Enlightenment, the common language of geographical maps comprised the cosmographical coordinates of latitude and longitude (Edney 1994a; Andrews 1996, 3–4). The new idealization of cartographic language in the nineteenth century was grounded in the apparently pictorial nature of all maps. The idea of cartographic language was further complicated by the use of map scale as a guarantor of each map's proportionality to the world and of its pictorialness, which has led some commentators to consider projective coordinate systems as constituting an ontological basis for cartographic language.

Map historians have also interpreted certain early modern observations—that the power and attraction of maps lay in the way in which maps present the world "out there" to the reader "in here"—as indicating an early modern origin for cartography. For example, as Georg Braun opined in the preface to volume 3 (1581) of his huge collection of city views and maps, the *Civitates orbis terrarum*:

> What could be more pleasant than, in one's own home far from all danger, to gaze in these books at the universal form of the earth . . . adorned with the splendour of cities and fortresses and, by looking at the pictures and reading the texts

---

*"pour ce que ces représentations ici sont, et plus grandes et plus semblables: la grandeur donne une plus grande distinction aux parties et la plus grande similitude une bien plus grande facilité à concevoir."

accompanying them, to acquire knowledge which could scarcely be had but by long and difficult journeys? (Quoted by Skelton 1966, vii)

Robert Burton, in his *The Anatomy of Melancholy* (1621), again suggested that the function of maps to make the world visible and legible applied to maps at all resolutions:

> Methinks it would please any man to look upon a geographical map . . . chorographical, topographical delineations, to behold, as it were, all the remote provinces, towns, cities of the world, and never to go forth of the limits of his study. (Quoted by Skelton 1966, vii)

Several scholars have made the basic mistake of confusing this widespread early modern appreciation of the visual impression that maps instill in the viewer—the twin beliefs that one can see the world "at a glance" and can do so without the effort and labor of actual travel—with the modern conviction that maps are *produced* by the same potent vision. One result of this mistake is the otherwise unwarranted assertion that modern map making *is* "the achievement of seventeenth-century engineering culture" (Weibel 2014, 450–54; also Harvey 1989, 240–59; Kaplan 2018, 34–35).

Christopher Packe's innovative and unique geomorphological map of eastern Kent, published in 1743 (fig. 4.1), does seem to prefigure the modern argument that all maps provide "God's-eye views" or "views from above." Packe explained that his self-consciously titled "philosophico-chorographical chart" constituted a "landscape" or "portrait" constructed with a perspectival "Eye" that was "here, every where present by turns" and that its character was manifestly different from that of contemporary geographical maps structured by latitude and longitude (Packe 1737, 4 and 15; Packe 1743, 3–4; see Edney 2012b, 433; Charlesworth 2019). However, Packe's particular innovation of representing landscape by means of watercourses would not be adopted by any other map maker; only in 1799 did Johann Georg Lehmann advance a complementary approach that used hachures to comprehensively delineate the shape of the land. Packe's rhetoric was not echoed by commentators before the 1830s, at the earliest. Despite his insight, Packe's work cannot be said to have influenced later developments.

In contrast to Packe's proto-topographical vision, early modern geographical maps were subject to a quite different visual regime, one that was understood to support reasoned understanding. When early modern commentators referred to geography as the eye of history, "eye" was not only literal but also figurative; history's other eye was chronology, in which lists of dates and events permitted historians to structure their narratives temporally (Mayhew 2003; Grafton 2007, especially 28; Edney 2019c). That is to say, pre-1800 sentiments about maps as tools

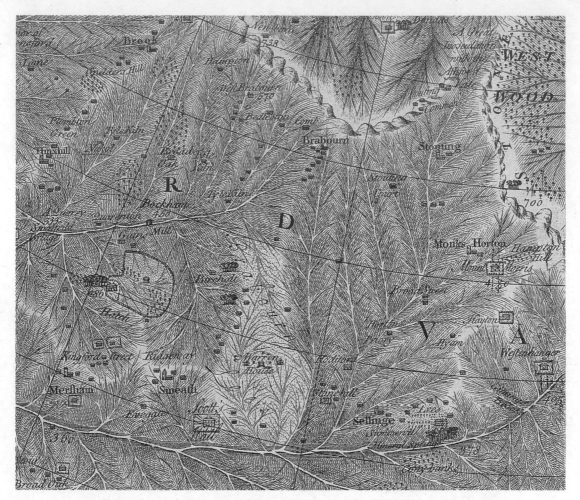

FIGURE 4.1. Detail of Christopher Packe, *A New Philosophico-Chorographical Chart of East-Kent* (London, 1743). Drawing an analogy between water flow (the hydrologic cycle) and blood circulation, Packe mapped the rivulets and rivers of eastern Kent as if they were veins and arteries; that is, the light areas represent ridges. Copper engraving; detail is approximately 20 × 21 cm. Courtesy of the Yale Center for British Art, Paul Mellon Collection, Yale University, New Haven (Rare Books and Manuscripts, G5753 K4 P3 1743+ Oversize).

of visualization varied according to the mode of mapping. But after 1800, the expanding institutional capacities of Western polities and new technologies together underpinned the formation of an idealized and universal visual regime that was deemed to be characteristic of *all* maps.

Systematic Mapping

Territorial knowledge was a key element in the increasing centralization of states in Europe in the eighteenth century. The different characters of those states gave rise to markedly different institutional characters for their territorial surveys: failed

FIGURE 4.2. Paris and its environs, from César-François Cassini de Thury's *Carte générale et particulière de la France* (Paris, 1762), being both the first sheet (*nombre* 1) in the numbered sequence of 182 sheets covering France, and the first to be published (*feuille* 1). Produced to a scale of 1 *ligne* (2.25 mm) to 100 *toises* (194.904 m), each sheet covered an area of 25,000 by 40,000 *toises* (48.725 x 77.96 km) and measured 59 × 90 cm. This sheet was originally published in 1757. Courtesy of the David Rumsey Collection (5694019); www.davidrumsey.com.

efforts in Spain and Portugal; in Britain, private, commercial surveys of England's counties but military surveys of large parts of Scotland and Ireland; strictly military surveys of each province belonging to the Austrian Habsburgs; the quasi-state survey of all of France, albeit with remoter provinces being surveyed under local authority; and so on. Of these various surveys, that for César-François Cassini de Thury's *Carte générale et particulière de la France* set the technical model for extensive territorial surveys carried on after 1790. It featured a coherent and comprehensive triangulation to serve as a geometrical foundation, permitting a detailed topographical survey to be carried over geographical regions, together with a regular division of the country, once projected onto a plane, into standardized sheets (fig. 4.2). The precise implications of this geometrical arrangement are introduced below, and discussed in more detail in chapter 5. The salient point for now is that the *Carte générale et particulière* proved the feasibility of a particular mechanism whereby centralized states might create many topographical maps that would fit neatly together to make a single map of an entire country. The proliferation of these systematic surveys and the eventual integration of their maps into many elements of governmental, military, and public life proved a major factor in the formation of the ideal.

MODERN SYSTEMATIC SURVEYS

ontology / pictorialness / observation / singularity and universality.   The kind of systematic survey undertaken for the *Carte générale et particulière de la France* would begin to be emulated by other European states in the second half of the eighteenth century (Kretschmer, Dörflinger, and Wawrik 1986; Edney and Pedley 2019). Some of the eighteenth-century attempts either failed to get off the ground—for example, the unimplemented proposal to map Portugal in 1788—or ran into financial or staffing problems, like Giovanni Antonio Rizzi Zannoni's survey of Padua (1773–80). But many were successful, if generally laborious and protracted, such as the survey of Denmark by the Videnskabernes Selskab in 1761–1805; the comte de Ferraris's survey of the Austrian Netherlands (1771–77); Rizzi Zannoni's survey of Naples (1780–1812); and Johann Gotttlieb Friedrich von Bohnenberger and Ignaz Ambros von Amman's survey of Swabia (1793–1828).

The Revolutionary and Napoleonic Wars prompted a more concerted effort by general staffs to remap Europe for planning and conducting modern warfare. French military engineers began numerous surveys of conquered territories in Italy and Germany, notably of Bavaria, the Rhineland, and the Papal States; many of these surveys would be continued and completed after 1815. Other Italian and German territories were mapped by other combatants: the Austrians mapped Venetia (1798–1805; Rossi 2007) and then began a thorough mapping of its extensive territories in the *Franziszeische Aufnahme* (1806–69); Prussian engineers mapped Saxony (1780–1825) and Westphalia (1796–1805). Cornelis Rudolphus Theodorus Krayenhoff undertook a detailed survey of the Batavian Republic in 1801–23. In the United Kingdom, the threat of French invasion led the Board of Ordnance, the institution responsible for military engineers and the artillery, to map the southeastern county of Kent; the survey began in 1790 and used William Roy's Greenwich–Paris triangulation as a foundation. This work gave rise to both a statewide geodetic triangulation and a series of surveys of counties, which by 1810 was starting to be called the "Ordnance Survey," but which did not develop the character of a coherent, statewide, territorial survey until 1824, when the entire establishment was transferred to Dublin to map Ireland for both cadastral and military purposes (Adams 1994; Oliver 2014, 66–102, especially 72–73).

The rapid adoption of systematic territorial mapping was marked by the coinage of a new term, in order to conceptualize the hierarchical practices involved: "triangulation." In 1802, a summary account in an official journal for French military engineers of those practices referred to the "general network composed of triangles of the first order" (Anon. 1802, 51). But shortly thereafter, a further article in the same journal began by referring back to the first account as providing "the theory of grand or primary triangulation" (Anon. 1802–3, 1).*

The separation of higher order from lower order triangulations was reinforced by the manner in which many of the uncertainties implicit in the prosecution of geodetic triangulations were solved in the early nineteenth century. Through the Enlightenment, geodesists had no consistent method to manage the errors that they knew permeated their work and instead relied on complex systems of redundant observations in order to double-check their work (Edney 2019a). The problem was solved independently by Adrien-Marie Legendre and Carl Friedrich Gauss, who each developed the method of least-squares analysis to model and control the distribution of error within large observational systems. Legendre applied the technique to his analysis of comet orbits, published in 1806. Gauss used it to adjust geodetic triangulations, beginning in 1799 with a commentary on the Survey for the Meter, between Dunkirk and Barcelona, and continuing with his own experimental triangulation of a hundred stations around Braunschweig in 1803–7. By 1810, the technique was being adopted by other astronomers and geodesists (Galle 1924, especially 9, 13–14; Gerardy 1977; Dutka 1995; Sheynin 1994, 1999, 2001; see also Porter 1986; Stigler 1986). Gauss further established the analytical tools needed for computing a high-level triangulation across the non-Euclidean surface of an ellipsoid. With these difficult issues resolved, everyday detailed mapping seemed to be grounded on a solid, certain foundation. With the primary triangulation complete and its errors properly distributed, secondary triangulations and detailed topographical surveys could be undertaken with simpler instruments and less skilled personnel, in the firm knowledge that their work would be rigorously controlled by the primary survey.

After 1815, state centralization and industrialization gave a further bureaucratic and statistical edge to the movement to systematically survey Europe. In addition to all the triangulation-based territorial surveys begun before 1815, and continued thereafter, new surveys proliferated after 1815: the French general staff's *Carte de l'état-major* (1818–66); the British survey of Ireland (1824–46); the Austro-Hungarian surveys of Tuscany (1817–27) and the Papal States (1841–43); Belgium's cadastral survey (1830–44); the Netherlands' new military survey (1830–55); and so on (Comstock 1876; Wheeler 1885; Stavenhagen 1904; Nadal and Urteaga 1990; Kretschmer, Dörflinger, and Wawrik 1986; Kain forthcoming). The practice extended to the United States after 1865. By the 1880s, it was commonly accepted among the chief civil and military officials of the industrializing world that there should be a standard, detailed map of each country, and statewide systematic

---

*Respectively: "canevas général se compose des triangles du premier ordre"; "théorie de la grande triangulation ou triangulation primaire."

surveys became organized as permanent agencies of government. Moreover, it seemed desirable to some in the federal U.S. government that the disparate mapping agencies that had developed for different kinds of territorial mapping—cadastral, topographical, and hydrographical—should be combined into a single agency that might truly unify all the surveys of mensuration, although this *desideratum* could not actually ever be implemented (Edney 1986).

Parallel to the systematic territorial surveys was the reorganization of coastal mapping and the establishment of modern hydrography. The rise of marine trade, and of marine warfare, meant that European states became increasingly interested after 1750 in the highly detailed mapping of their entire coasts and coastal waters. The original push was exemplified by the British surveys along the coasts of North America after 1763, published by J. F. W. Des Barres as the *Atlantic Neptune* (1774–82) (Hornsby 2011; Edelson 2017; Johnson 2017). By the end of the century, the French were using detailed coastal triangulations, based on the main national survey by the Académie des sciences, to control detailed surveys of swathes of land and water to either side of the coast (Chapuis 1999). The scheme was progressively adopted around Europe's coasts, and also in the fledgling United States, where the laborious trigonometrical work took decades to complete and caused no end of political problems for the U.S. Coast [and Geodetic] Survey, founded as the Survey of the Coast in 1807 (Wheeler 1885, 497–538; Edney 1986).

The new territorial surveys went hand in hand with a political reconception of the nature of the territory being mapped (Winichakul 1994; Elden 2013; Di Fiore 2017). Europe's *ancien régime* polities depended as much on nonterritorial structures of authority, whether patrimonial, feudal, or jurisdictional, as on direct territorial control. The result was the proliferation of spaces whose political and territorial status were often ambiguous. The reconfigurations of European polities prompted by Napoleon, including the collapse and final dissolution of the Holy Roman Empire in 1801–6, led the Congress of Vienna to develop a more strictly spatial understanding of territory that would be implemented throughout the nineteenth century. Instead of allocating manors, parishes, or communes to one side or another, nineteenth-century boundary commissions consistently sought to delimit precise lines in the landscape. Territorial surveys at once enabled and were enabled by this sea change in political relations. Early modern boundary settlements had featured, to some extent, precise delineations (Sahlins 1989)—and so stand as another instance in which modern mapping practices were prefigured by early modern ones—but post-1800 practices were applied in a far more systematic and philosophically coherent manner (e.g., Chester 2009).

The increase after 1790 in systematic, triangulation-based, statewide surveys promoted the use of projective geometries that seemed to combine plane and cosmographical geometries and to permit easy interchange between them. (Chapter 5 explains this convergence in detail.) All these new surveys fostered the conviction that mapping necessarily creates such interchangeable sets of coordinates. Indeed, it was these surveys, combining the topographical mapping of landscapes and the hydrographic mapping of oceans, at least close to shore, that gave rise to the new sense of unity of process that would be given the new label of "cartography."

In practice, as ever, it took a great deal of time and effort to implement this new technology, as several detailed, empirical studies indicate (especially, Skelton 1958; Scharfe 1972; Chapuis 1999; Godlewska 1999; Blais and Laboulais-Lesage 2006; Edney 2017d). The Enlightenment model of geographical mapping continued in use, together with its claim to be able to accommodate chorographical surveys. Such detailed surveys were just another source to be incorporated within the geographical framework of meridians and parallels, so all that was needed was to make the geographical maps sufficiently large to be able to show the fine details (Edney 2019b). Early in the nineteenth century, for example, Aaron Arrowsmith made his reputation with his huge, multisheet maps that he updated frequently with new surveys; by 1822, even his map of India in 9 elephant-sized folios proved insufficient, and he proposed a 102-sheet map series at 4 miles to 1 inch (fig. 4.3). However, even as he did so, it was becoming apparent that the system of fitting chorographical surveys to the geographical framework, using a few places whose latitude and longitude had been independently determined, was insufficiently rigorous. While the British committed in the 1830s to a program of triangulation-based surveys, those surveys took so long to complete that most of British India continued to be mapped piecemeal (Edney 1991; Edney 1997, especially 325–40). Yet, by the twentieth century, the continued commitment to the principles of systematic, triangulation-based surveys led to the normalization of the technology and the projective geometries it produced (Rankin 2016).

THE PUBLIC SPHERE AND DEMOCRACY

materiality / discipline / publicity / morality. A major innovation of the new wave of systematic territorial surveys initiated by the *Carte générale et particulière de la France* was that the resultant maps were published. They were not just printed as a means to distribute the maps to lower level officials or to eliminate the introduction of errors when copying the maps in manuscript, they were printed to be actively sold on the open market. The strictly military surveys of the eighteenth century, such as William Roy's survey of Scotland or the Austrian *Josephinische Landesaufnahme*, had remained in manuscript and were used by only a limited coterie of senior officials. But the later territorial surveys were never strictly military in their functions. Like the state-supported commercial venture of the *Carte générale et particulière*, the later surveys also provided information to civil branches of government and to the emergent public sphere. Indeed, the shift to

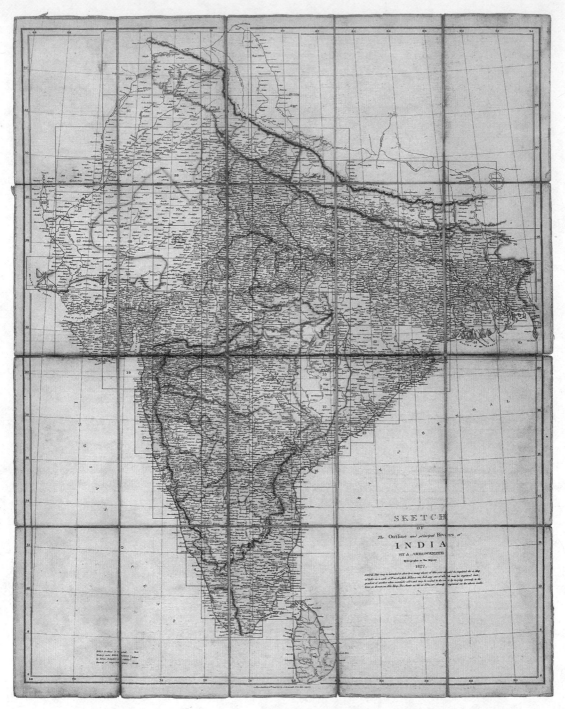

FIGURE 4.3. Aaron Arrowsmith, *Sketch of the Outline and Principal Rivers of India* (London, 1822), indicating the sheet lines for a proposed 102-sheet map of India. Arrowsmith published the first 16 sheets as the *Atlas of South India* (London, 1822). Size of the original, 82 × 64 cm. Courtesy of the David Rumsey Collection, David Rumsey Map Center, Stanford Libraries (G7651.C3 1822 .A7); purl.stanford.edu /wx898bn7470.

print in territorial mapping serves as a marker of the relative degree of openness of political debate and the growth of democratic institutions. In England, where public debate had begun as early as the 1640s, the Board of Ordnance built on a tradition of eighteenth-century commercial topographical mapping when it published its own first map of an English county in 1801. By contrast, the autocratic Austro-Hungarian empire kept the *Franziszeische Aufnahme* (1806–69) in manuscript and did not start printing its official topographical maps until the *Franzisco-Josephinische Landesaufnahme* (1869–87).

The causes and patterns of the development of public discourse are complex, and are inevitably the subject of an extensive literature that I cannot explore here (but see especially Habermas 1989 [1962], Warner 1990; Calhoun 1992; Broman 1998; Mah 2000; Melton 2001; Withers 2001). The salient point is that people of the "middling sort"—professionals, merchants, and lesser gentry—who were otherwise excluded from participation in the royal courts, including parliaments and courts of law, progressively claimed equality with social and political elites, and therefore claimed the right to comment on and to shape the formulation of state policy, cultural production, and economic activities. They did so by debating policy—everything from how to worship to how to write poetry—in several new spaces of sociability, from London coffeehouses and Parisian salons to new forms of print, notably newspapers and topical pamphlets. The goal was to construct and refine a "public opinion" that would influence and regulate the activities both of civil society, the arena of commodity exchange and labor, and of the state, the arena of authority and sovereignty. This public clamor produced political change, beginning with the American Revolution—whose participants identified the new phenomenon of print discourse as the key enabler of revolutionary thought and effort—and continuing through the modern era with the nineteenth-century development of the modern industrial democracies. In the process, the social scope of the public was forcefully widened from the social elites and the middling sort, who had long sought to exclude their own social inferiors from public discourse, to embrace the laboring classes and women as well.

The rhetorical heart of the public sphere was the disregard of its participants' social rank and status. Within the apparently open and newly universal arena of print—as Daniel Defoe had exclaimed, "Preaching of Sermons is speaking to a few of Mankind; printing of books is talking to the *whole world*" (1704, unpaginated preface; emphasis added)—the public could engage in the impartial evaluation of arguments according to their logic and evidence, not according to the status of those who made them. Printed political discourse was thus founded on a "principle of negativity" by which authors negated their own individuality in order to stand forth as generic, virtuous citizens of the public sphere (Warner 1990, 42–43).

Moreover, the public sphere required and demanded publicity, which is to say to make public knowledge that is otherwise restricted to either civil society or the state. How else could good policy be formulated, if information was not shared? The result was a steady rise in official publications, from the trade figures and tax receipts that underpinned the later eighteenth-century work of William Playfair in Britain or August Friedrich Wilhelm Crome in Germany to the eventual publication of proceedings of the British Parliament after 1803 (Brewer 1990, 221–49; Klein 2001; Nikolow 2001). The practice of official publication of useful information only intensified after 1800. Making the results of official territorial surveys of land and sea available was just like the publication of census results: they *had* to be published, for the good of the public.

While having clear origins in the long eighteenth century, the public demand for information took on new significance in the nineteenth century. The late eighteenth century saw the formation of new national museums and libraries to serve as the storehouses of knowledge, notably the institutions that are today known as the British Library, the Bibliothèque nationale de France, the Library of Congress, and so on. These major institutions were more open than the older, private libraries, but access to them still remained restricted. After 1800, however, the number and range of public libraries exploded across Europe, as governments, philanthropists (Cain 1994), and private groups sought to increase public access to information and culture, and especially to geographical publications.

Herein lie the origins of several of the ideal's preconceptions. First, the expectation that maps of all sorts should be printed for public distribution, a belief that did not depend on, and paid no attention to, the actual effectiveness of the marketplace for disseminating printed materials. Second, the expectation that the provision of information to the public should be morally altruistic, because the public demanded and required correct and unbiased information for the proper formulation of policy. There was, of course, the need to ensure the accuracy of instrumental maps, so that marine charts did not lead mariners astray, for example, but a fundamental principle was that more intellectual maps must be as correct and as up to date as possible so as not to mislead the public's intellectual understanding of the world. And, third, the expectation that maps were to be evaluated for their correctness and propriety, just like any other informational text, and they were accordingly to be subject to public review, debate, and discipline.

THE NEOLOGISM, "CARTOGRAPHY"

ontology / singularity and universality. The manner in which systematic surveys seemingly unified all mapping—whether terrestrial or marine, general or particular—gave rise to the key neologism of "cartography." Various claims that the term "cartography" was used as early as the sixteenth century have been disproven (Harley 1987, 12n93). The word was actually coined in about 1790 and again, apparently independently in 1808, before taking root in the 1820s and flourishing in the 1830s

(Kingston 2006; Van der Krogt 2006, 2015).* A history of the neologism can now be hazarded, exploring how "cartography" initially bore varied meanings before eventually gaining a degree of stability.

The word "cartography" is an etymological hybrid that combines classical and modern elements: "-graphy" derives from the classical Greek -γραφία, "writing or describing"; "cart-" from the late medieval Latin *carta* or early modern French *carte*. *Carta* began to be applied to spatial images on paper in thirteenth-century Italy; this precise usage slowly spread through the rest of Europe, in Latin (as in Martin Waldseemüller's great 1516 *Carta marina*), in Italian, and in other western vernaculars as *carte* / *Karte* / *chart*. By the later 1600s, *carta* had largely supplanted other Latin-derived terms that had long been applied to spatial images, such as *descriptio*, *tabula*, or *typus*.

It has been argued that since both elements of "cartography" derive ultimately from classical Greek, the term is not in fact a hybrid (Harley and Woodward 1987, xvii n13; Van der Krogt 2015, 126). The late medieval Latin *carta* had stemmed from the classical Latin term *carta* or *charta* (referring to a leaf of paper or papyrus), which had in turn derived from the classical Greek χαρτης (referring to a papyrus leaf). However, it is clear from the context of its coinage and acceptance that the term "cartography" was intended to relate to maps, not to sheets of papyrus. The relationship is made still clearer by some usage in nineteenth-century English and German, which specified "chartography" as relating specifically to *charts* (Van der Krogt 2015, 126n19). Thus, "cartography" was indeed formed by the active combination of classical and modern elements.

Combining "cart-" and "-graphy," the neologism reads as if it had been derived directly from some supposed but actually nonexistent Greek word, χαρτογραφία. That is, "cartography" was actively modeled on the Classical Greek word "geography" (γεωγραφία, from γεω- ["geo-," the combining form of γη, "earth"] and -γραφία). The parallel with "geography" is evident in the medial *o* of "cartography," which should properly have been *a* in line with the declension of *carta*. By comparison, a correctly formed hybrid is "stratigraphy," derived from the Greek "-graphy" and the Latin *stratus*, "layer" (see also Liberman 2014).

Any freshly minted neologism is open to semantic interpretation as users grapple with its novelty. "Cartography" has had an especially wide scope for ambiguity and flexible application because the manner of its construction offered two overlapping suites of potential meanings that did not necessarily align with one another.

First, there were the shades of meaning granted by varying interpretations of

---

*I presume that Scharfe (1997, 26) made the same argument as Van der Krogt when he identified the "semantic independence of 'cartography' (from about 1829–39) from geography, topography and geodesy," although he did not explain just what he meant.

"-graphy." The fundamental meaning of the original Greek -γραφία was "writing." For example, βιβλιογραφία, "bibliography," originally meant simply the writing of books. But in modern usage, the suffix has acquired several further connotations such as "writing *on*," as in "lithography" (writing on stone); "writing *with*," as in "photography" (writing with light); and "writing *about*," which has generally been used for names of traditionally descriptive sciences such as "bibliography" and "geography." In this respect, "cartography" offered a range of potential meanings to those who sought to determine its meaning solely from its etymology: writing maps (i.e., making maps); writing *with* maps (i.e., making maps for other endeavors); and writing *about* maps (i.e., describing maps, whether the study of individual maps or the descriptive study of maps generally).

Second, having been coined in parallel with "geography," "cartography" also acquired the three primary usages that had, by the early 1800s, accrued to the older term: as the intellectual endeavor of knowing the world, or "geography"; as that endeavor's subject matter, or "*the* geography" (as in "the geography of England"); and as the product of that endeavor, as the book or map that describes the placement and internal arrangement of a region, which is to say, "*a* geography." These semantic extensions accordingly offered "cartography" still further range: cartography, signifying the intellectual endeavor of knowing about maps, which is largely coincident with "writing about" maps; *the* cartography, meaning the idealized archive of spatial knowledge of a region, as manifested in a canon of maps, as in "the cartography of England"; and *a* cartography, meaning an account of "the cartography," which today would be termed a cartobibliography.

The early history of the word "cartography" reveals that the neologism was indeed deployed with several of these possible meanings, and their shades have persisted. In other words, not only the idealization of mapping but also its apparently universal label developed around multifaceted and potentially conflicting concepts.

The earliest reliably known instance of "cartography" occurred in a manuscript petition to the French foreign ministry, written between 1787 and 1791, by Nicolas-Antoine Queuxdame, known as Tessier (Kingston 2006; Van der Krogt 2015). Tessier was then completing an inventory of the huge map collection assembled by the geographer Jean Baptiste Bourguignon d'Anville, which the ministry had acquired on d'Anville's death in 1782 (Heffernan 2014, 10–12). He proposed to expand the inventory into a general listing of all known maps, which he called "a comprehensive gazetteer or, if one can use this expression, a universal Cartography."* In qualifying his use of the neologism ("if one can use this expression"), Tessier seems to have suggested that he had heard the term from someone else but that it was probably unfamiliar to ministry officials. Tessier's project ("*a* cartography") was a

---

*"une nomenclature géographique ou si l'on peut employer cette expression, une Cartographie universelle."

description of maps, analogous to "a geography." In this context, the neologism was a rhetorical flourish that, like the petition itself, had no effect.

Some commentators, when faced with the neologism after it had entered general circulation in the 1820s, would similarly construe "cartography" to mean the endeavor of studying and describing maps. At least one nineteenth-century lexicographer turned to etymological principles when faced with the task of explaining the new coinage, defining "*cartography*" as both "a description, or an account of maps and charts" and "illustration by maps or charts," which is to say, both writing *about* and writing *with* maps (Worcester 1849, art. "Cartography"). The practice of listing maps in inventories and catalogs continued under the guise of "historical cartography." Even with Herbert Fordham's (1914) coinage of "cartobibliography" (see Blakemore and Harley 1980, 37; Van der Krogt 2015, 140), the naive use of "*a* cartography" in the sense of a catalog demonstrably persisted in the nineteenth and twentieth centuries (Bom HGz 1962 [1885]; Wagner 1932).*

It has been argued that the viscount of Santarém had specifically coined "cartography" to mean "the study of maps," which is to say writing *about* maps (Harley and Woodward 1987, xvii; Harley 1987, 12). However, this interpretation is unwarranted: when Santarém had claimed in 1839 to have "invented" the word *cartographia* (Santarém 1906, 30), he was in fact rehearsing the more generally accepted meaning for the word that had already been propagated in Paris (Van der Krogt 2015, 127). In this respect, we can no longer follow Armando Cortesão (1935, 2: 365–66; 1969–71, 1: 4–5) in crediting Santarém with creating the neologism.

Actual responsibility for creating and popularizing the neologism in the early nineteenth century almost certainly lay with the exiled Danish geographer Conrad Malte-Brun. He had used *chartographie* in 1808, in reference to a proposal for a 204-sheet map of Germany (Bertuch 1807). Such a map of Germany—drawn "at a scale large enough to contain all the interesting details of topography" so as to be of use to geographers, travelers, administrators, and the military—would fill "this lacuna in cartography" (Malte-Brun 1808, 264).† This usage embraced both the structured practice of making maps and the resultant map archive (*the* cartography).

Malte-Brun's coinage eventually found fertile ground among the members of the Société de géographie, founded in Paris in 1821 as part of the era's general

---

*I encountered an attempt at a new coinage of this meaning of "cartography" in a letter from John S. Putnam of Northwestern University Press to Lawrence W. Towner (President, Newberry Library), 28 January 1966, Newberry Library Archives Group 07/07 (Hermon Dunlap Smith Center), box "Early History, Proposals, Events," Folder 1.

†"Depuis long-temps le besoin d'une Carte de l'Allemagne, dessinée sur une échelle suffisamment grande pour contenir tous les détails intéressans de la topographie, à été senti très-vivement non seulement par les géographes et les voyageurs, mais aussi par les autorités constituées, et surtout par les militaires.... La célèbre maison Bertuch ... entreprend aujourd'hui de remplir cette lacune dans la chartographie."

reconfiguration of geographical practice (see below). In his eight-volume world geography, Malte-Brun used *géographes* for geographical map makers, except for two instances in the sixth (1826) volume, when he used *cartographes* with a negative and perhaps sarcastic intent, complaining that "cartographers since [Vincenzo] Coronelli" had ignored certain districts of Bosnia and that "cartographers" had perpetuated a meaningless toponym (Malte-Brun 1810–29, 6: 230 and 6: 257n1). Malte-Brun also co-authored a March 1826 report to the society on the travels through North Africa by Jean Raymond Pacho, which held out—quite seriously and not at all facetiously—the prospect of a "cartographic criticism" of source materials (Barbié du Bocage, Joubert, and Malte-Brun 1826, 99). Furthermore, a caption to an archaeological plan of Cyrene in Pacho's report noted that relief was shown by shading, "according to the method adopted for cartography" (Pacho 1827, 363).* Finally, Philippe François de La Renaudière (1828, 63) briefly reviewed the history of mapping in his history of geography and observed how

> Ortelius finally put some order into geography. He first separated ancient geography from the modern. He did much for both of them, and put erudition even into cartography.†

These early occurrences of "cartography" all referred to an endeavor, but their isolation suggests the term's novelty. Malte-Brun's usage of *cartographes* was not, for example, continued within either of the two "improved" editions of his geography that were translated into English and published in the United States (Malte-Brun 1824–29, 1827–32).

Although no statements accompanied these early occurrences about how and why the neologism had been coined, each occurrence nonetheless indicates that, for their authors, "cartography" already connoted the detailed, conscientious, critical, and formal endeavor of preparing all kinds of maps from the new kinds of systematic surveys then in development. These occurrences variously used "cartography" to refer equally to fine-resolution topographical mapping, to the organized and "erudite" mapping of the world at much coarser resolutions, and to the intersection of such mapping activities in the remarkable six-volume *Atlas universel* (1827) by Philippe Vandermaelen (Van der Krogt 2015, 131–32; see fig. 5.21). Together, these early occurrences construed "cartography" to be the singular, normative, and thoroughly scientific endeavor of observing and measuring the world and making maps, regardless of their resolution.

The idealization of "cartography" as embracing all mapping processes and the

---

*"selon la méthode adoptée pour la cartographie."

†"Ortélius mit enfin un peu d'ordre dans la géographie. Le premier il sépara la géographie ancienne de la géographie moderne. Il fit beaucoup pour toutes deux, et porta l'érudition jusque dans la cartographie."

priority of the word's coinage by scholars working in France are both evident from an early, isolated instance of "cartography" in English (Van der Krogt 2015, 135–36). The exiled Prussian polymath Francis [Franz] Lieber, who had emigrated to the United States in 1827, prepared a detailed and ambitious curriculum for a new Philadelphia orphanage. Within the section on the five different kinds of "drawing" that the boys were to be taught, he specified:

> 3. Drawing of maps or *chartography*, (at least I believe we might use this word, formed after the French *cartographie*, which comprises the drawing of geographical and topographical maps, charts, and all the drawing of mensuration). (Lieber 1834, 98; original emphasis)

A passing thought in a much larger work, Lieber's statement presents an understanding of "cartography" that seems already well digested. Lieber enumerated what had hitherto been understood as distinct modes of mapping—geography, topography, charting, and property and engineering mapping (i.e., "all the drawing of mensuration")—as parts of a common endeavor. In doing so, he set aside any concern for the significant practical differences in producing and consuming these different kinds of maps. Lieber thus enunciated the core of the ideal: cartography is the universal endeavor of the direct and coherent measurement of the earth's curved surface and its reduction to a plane image.

Even so, uncertainty remained after 1830 over the precise meaning of the term and of the idealization for which it stood. While the history of cartography developed after 1830 as the history of the idealized endeavor (see below), a core element of the new field acquired the name "historical cartography." This particular element comprised the arrangement of maps of a given region in chronological sequence and the assessment of the importance of each individual map according to how well it improved or enhanced the region's geographical archive. The term relied on variant meanings of the neologism, as "*the* cartography" of the archive of spatial knowledge of a region and as "writing *about* maps." The semantic stability of "historical cartography" was further destabilized by its use in reference to the mapping of past times, a usage that relies on the idea of cartography as "writing *with* maps." Finally, the growth of academic cartography after 1950 was accompanied by definitions of the field as not only that of making maps—all maps—but also of the study of map making, which is to say cartography as "writing *about* maps." Such semantic flexibility has consolidated the ideal of cartography by enfolding the analysis of maps in the same rhetoric of rationality and science as the practice of map making.

Regardless of the precise nuance adopted, definitions of "cartography" have promoted the validity of the generic category of phenomena called "maps." Lieber's definition exemplifies how, in the Anglophone tradition at least, definitions of

119

"cartography" have consistently followed a two-stage process: first, a declaration, usually brief, that cartography is the making of normative maps; second, a much longer explication of what kinds of map are indeed normal. The long-standing practice of enumerating the several different kinds of normative map culminated in the tautology perpetrated by the newly formed British Cartographic Society in 1964 and adopted by the International Cartographic Association in 1973. Disciplinary concerns hid the simplicity of the initial declaration behind some rather baroque specifications:

> Cartography is the art, science and technology of making all kinds of maps, together with the study of maps as scientific documents and works of art.

The second stage was a lengthy enumeration:

> In this context [normative] maps may be regarded as including all types of [geographical] maps, plans, charts and sections, three dimensional models and globes representing the earth or any heavenly body at any scale. (Anon. 1964; Meynen 1973, 1; see Maling 1991)

But as academic cartography developed further, such enumerations were found to lack the academic rigor needed to codify "the common properties shared by all maps, which set them apart from artefacts which are not maps" (Visvalingham 1989, 26). In seeking that rigor, map scholars have sought only to define the normative map, and in doing so, they have not questioned the idealized concept of cartography.

Mathematics and Rationality, Empires and States

The modern sentiment that maps are necessarily "scientific" rests not only on the projective geometry promoted by state-sponsored territorial mappings, but also on developments in politics and mathematics. As a result, cartography was construed as an innately rational endeavor. This new rationality might look similar to that of the Enlightenment *philosophes*, but it was formed from wholly new cloth.

MODERN IMPERIALISM AND SOCIOLOGY

individuality / observation / singularity and universality. With the rapid intensification after 1800 of European overseas imperialism, and with the territorial expansion of the United States and the newly independent countries in Latin America, cartography became an especially culturally resonant activity. As the ideal developed, cartography was specifically construed as a strictly *Western* phenomenon, as a primary marker differentiating the imperial Self (rational and liberal) from the colonized Other (irrational and despotic). This difference was apparent early on: one of the first uses of "cartographers" was made by Karl Ritter in reference to British surveyors in South Asia (Edney 2009, 42). And it has persisted. When Armando Cortesão (1969–71, 1: 4) stated that the importance of the history of cartography stemmed from the manner in which "historico-geographical studies are an indispensable background to the history of civilization, which is the highest stage in the history of mankind," he implicitly referenced the nineteenth-century ideology of unilinear cultural development whose stages were mapped onto ethnic categories to construct a racist hierarchy of African (black) savagery, Asian (yellow) barbarism, and Western (white) civilization.

The practice of cartography was held to distinguish rational, liberal, and moral Europeans (and their North American descendants) from irrational, despotic, and amoral peoples on other continents; in this formulation, Europeans did cartography, other peoples *could* not. Indeed, nineteenth-century commentators such as Alexander von Humboldt (1836–39, 1: viii–ix, 1: 1–3) and Baron Charles Athanase Walckenaer (1835b, 638) held that the Renaissance discovery of the New World led to the dissolution of the medieval worldview and established modern rationality and morality, but only for Westerners. The ideal of cartography accordingly contributed significantly to self-justifications by Western nations that their imperial activities would bring scientific modernity to the benighted others (Edney 2009, 41–43; Crespo and Fernández 2011, 407–8; see, e.g., Ramaswamy 2017).

In the twentieth century, Jean Piaget would enshrine this sense of difference in his implicitly racist theories of cognitive development. Piaget tracked the development of individual cognition, in part through cognitive concepts of space, from the child's nonscalar, topological sense of space (defined by routes and their interconnections, or nodes) to the scalar, topographical sense enshrined in the ideal of cartography. But his evidence for people working within a topological sense of space drew as much on ethnographic studies of "childlike" adult indigenes as it did on studies of children (Blaut 1993, 99–101; Shweder 1985). Moreover, Piaget revealed little appreciation for the ways in which Western adults also function topologically. Two points are clear here: first, the presumption of a racial divide in cognitive development; second, the confusion of "cognitive mapping," common to all adult members of *Homo sapiens*, with socially mediated acts of representing spatial relationships (Wood 1993).

Moreover, the same contrast of the imperial Self to the colonized Other has been applied within male-dominated Western culture. Men have placed women in the role otherwise filled by the colonized Other, which is to say that men are construed as rational and logical, women as intuitive and emotional. Because they are Western, white women could be afforded some scope for rational thought and action, but until recent decades they have been permitted this only if they also forgo the primary practices considered as "female" (child-rearing, home-keeping, and so on). Even so, present-day academia and knowledge-based industries remain heavily gender-biased.

Men are thus presumed to think in terms of topographical, coordinate, and scalar space, women in terms of topological space. Gendered differences in spatial abilities, including mapping, have been widely documented (Huynh, Doherty, and Sharpe 2010, 272–73). Yet only recently have scholars recognized the difficulty in such psychological studies of controlling for "nurture," which does seem to have an effect on the spatial ability of men and women alike (Hoffman, Gneezy, and List 2011). Indeed, evolutionary biologists now "wonder not whether, but *why*, sex" should even be presumed to be the sole or even primary determinant of "male and female brains, and male and female natures" (Fine 2017, 88).

ANALYTICAL DEVELOPMENT OF MAP PROJECTIONS

ontology / singularity and universality.   Early modern cosmographers generated a profusion of different ways to depict the whole world and its regions (Snyder 1993; Shirley 2001). Nonetheless, they favored those that had basic properties, either equidistance or conformality, and that could be easily drawn with straight lines and arcs of circles (Morrison and Wintle 2019). This pragmatic bias in geographical and world mapping began to be displaced in 1772, when Johann Heinrich Lambert published the results of his application of the calculus to the analytical creation of new projections with specific properties (Lambert 2011 [1772]). While his own newly designed projections were little used until the twentieth century, Lambert's work sparked a new concern among geographers and mathematicians to perfect the mathematical underpinnings of all maps. Marie Armand Pascal d'Avezac de Castera-Macaya (1863, 138–50) classified all map projections by their mathematical structure—azimuthal, conic, cylindrical, pseudo-conic, pseudo-cylindrical, and so on—which has promoted the conviction that all map projections are ineluctably mathematical transformations and were always considered so (see Watson 2008).

The new projections advanced in the nineteenth century were defined as much for the ellipsoid as for the sphere, and statistical techniques were developed to precisely model the complex geometries inherent in any mapping of the earth's curved surface onto a flat map (Snyder 1993, especially 76–94). The general result was the extension of the projective geometries of systematic surveys to all modes and resolutions. The *Carte générale et particulière de la France* had used the transverse *plate carrée* (or equirectangular) projection—a projection so simple it involved barely any calculations—to define the key points of the survey and had then worked in projective coordinates. Lambert's analytical approach eventually led, over the course of the nineteenth century, to the establishment of equations for *every* projection: $x(\phi, \lambda)$ and $y(\phi, \lambda)$. Projections came to be defined not by their aesthetic and the methods of their construction but by scale-independent mathematical formulas and abstract properties.

Three particular projections were widely adopted by both geographers and systematic surveyors after 1900. Lambert's own conformal conical projection with two standard parallels has been commonly used by geographers in mapping areas of east-west extent, such as the United States, and also, in narrow latitudinal bands, as the basis of systematic topographical and cadastral surveys. Since 1800, the conformal Mercator projection has been widely adopted for systematic hydrographic surveys and official marine charting, and also for some geographical mapping. Lambert had defined its transverse aspect for coarse-resolution maps of north-south extent, and Gauss had then redefined that aspect for the spheroid; after 1930, the transverse Mercator was adopted, in narrow longitudinal bands, for systematic topographical and cadastral surveys. That the same projection formulas can be used to translate between cosmographical (ϕ, λ) and projected coordinates ($x[\phi, \lambda]$, $y[\phi, \lambda]$) for both fine-resolution and coarse-resolution mapping has only reinforced the conviction that "the map" depends on a universal ontology.

THE CHANGING STRUCTURES OF SCIENCE

ontology / observation / singularity and universality.   The disciplinary structures of natural philosophy underwent substantial change in the late eighteenth and early nineteenth centuries. The overall narrative is well established: increasing specialization and professionalization led to specific groups of scholars identifying not only as natural philosophers but also more specifically as "astronomers," "mathematicians," "geologists," and so on. The trend is apparent in the establishment of specialized societies, such as the Geological Society of London (1807), the Royal Astronomical Society (1820), and the Zoological Society of London (1826). The reconfigurations in the field of "geography" (generally, see Livingstone 1992) had special importance for the contemporary status of map making.

"Geography" began to change as a field of study late in the eighteenth century. The framing devices for spatial knowledge deployed throughout the early modern era—cosmography, geography, chorography, and topography—began to be displaced by a thematic and scaleless distinction between human and physical subject matters. The shift is evident as early as Johann Cristoph Gatterer's (1775) outline of geographical knowledge (see Lüdde 1849, 10 [item 33]). It would be codified by Immanuel Kant's division, at the very end of the eighteenth century, of the overall discipline of "world-knowledge" into two parts: physical geography and anthropology (Wilson 2011).

As geography became less concerned with describing the world and became more focused on systematic accounts, the field progressively dissociated itself from, and ceased to be synonymous with, coarser resolution mapping. This process was helped by the apparent solution in the eighteenth century of the intellectual problems of geographical mapping, such that coarser resolution mapping was no longer intellectually challenging. Rather than being concerned with how to map the

123

world, geographers were faced with a world map that was full of blank spaces. The endeavor of geography increasingly emphasized the active and manly field science of exploring and cataloging the world's features and their organization. For example, the Société de géographie, founded in Paris in 1821, aimed specifically to promote and publish the results of new geographical explorations and discoveries in order to fill up the world map. Although the new society understood its tasks to include map production, its role in this regard was only to have them engraved ("*les faire graver*") rather than to actively compile them anew (Godlewska 1999, especially 130; see also Godlewska 1989). The society's members reconfigured the history of geography, previously dominated by the history of coarse-resolution mapping, as the history of European exploration (e.g., La Renaudière 1828). Similarly in Britain, the Royal Geographical Society, created in 1830, aimed to promote exploration, and its first secretary, Captain Alexander Maconochie, briefly held a chair in geography at the newly created University of London (1833–36).

High-level, geodetic surveying was already its own discipline, aligned with astronomy and geophysics, and had grown far removed from the realm of everyday mapping. Now the reconfiguration of geography as being largely concerned with social and physical environments cast loose the supposedly scaleless and mathematical practices of map making, permitting them to be recast as the rational, fundamental technology of "cartography."

There is a close parallel between the formation of the concepts of "cartography" and of singular "Science," which would similarly be "asserted as a component of European cultural hegemony in the nineteenth and early twentieth centuries" (Golinski 2012, 20). Even as the disparate kinds of mapping were being brought together into the common framework and supposed singular process of cartography, so too was a singular concept of science being articulated in reaction to the increasing specialization and institutional fragmentation of the natural sciences. In particular, in his *Cours de philosophie positive* (1830–42), Auguste Comte argued that "all sciences are 'branches of one Science, to be investigated on one and the same Method.'" Science is science not because of what it studies but because of *how* it studies. Comte grounded his argument in a historical narrative of the successful application of an abstract and universal scientific method to different realms of natural and social philosophy: in chronological sequence, to mathematics, astronomy, physics, chemistry, physiology, and finally sociology, which in the early nineteenth century was still in its infancy as a science (Golinski 2012, especially 23, quoting Lewes 1853, 10). One result of this historical justification would be the creation of the concept of Europe's scientific revolution (Shapin 1996). In Britain, this reconceptualization of Science was championed by William Whewell who, among other things, coined the term "scientist" in 1833 to describe someone who studies Science (Ross 1962, 71–72). It is significant, in this respect, that the incarnations of "cartography" in this period seem to have been mostly as *cartographe*, not *cartographie.* The unity of both endeavors—science and cartography—lies in their common processes by which disciplined people—scientists and cartographers— investigate the world. This parallel points to a *Zeitgeist* that needs further explication, at least in terms of the development of the ideal of cartography.

There was also a revealingly gendered aspect to the coinage of "scientist." As has often been noted, the specific occasion of the term's first appearance in print was Whewell's review of Mary Somerville's interdisciplinary text, *On the Connexion of the Physical Sciences* (1834). Whewell obviously could not refer to Somerville as a "man of science," and he also sought a term that would capture the supposed manner in which women, "if they theorize," do so with "a clearness of perception," without thought for practical application and without any conflict between theory and practice, that permits them to see the unity of scientific method in a manner impossible for men (Whewell 1834, 65; see Neeley 2001, 3). Such arguments insist on the individuality of cognition and therefore of scientific and cartographic behavior.

THE HISTORY OF CARTOGRAPHY

individuality / materiality / discipline / singularity and universality.   The concept of cartography, like the new singular concept of science, was developed and elaborated over the course of the nineteenth century in concert with the establishment and growth of an equally idealized field of study: the "history of cartography." Scholars had been interested in early maps before 1800, of course. Some historians and antiquaries had looked at early maps as sources of evidence for their studies of the ancient and medieval worlds, and some professional map makers had written histories of the development of their work in order to position themselves at the forefront of critical practice (Skelton 1972, 62–73; Harley 1987, 7–12; Edney 2019c; Withers 2019a). Such studies continued to be pursued by communities of substantive and internal map historians throughout the nineteenth and twentieth centuries (Edney 2012a, 2012b, 2014a, 2014b, 2015b). But after 1830 there coalesced a new, international community of scholars with an explicit interest in early geographical maps and, for the first time, early marine charts. An initially small coterie had developed by 1860 into a combination of map librarians, historians of geography and empire, and the collectors and dealers in rare maps and books, all of whom understood themselves to be "historians of cartography" (Cortesão 1969–71, I: 1– 70; Skelton 1972, 70–102; Blakemore and Harley 1980, 14–44; Harley 1987, 12–23).

The particular stimulus for the new field of study was the new drive to explore and exploit the remaining empty spaces on the world map, which led some scholars to revisit the history of European exploration and discovery. For a small community of scholars centered in Paris in the 1830s and 1840s—shortly after their colleagues captured the new idealization in the neologism "cartography"—the first great era of European expansion in the Renaissance offered a compelling demonstration of the human capacity for scientific and moral progress, as manifested in the progressive improvement in geographic knowledge. The fifteenth and early sixteenth centuries formed a watershed for humanity: the germ of a scientific spirit, only sporadically expressed during the Middle Ages, now flourished and led to profound changes, not only in knowledge of the material world but also in human thought, reason, and morals. For Alexander von Humboldt, Edme François Jomard, the exiled viscount of Santarém, and others, early maps gave almost visceral access to this compelling narrative of human cognitive development, of the progressive extirpation of error, and also of the positive benefits of pursuing geographical conceptions that ultimately proved mistaken (especially Humboldt 1836–39, 1: viii–ix and 1: 1–3; see Ette 2010; Walckenaer 1835a, 1835b). Their primary tasks were to locate early maps and charts, to critically assess their relevancy for the histories of geographical conceptions, geographical exploration, and human (i.e., Western) civilization, and to make qualified maps available to other scholars in facsimile atlases (Godlewska 1995).

As the field continued to grow, being especially motivated in the 1890s by the quadricentennials of the voyages of Vasco da Gama and Christopher Columbus (Wright 1945, 505), historians of cartography sought to narrate the growth of human/Western civilization. The history of cartography was a means to trace the development of the literal worldview. Cartography was understood as a universal endeavor "to which all the chief races have contributed" as part of "man's appointed . . . task of subduing the earth and gaining a fuller and securer life, and increased control over the still unused resources of our bountiful earth" (Gregory 1917, 64–65). The drive to observe and map the world thus appears to be universal, yet only Western civilization—coincident with the Western mind—has had the technological and scientific abilities to bring cartography to fruition.

Historical narratives have variously placed the origins of that achievement in multiple moments. But the origin stories are overly precise because they all ignore their ostensible subject—the totality of all mapping practices—and instead address just one or two particular mapping modes. Arguments that cartography originated in the exertion of territorial control after imperial expansion by the ancient Egyptians, by Alexander the Great, or by the Romans have focused exclusively on chorographical mapping; that it originated with Claudius Ptolemy's *Geography* in second-century CE Hellenistic Egypt, on cosmographical and geographical mapping; that it originated with the modern-looking outlines of medieval marine maps of the Mediterranean, on marine mapping; and that it originated with the general geometricization of life in early modern Europe, on topographical and chorographical mapping. Moreover, each origin story places the beginning of cartography in a relatively limited chronological period. All such accounts fundamentally fail because they conceive of cartography as an actual endeavor and not as an idealized and invalid representation of how people have actually gone about making different sorts of spatial representations.

126

Historians of cartography self-consciously limited themselves to the history of *early* cartography, which is to say, to the periods before 1800 that led in a triumphal narrative of ineluctable progress to the achievements of modern cartography itself, as emphasized in the titles of several important nineteenth-century works (Daly 1879; Nordenskiöld 1889; Harrisse 1892). The precise formulation of the progressive narrative inevitably varied over time, not least when the substantial technological strides in mapping made during World War II prompted Lloyd Brown (1949) and Gerald Crone (1953) to advance new general histories that gave substantial attention to modern, territorial surveys. Nationalistic tendencies further emphasized cartography's various "golden ages," when different European societies sequentially "took the lead" in assembling the world/archive and disseminating the results: fifteenth- and sixteenth-century Italy; sixteenth- and seventeenth-century Netherlands; eighteenth-century France and Britain; nineteenth-century Britain and Germany.

But the overall story remained the same: the steady accretion of geographical knowledge and then, after about 1800, of territorial knowledge, all marking the advance of Western civilization. In all of this, cartography stood as a single endeavor—that of knowing the world—that manifested the state of science and scientific inquiry in each period. It had advanced in the Hellenistic era, been brought low in the Middle Ages, recovered in the Renaissance, improved significantly in the Enlightenment, to the point where "science claimed cartography" and extirpated artistic elements (Rees 1980, 60; fig. 4.4), and thereafter steadily perfected itself in the modern age. Throughout, scholars have blurred the distinctions originally drawn in the mid-nineteenth century between the "history of cartography" as the history of the endeavor and "historical cartography" as the history of the map coverage of particular regions. The history of all mapping has thus been boiled down into a narrative of unalloyed Western progress in both the quantity and the quality of spatial data, often presented as series of "firsts," either of geographical features (e.g., the first map to show X) or improved techniques (e.g., the first map of a region to use triangulation).

RATIONALIZING LINEAR MEASURES

ontology / observation / singularity and universality.   The idea of a comprehensive and universal geographical archive was sustained by the development and widespread acceptance of the rational metric system of measures. Traditional acts of measurement could only ever be mundane and embodied, in that they all mediated between the human body and either things or the wider world. Customary measures evolved within specific technologies for different arenas of human activity—commerce, agriculture, construction, travel, and so forth—and were expressed through localized practices. They accordingly proved resistant to regulation by Europe's monarchs. Each measure varied considerably between countries and

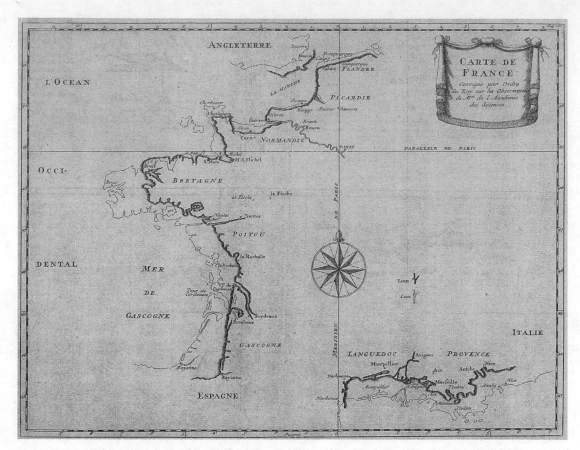

FIGURE 4.4. The emblem of cartography's supposed scientific reformation in the Enlightenment: Jean Picard and Philippe de la Hire, *Carte de France corrigee par ordre du Roy* (Paris, 1693; reprinted, 1729). Originally prepared in the early 1680s, this map contrasted the thin line of the "old" coastline of France—copied from a map presented by Guillaume Sanson to the dauphin in 1679—to the thick, shaded line of the coastline as "corrected" by longitude determinations by members of the Académie des sciences, who had observed eclipses of Jupiter's satellites. Since Christian Sandler (1905), historians have repeatedly reproduced this map as a marker of the supposed Enlightenment transformation of cartography from "an art" to "a science" (whatever those terms might actually mean). Copper engraving, 27 × 35 cm. Courtesy of the Osher Map Library and Smith Center for Cartographic Education, University of Southern Maine (Osher Collection); www.oshermaps.org/map/622.0001.

even between districts. In England, for example, monarchs might have defined a mile to be 1,760 statute yards (1,609 m), but local usage perpetuated both shorter and longer miles (1,524–2,286 m). And while statutes decreed that agricultural land had to be measured with a standard rod 16.5 feet (5.03 m) in length, in practice the actual wooden rods varied in length from 9 to 28 feet (2.74–8.53 m) according to the district and the kind of land being measured, whether arable, pastoral, woodland, or waste (Zupko 1977, 1985).

Rationalization of measures began in medieval Europe with various attempts to standardize the weights of coins. It continued in the sixteenth and seventeenth centuries with the formation of the science of artillery, which required standards for both weight (of cannonballs) and magnitude (of balls and gun bores) (Armoghate 2001). Natural philosophers began to propose general systems for standardized linear measures in the later seventeenth century, using as a foundation either the length of a seconds pendulum or the earth's size, but they could not overcome basic technological problems. Linear measures remained approximate, hindering the ability of geographers to collate information from multiple sources.

The rationalizing zeal of revolutionary France eventually succeeded in creating a universal system of measures: the metric system was initially promulgated in 1793, and customary units were officially abolished across France in 1799 (Zupko 1990, 75–105, 135–69). Furthermore, the desire for a *true* universal standard continued in the search for ancient, *ur*-measures that had only been corrupted and variegated over time. French scholars such as Edme Jomard thought that Greek measures had been derived, in a manner akin to the new meter, from the length of a degree as determined by the ancient Egyptians, a measure that had also been encoded in the dimensions of the great pyramids at Giza (Jomard 1809; see Godlewska 1999, 138n24); credence continues to be given to Jomard's calculations (e.g., Bernal 1987, 184–85). The British engineer Thomas Best Jervis (1836a, 1836b) claimed to have determined the ancient "Primitive Universal Standard" and argued that this should be the basis of a single, India-wide system of weights, measures, and coinage. British pyramidologists resurrected the supposed ancient Egyptian measures in the 1860s, in the midst of a long-running debate over whether to adopt the metric system in the United Kingdom (Schaffer 1997; O'Gorman 2003; Withers 2017, 17).

Unlike the revolutionary calendar, parts of the metric system survived the Bourbon restoration and formed the seeds from which the whole system would eventually be adopted across continental Europe. Acceptance of the new system was not straightforward: retention of customary units became one element in the opposition of local elites and entrenched customs to the increasing centralization of European states, in France (Weber 1976, 30–33) as well as in less well-organized states (e.g., Branco 2005). It is fair to say, however, that metric measures were generally accepted by participants in continental Europe's main economic markets by the 1840s, although customary units lingered in remoter, rural areas (Kennelly 1928).

The metric system implemented a Platonic ideal of measurement: metric measures were defined not by the relationship they created between human bodies and the things or the world being measured, but by their relationship only to other measures. Metric measurement thus became a generic and universal act. And if measurement was generic and universal, so was the corpus of information it produced. An abstracted system of measures that was neither mundane nor embodied could be applied to any type of territory, regardless of extent and resolution.

Moreover, as detailed in chapter 5, universal linear measures led older expressions of the geometrical correspondence of maps to the world, all of which related smaller to larger customary units (e.g., 1 inch to 1 mile), to be displaced by the numerical ratio that relates the same abstract measure on the map to the same abstract measure on the ground (e.g., 1:50,000). Because either side of the numerical ratio is expressed in the same units, those units effectively cancel each other out, and the ratio becomes a truly abstract expression: independent of any measure, it can be applied to any kind of map, with any degree of resolution, of any part of the world. Indeed, the numerical ratio itself serves as a measure of the degree to which a particular map apparently generalizes, or is reduced from, the archive. Finally, the metric system encouraged elegantly abstract numerical ratios, in which denominators reflect the multiples of tens on which the system as a whole is grounded (e.g., 1:1,000, 1:10,000, or 1:100,000). Of course, it took time for such change to be put into practice. After failing to implement such a neat map scale with the *Carte de l'état major*, the French began a new territorial survey in the 1880s at the rounded map scale of 1:50,000 (Berthaut 1898–99; Huguenin 1948).

We can see the extent to which the metric system has construed the geographical archive to be universal and independent of particular measures by considering those parts of the world where the metric system was adopted late, or not at all. In 1824, the British created the imperial system of measures, officially extending the existing system of statute measures to the rest of the British empire, in order to compete with the increasingly popular French system (Zupko 1990, 176–80). Map scales continued to be defined with customary measures, often simultaneously expressed with numerical ratios and scales. The earliest specification of numerical ratio on Ordnance Survey maps seems to have been on its series of urban maps at 5 feet to 1 mile (1:1,056), undertaken in the first half of the 1850s (see fig. 5.20). When the Ordnance Survey finally metricized in the 1970s, it settled on 1:50,000 for general topographical map coverage, but at the same time still insisted on adding to the maps a verbal explanation of the ratio—2 centimeters to 1 kilometer—in the manner of older, premetric measures.

The U.S. adherence to linear measures based on the older British system has led to rather idiosyncratic map scales for federal base mapping. Henry Gannett's advocacy of the International Map of the World at 1:1,000,000 led him in the 1880s to promote base mapping by the U.S. Geological Survey at 1:62,500, as the fraction (one-sixteenth) of 1:1,000,000 that was closest to 1:63,360 (1 inch to 1 mile). In the twentieth century, first the Tennessee Valley Authority in the 1930s and then, after 1945, the Geological Survey adopted a map scale commonly used by American engineers, of 1 inch to 2,000 feet (1:24,000), for the systematic remapping of the country. In this last example, at least, the numerical ratio remains the mathematical equivalent of an otherwise customary expression of correspondence.

ontology / singularity and universality. The extension of systematic surveys after 1790 added yet more complications to the already complex issue of counting longitude. Within the context of cosmographical geometry, geographers displayed longitude on their maps from various first meridians. Eighteenth-century French and Dutch geographers had abandoned many of the first meridians used on Renaissance maps, and had focused on Ferro and Tenerife, but geographers in Europe's cultural margins advocated the use of local meridians, such as those through London or Uppsala, for zero meridians. Then, the perfection and adoption of astronomical methods to determine longitude in the field promoted the use of longitudes calculated with respect to the observatories in Greenwich (using tables for lunar distances) or Paris (using tables for Jupiter's satellites). Yet other observatories recalculated those tables to determine longitudes with respect to other points: the Dutch, for example, used the peak of Tenerife, the Spanish the observatory at Cadiz. The new chronometers measured longitude from each ship's port of origin. And adding further to this proliferation, each new systematic survey after 1790 featured the trigonometrical calculation of the longitude of each triangulation station from a central meridian, such as those through the observatories of Paris or Pulkova (Edney 2019d).

The multiplicity of zero or "prime" meridians was only occasionally the subject of commentary before 1800. But thereafter the extension of systematic surveys and the increasing integration of cosmographical and projective geometries led to the increasing impression that the situation was chaotic and had to be at least simplified and preferably settled (Withers 2017). After all, construction of a single archive of spatial knowledge by projective coordinates ($x[\phi, \lambda]$, $y[\phi, \lambda]$) requires a coherent and uniform λ. If one thing demonstrates how the development of the ideal of cartography was a lengthy process, it is that the desire for a common zero meridian did not immediately lead to the adoption of one. While mapping practices remained largely distinct, the multiplicity of zero meridians did not cause problems, as long as practice was consistent within each mode.

What moved the nations of the industrialized world to finally identify a single prime meridian was the adoption of new technologies of transportation and communication. Long-distance railroads, such as the Union Pacific Railroad from St. Louis to San Francisco (completed in 1869), and telegraphs, such as the first transatlantic telegraph (opened in August 1858), promoted the need to define standard time to coordinate operations. Eventually, in October 1884, the U.S. government convened an international conference in Washington, DC. Given the prevalent use of marine charts and observational tables based on the meridian of the Greenwich Observatory, stemming from the British empire's economic and naval might, the conference adopted the Greenwich meridian as defining "standard mean time" (Withers 2017).

As a side effect, the new standard time was taken as implying a standard zero for longitude, but the adoption of this prime meridian was still not required for mapping. The French continued to use Paris as the prime meridian for official mapping, while the U.S. Congress did not overturn the 1850 act requiring U.S. federal land maps to use the meridian of the Naval Observatory in Washington, DC, until 1912. But the voluntary adoption of Greenwich as *the* prime meridian for all mapping nonetheless gave a new degree of unity and certainty to the spatial archive.

SET THEORY AND THE TRANSFORMATIONAL MAP

ontology / observation / singularity and universality. The idealization of projective geometry was consolidated with mathematicians' development of set theory in the 1870s. The ideal's insistence that all maps are images in a two-dimensional plane of a three-dimensional figure, converted by means of a projection, now acquired a purely mathematical formulation.

A key element of set theory is the act of transformation between two sets. Carl Friedrich Gauss (1847) had analytically defined one such transformation, that of the earth to the plane in a map projection, which he called an *Abbildung* (picture, representation) or "imaging." In the 1870s, Richard Dedekind formulated a general treatment for such transformations, from one dimensionality to any other, and in doing so appropriated Gauss's term for the specific transformation. For Dedekind, an *Abbildung* was any transformation between sets in which the second, transformed set possessed the same structure as the first (Sieg and Schlimm 2005). Influenced by Gauss's initial special case, Anglophone mathematicians quickly adopted a different term for such structure-preserving transformations: "mapping" or just "map" (Peirce 1931–58, 3: 388, ¶609 [1911]).

In line with this new concept, one British encyclopedia soon recast the definition of "map" to be, fundamentally, a mathematical transformation from three to two dimensions and, only secondarily, the graphic image of the earth:

> Map. A representation of the surface of a sphere, or a portion of a sphere on a plane. The name however is commonly applied to those plane drawings which represent the form, extent, position and other particulars of the various countries of the earth. (Anon. ca. 1885, quoted by Andrews 1998, no. 58)

And in 1903, Charles Sanders Peirce, the American logician, mathematician, and geodesist, defined a map in terms of pure mathematics within a tract on semantic graphs:

> A map of the simplest kind represents all the points of one surface by corresponding points of another surface in such a manner as to preserve the con-

tinuity unbroken, however great may be the distortion. (Peirce 1931–58, 4: 400, ¶513 [1903])

The new perspective offered by set theory greatly reinforced and extended the ideal. It certainly influenced Josiah Royce and perhaps Lewis Carroll in their development of the idea of a perfect transformation and therefore of the idea of the map at 1:1 (see chapter 2). Among art theorists, it helps explain the apparently unproblematic nature of photography; Ernst Gombrich (1975, 123), for example, stated that standard attitudes held that the photograph "maps the optical world by mapping the visual sensations which correspond to it."

Peirce, however, complicated the logic (Eisele 1963). *Contra* Royce, he pointed out that a celestial map would not show itself, suggesting that the map-as-imaging was more complex than territorial mapping would otherwise suggest. Overall, Peirce seems to have found the idea of a map at 1:1 to be trite, and he was far more interested in the complex mathematics underpinning actual maps. For Peirce (1931–58, 3: 388, ¶609 [1911], and 4: 400, ¶513 [1903]), the quintessential map was the Mercator projection, which reaches to infinity at either pole and which can be endlessly repeated so as to stretch infinitely along the equator; in this case, the transformation was not one-to-one (one point on the world to one point on the map) but one-to-many (one point on the world repeated many times on the map). This argument has perhaps influenced those commentators who, despite the fact that discursive constraints limit the extent of projected spaces, have nonetheless insisted that the projective geometry of $(x[\phi, \lambda], y[\phi, \lambda])$ coordinates implies an infinite Cartesian (x, y) plane.

In the twentieth century, the mathematical "map" has recursively supported the conviction that cartography is necessarily grounded in measurement and science, each map being constituted from a transformation from three dimensions to two. In about 1992, for example, there was a telling exchange on Jeremy Crampton's short-lived "ingrafx" listserv. Someone asked about Arno Peters's political arguments concerning map projections, and in the ensuing discussion, an academic cartographer declared that "a map projection is simply a transformation from three dimensions (the world) to two (the map); politics has nothing to do with it."* Some academic cartographers have even explicitly construed Peirce's generic mathematical definition to refer specifically to cartography's normative map (Andrews 1998, no. 73; Mastronunzio and Dai Prà 2016, 183–84). This confusion of terms has underpinned the arguments by academic cartographers that mapping is

---

*Unfortunately, this particular thread seems not to have been archived at groups.google.com/forum /#!forum/bit.listserv.ingrafx. I am therefore quoting from memory, and I might well be imprecise, so I am not naming the individual being quoted.

innately transformational, both geometrically and cognitively (Robinson 1965, 35; Morrison 1976; Tobler 1979; see Robinson 1979, and Visvalingham 1989).*

Seeing the World

Beginning in the Renaissance, the mapping of places and of regions (chorography) on plane geometries was thoroughly intertwined with the preparation of perspective views of landscapes, fortifications, and urban places, as can be seen in a remarkable, recent exhibition catalog (Gehring and Weibel 2014; see Edney 2007b, 121–30). Observation—that is, vision actively guided by reason—produced geometrically structured "realistic" and "mimetic" diagrams, views, and plans of the parts of the world, from small objects and plants to buildings and fortresses to entire regions. Such work is bound up in social power relations, especially in imperial situations (see, e.g., Mitchell 1994; Edney 1997, 46–76; Crowley 2011). The widespread adoption of systematic territorial surveys after 1790, with their increasingly precise attempts at delineating landscapes with hachures, contours, and shaded relief (Imhof 1982 [1965]), served only to intensify the conviction that both scientific illustration and landscape imagery were mimetic and underpinned the act of mapping, turning the territorial plan into an apparently unmediated re-creation of the world (Valerio 2007; see Edney 2009).

Moreover, the first half of the nineteenth century was marked by the flourishing of an array of new forms of imagery and of visual technologies, which complemented the ideas of rational geometry and archive. They have contributed to the idealization of the cartographic act of seeing, whether seeing the world in the process of making maps, or seeing the world through maps in the act of reading. They underpin the modern conviction that the key function of all maps is to make visible the invisible. In this respect, the visuality of the ideal seems to date back to the Renaissance and the scientific revolution, when European natural philosophers began to make tools, such as thermometers, that gave visual expression to nonvisible phenomena, such as heat, and so permitted their measurement (Latour 1990). But, as with the idealized rationality of the new cartography generally, the common understanding of the nature of maps and mapping depends on the specifically nineteenth-century rationalization of vision and observation.

PANORAMIC VISION AND EARLY AVIATION

pictorialness / observation / singularity and universality. Cultural assumptions about the nature of vision altered dramatically in the first half of the nineteenth century

---

*The idea that maps are cognitive transformations is further reinforced by assertions by some semioticians and logicians, following Alfred Korzybski (1933; see chapter 2) and Peircean semiotics, that, in their representation of the terrain, normative maps exemplify the iconic sign (e.g., Kotarbińska 1957, 111, 142–43; Pietkiewicz 1968, 274–75). This suggestive connection demands further analysis and elucidation.

(Crary 1990; Jay 1993). Of particular importance was the development of the panorama, the first being built in London in 1788. Panoramas were not just extralarge versions of the perspective views popular since the Renaissance; they were huge works that presented a full, 360-degree experience. Within a large circular structure, perhaps entered through a dark tunnel to ensure that the view would be sudden and dramatic, viewers could examine huge, encircling views lit from above (fig. 4.5). An alternative form was the large, three-dimensional urban relief model, or "panstereorama" (Ellis 2018). The introduction of passenger railway travel led to the production of a cheaper and more accessible form of panorama, mounted on rollers so as to mimic the constant unfurling of landscape seen from a train (Oettermann 1997; Comment 1999; Brandenberger 2002; Bigg 2007; Della Dora 2007; Charlesworth 2008, 1–34; Byerly 2013, 29–82).

There is a pronounced similarity between the panorama and the technology of the panopticon proposed by Jeremy Bentham in 1798: "In the panorama, the world is presented as a form of totality; nothing seems hidden; the spectator, looking down upon a vast scene from its center, appears to preside over all visibility" (Wallach 2005, 111). *Seems* is the key word here. The panorama gives the impression of a complete and empowered view, but not the actuality, which must still be visually consumed in fragments. Such constructions "promised a synthesis and condensation of an entire landscape that would allow the viewer to comprehend and consume it," not all at once (as the early modern formula defined the conceptual power of both geographical maps and urban views) but through repeated examination and investigation (Byerly 2007, 151). That is, the panoramas promoted a "totalizing" and "knowing" fiction, according to which distant places could be visually encompassed and intellectually comprehended. This grand perspective was founded in the desire of metropolitan audiences to view, in a genteel and safe manner, exotic landscapes (Ziter 2003, 22–53; Oleksijczuk 2011) and spectacular cityscapes, and would eventually be applied to industrial cities to allay middle-class anxieties about the burgeoning urban poor and the otherwise inaccessible and uncontrollable slums (Kasson 1990).

The panorama sought, in other words, to make real the centuries-old fixation on the "Apollonian view," as Denis Cosgrove (2001) called it, the view from above. It was inevitable that balloonists quickly adapted the conceit of the panorama to their aerial views when they began to achieve that perspective previously reserved for the gods (Kaplan 2018). The first courageous adventurers to cut loose from the earth's surface in hot-air balloons had emphasized the difficulty of seeing the landscape over which they floated, because clouds obscured the ground. For example, Thomas Baldwin included in his *Airopaidia* of 1786 two maps—a colored "prospect" and a monochrome "explanatory print"—of a 1785 balloon trip between Cheshire and Lancashire; in both the maps, the ground was largely obscured by clouds (Brownstein 2013; Thébaud-Sorger 2013; Verdier 2015, 306–8;

The geometrical Ascent to the Galleries.
in the Coloseum, Regents Park.

E IV.

Published June 1829, by R. A. Ackermann & Co. 96 Strand.

FIGURE 4.5. *The geometrical Ascent to the Galleries in the Coloseum, Regent's Park* (Anon. 1829, pl. 4). Thomas Hornor's great panorama of London, built in 1829, depicted the city as if seen from the top of St. Paul's Cathedral; visitors entered through an enclosed passageway and then rose on a screw-driven elevator to viewing galleries at the top of a central tower from which they could view the panorama in a manner directly akin to Jeremy Bentham's panopticon. Colored aquatint, 32 × 24 cm. Courtesy of the Yale Center for British Art, Paul Mellon Collection, Yale University, New Haven (Rare Books and Manuscripts, Folio A D 16).

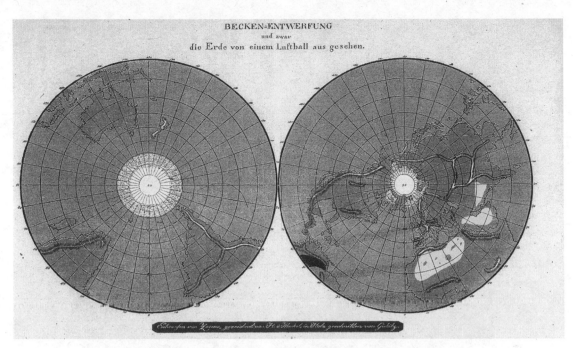

FIGURE 4.6. The world "as seen from a balloon." An early instance of the claim that even coarse-resolution world maps were grounded in direct observation. August Zeune, *Becken-Entwerfung und zwar die Erde von einem Luftball aus gesehen*, extracted from August Zeune (1833), but apparently first printed in Zeune (1811). Hand-colored woodcut, in blue (ocean), green (land), and yellow (desert), leaving white (polar ice); each hemisphere is 12 cm in diameter; azimuthal stereographic projection. Courtesy of Universitätsbibliothek, Bern. (MUE Ryh 1103:38).

Ford 2016a; Ford 2016b; Kaplan 2018, 69–70, 81–96). As ballooning became ever more popular, a newly empowered vision made the problematic clouds disappear and turned the earth into a horizontal panorama (fig. 4.6). As one intrepid aeronaut wrote in 1852:

> And here began that peculiar panoramic effect which is the distinguishing feature of a view from a balloon, and which arises from the utter absence of all sense of motion in the machine itself. The earth appeared literally to consist of a long series of scenes, which were being continually drawn along under you, as if it were a diorama beheld flat upon the ground, and gave you almost the notion that the world was an endless landscape stretched upon rollers, which some invisible sprites were revolving for our especial amusement. (Mayhew 1852)

Descriptions of this aerial perspective also shifted from panorama to map (Tucker 1996; Dorrian and Pousin 2013), as noted in 1851:

> We looked down on the country as we passed rapidly over it; first, the suburbs of London, and then, in succession, villages, woods, fields, and houses,

THE IDEAL OF CARTOGRAPHY EMERGES

distinguishing still life from what was in motion, having before us, as it were, a varying animated map. (Burgoyne 1851, 531, quoted by Byerly 2007, 159)

It became possible to depict large swathes of land as if seen from a high-flying balloon, as in a map of South Asia in 1857, whose graphic complexity compensated for an actually low level of geographical content (fig. 4.7).

Such imagery and commentary helped establish the topographical map produced from direct observation and measurement as the default option for the normative "map." And, just as individuals might view the panorama to make them believe that they understood exotic places, whether in the empire or at home, so they could think that by examining detailed topographical maps, they might understand and comprehend distant and local landscapes. The scalelessness of cartography was, in this context, demonstrated by nineteenth-century georamas in Europe (Javorsky 1990; Oettermann 1997, 90–93; Besse 2003, 169–239; Lightman 2012). These great spheres permitted individuals to view, in the manner of a 360-degree panorama, the entire globe depicted on their inner, concave surface, treating the coarse-resolution map of the world as being as visible and as comprehensible as the fine-resolution map of place (fig. 4.8). A related phenomenon in the United States was the growing popularity through the early nineteenth century of extra-large wall maps—some over 2 meters in height—that were published for public consumption. These works were large, yet they featured small type, because they were constructed from elements designed to be used in regularly sized atlases and books, and so required repeated and close examination of all their parts (Brückner 2017, 117–239).

PHOTOGRAPHY

pictorialness / observation / singularity and universality.   Photography developed rapidly in the 1830s and, as the silver-halide process was refined, photographs became naturalized as unproblematic images of nature and society, even to the point of becoming legally admissible as evidence. Only a few photographs were judged, according to their social context, to be products of human art. Regardless of the ways in which they reproduced the world in monochrome, the monocular vision of photography agreed with and reinforced existing Western conventions of single-point perspective. Once photographs were thoroughly naturalized, the camera became an observational machine—that is to say, something that works, empowering the one who wields it—further emphasizing the authoritative position of the surveyor and explorer (Laussedat 1891; see Tagg 1988; Ryan 1998; Schwartz and Ryan 2003).

All together, the instrumental vision of photography, the vast visual scope of the panorama, and the high vantage of the balloon view established by the 1860s that maps were innately panoptic views from above. While Christopher Packe had, in the early eighteenth century, understood his pioneering map to be a constructed

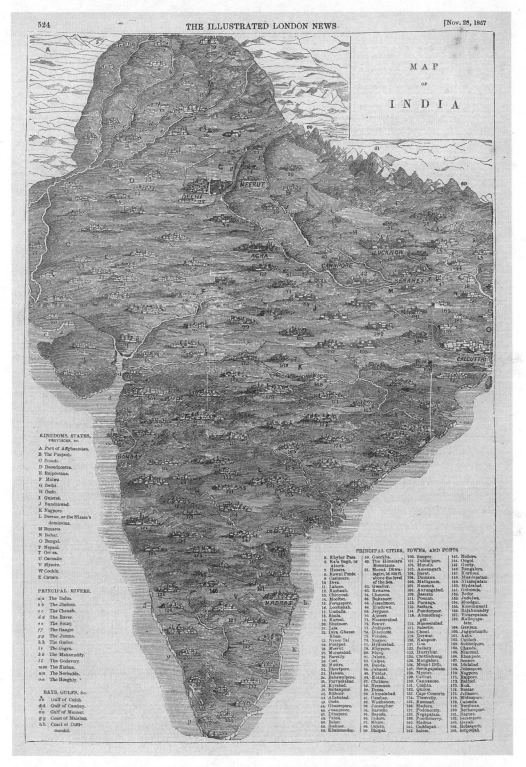

FIGURE 4.7. "Map of India," *Illustrated London News* 31, no. 889 (28 November 1857): 524. Published as a reference for the ongoing reports of the Sepoy Rebellion, this map was likely derived from Mary Read's large, five-color *India at a Glance. No. 2. Bird's Eye View of India, . . . from Cape Comorin to the Himalaya Mountains* (London: Read & Co., 12 September 1857). 36.5 × 24 cm. Courtesy of the Yale Center for British Art, Paul Mellon Collection, Yale University, New Haven.

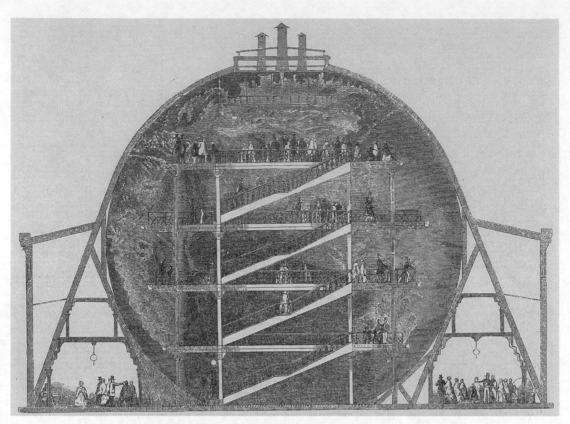

FIGURE 4.8. "Mr. Wyld's Model of the Earth. Sectional View," *Illustrated London News* 18, no. 491 (7 June 1851): 511. Created in conjunction with the Great Exhibition, James Wyld's "Great Globe" stood in Leicester Square, London, from 1851 to 1862. The globe was constructed to a scale of 6 inches to 1 degree of longitude, giving it an equatorial circumference of 180 feet (53 m). According to the account accompanying the image (p. 512), Wyld had "recollect[ed]" that only a limited part of the sphere can meet the eye at once" and so had realized that "by figuring the earth's surface on the interior instead of the exterior of his globe, the observer would be enabled to embrace the distribution of land and water, with the physical features of the Globe, at one view." Wood engraving, 16.5 × 23 cm. Courtesy of the Yale Center for British Art, Paul Mellon Collection, Yale University, New Haven.

view, made by his own artifice, the ideal of cartography developed the presumption that maps, even coarse-resolution regional maps, *are* "views from nowhere" or "God's-eye views." The twentieth-century development of aerial photography and then orbital satellites has only intensified this conviction (Propen 2009; Cosgrove and Fox 2010; Dorrian and Pousin 2013; Dyce 2013; Haffner 2013; Kaplan 2018). The ubiquitous NASA image of the "Blue Marble" from the Apollo 17 mission in 1972 (fig. 4.9) and its successor images have popularized the idea that one can see the whole terraqueous globe and its physical processes (Cosgrove 1997). The detailed imagery from satellites such as Landsat has in turn granted that universal image much finer resolution.

FIGURE 4.9. "The Blue Marble." This famous photograph was taken 7 December 1972 from Apollo 17, from an altitude of 21,750 nautical miles (40,280 km) and shows the full disk of the earth, without any shadows ("terminator"). Courtesy of U.S. National Atmospheric and Space Agency (AS17-148-22727); images-assets.nasa.gov/image/as17-148-22727/as17-148-22727.html.

The conceit was writ large in the remarkable and widely reproduced "satellite map of the earth" or *The Earth from Space*, by Tom Van Sant's GeoSphere project. The author's own claim is that this "is the first satellite map of Earth, showing the real world as it appears from space" and that until this image was published in 1990 "no one knew what the world, their home and place in space truly looked like" (Van Sant n.d.). Yet Van Sant actually composed the map from several thousand individual scenes, imaged between 1986 and 1989, all selected for lack of cloud and high sun (for limited shadows), all manipulated and painted by hand, to show "natural" colors of forests and deserts, and projected to image the entire world and not just the single hemisphere that is all that one can see, at most, from space (Wood 1992b, 48–69).

pictorialness / singularity and universality. An alternative line of development intensified cartography's pictorial claims: the application of the established practices of observing and measuring the gross features of a landscape to the mapping of spatial distributions of precisely identified elements. The mode of analytic mapping—generally but improperly known as "thematic mapping"*—visualizes the spatial distributions of social, cultural, and natural phenomena, together with manifestations of spatial processes, as part of the process of understanding and explaining those distributions and processes. The particular application of analytic mapping to the world's intangible, invisible, and unstable elements (to rephrase Pietkiewicz 1968, 272) built on and reinforced the observational claims that cartographers possess an empowered and innately scientific vision as they examine the world in increasingly minute detail. The post-1800 rise of analytic mapping, whether based on sampling or on comprehensive surveys, has been of profound importance for the natural and human sciences, and for the bureaucratic management of modern industrial states. The shrieking superlatives and exaggerated titles that have been foisted on some recent map books by the marketing departments of commercial publishers are actually almost warranted when it comes to analytic mapping: in making visible that which cannot otherwise be seen, analytic mapping has indeed "changed the world" (Winchester 2001; Johnson 2016).

Some analytic mapping had been undertaken in the eighteenth century in order to visualize and understand the distribution of phenomena, in what amounted to a preliminary turning away from studying cosmographical relationships to investigating the terraqueous globe (Porter 1980; Török 2019). After 1800, the rise of the census in the industrializing world and Alexander von Humboldt's advocacy of detailed, sustained, and systematic recording of the natural world together led analytic mapping to flourish as a fundamental component of the rapidly evolving natural and social sciences themselves and of their harnessing for political and statist purposes (fig. 4.10).

Of course, it is impossible to locate and track every thing. Analytic mapping entails first the sampling of the natural and human environments, and then the compilation of the results in sophisticated and effective visualizations. Analytic mapping was thus central to the inductive logic of nineteenth-century field science: by mapping out the spatial relationships of plants or peoples, it was possible to develop laws and conclusions, from Humboldt's own realization, by 1805, that botanic zones up the side of an Ecuadorean cordillera parallel those from the equator

---

*"Thematic mapping" is a catchall term for any map that has any kind of specialization or "theme" (from medieval maps of the apocalypse—as in Van Duzer and Dines 2016—to common-or-garden road maps), as opposed to "general purpose" territorial maps. All maps, though, can be said to have a theme; overtly special-purpose maps are part and parcel of their respective modes (e.g., road maps are part of regional mapping, forestry maps are part of place mapping). I thank Max Edelson for helping me settle on "analytic" as the appropriate label for the mode.

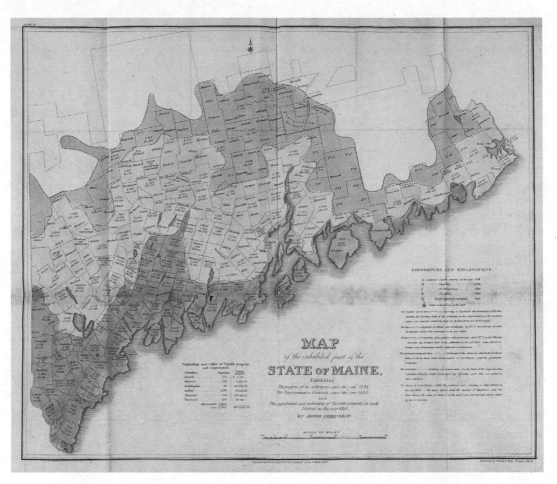

FIGURE 4.10. Moses Greenleaf, *Map of the Inhabited Part of the State of Maine Exhibiting the Progress of Its Settlement since the Year 1778* (Greenleaf 1829, map 6). This demographic map, included in what was perhaps the first analytic atlas produced in the United States, was part of Greenleaf's argument for the inevitability of the future development of the interior of Maine. Hand-colored copper engraving, 49.5 × 78.5 cm. Courtesy of the Osher Map Library and Smith Center for Cartographic Education, University of Southern Maine (Osher Collection); www.oshermaps.org/map/3564.0007.

to a pole (Humboldt and Bonpland 2009 [1805]). The history of analytic mapping has accordingly been studied in line with modern academic divisions between the natural and social sciences; between the mapping of physical phenomena by geologists, botanists, zoologists, and so on, and the mapping of social phenomena by economists, demographers, sociologists, and so forth. This distinction provides the basic structure of general histories of analytic mapping (Robinson 1982; Delaney 2012).

This sharp distinction was, however, not in evidence during the nineteenth and early twentieth centuries. Max Eckert (1921–25), for example, grouped demographic, ethnographic, and historical-political mapping together with botanical and zoological mapping under the category of "organic" maps. Analytic mapping

143

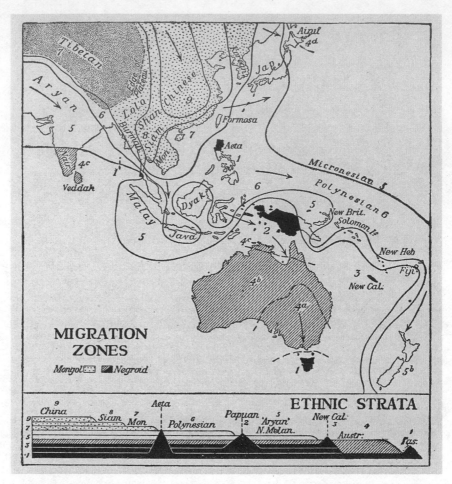

FIGURE 4.11. Using analytic maps to create and explain new theories. T. Griffith Taylor, *Migration Zones/Ethnic Strata* (Taylor 1928, 223), used a geological metaphor—which he called his "lava analogy"—to explain how different human races had repeatedly spread out from the evolutionary hub of central Asia, like successive lava flows. Each new "ethnic stratum" of supposedly more evolved and superior humans had swamped earlier populations, leaving them exposed only on the geographical margins. By modern standards, the facts and assumptions of this utterly racist argument are all wrong. Lithography from line drawing, 11 × 10 cm (image).

supported neo-Lamarckian ideas of environmental determinism and evolution that were deployed in justification of European imperialism (Livingstone 1984; Livingstone 1992, 177–259). Indeed, analytic mapping was central to the deductive logic of the environmental determinism that underpinned much anthropological and sociological thought (fig. 4.11; Winlow 2009; also Winlow 2006; Livingstone 2010). Further intersections of the social and the natural are also evident in the social and technical history of analytic mapping. Socially, analytic mapping is primarily a function of modern industrialization and the need of modern states to categorize, identify, and manage human, natural, and economic resources. Whether the surveys and mapping are undertaken directly by each state or indirectly by academics is irrelevant; both are integral to the ability of the modern state to function (Hannah 2000; Demeritt 2001). Technically, analytic mapping has deployed the same particular semiotic strategies for symbolizing and visualizing spatial phenomena, such as the isoline, graduated circle, or flow line, regardless of the phenomena being mapped. In considering the history of analytic mapping, some map historians have emphasized the development of these strategies (Robinson and Wallis 1967; Robinson 1971; MacEachren 1979; Friendly 2005; Friendly 2008a). An emphasis on the techniques of analytic mapping has assisted academic cartographers in their claims to pursue a discipline distinct from other natural and social sciences.

Analytic mapping has promoted the idealization of the inherently visual and scientific nature of cartography. It built on the ideal's elements of simplification and visualization by applying the visual nature of maps to things that are not visible; making the invisible both visible and measurable is, of course, a hallmark of modern science. The sense of analytic mapping as being necessarily scientific only developed as the practices and institutions of the natural and social sciences themselves were refined over the course of the nineteenth century (Hannah 2000; Palsky 2002; Gilbert 2004; Schulten 2012; Dunlop 2015; Hansen 2015).

The apparent interwovenness of the mapping of distributions with the mapping of locations has been reinforced and justified by the superimposition of a conceptual continuum onto a historical narrative. Conceptually, the continuum moves from general maps that record the locations of all geographical features ("base maps"), through "special purpose" maps that focus on just one or two classes of feature, such as maps associated with rail and road travel or the management of a particular resource, to analytic maps that map just one or two kinds of feature. Historically, special-purpose mapping is understood to have been the lineal precursor of analytic maps *per se* (Licka 1880). This narrative was crucial for postwar justifications of academic cartography (e.g., MacEachren 1979; Petchenik 1979; Castner 1980), and especially for Arthur Robinson's (1982) recasting of the general-/special-purpose divide into the "substantive" and "visual phases" of cartographic history (see Edney 2014a, 87), and it is still being rehearsed (e.g., Slocum and Kessler 2015, 1503–5). It must be recognized that the conjunction of the continuum with the narrative rests on graphic and linguistic confusions, both in the apparent similarity of map forms and in the way that special-purpose maps are deemed to address particular "themes"; special-purpose and analytic maps are products of markedly distinct spatial discourses and are not directly related.

Finally, within professional circles at least, their similar reliance upon statistical manipulation has further promoted the unity of analytic and locational mapping. Gauss developed least-squares analysis in large part to model unquantifiable errors within the huge number of observations involved in a geodetic triangulation; the analysis provides final values for the angles of each triangle, such that the sum of the squares of all errors across the triangulation is minimized. By the twentieth century, the major mapping agencies were further defining the accuracy of the content of their topographical maps in terms of descriptive statistics and probabilities, in the same kind of way that social scientists had defined the reliability of their censuses and other social survey data. That is to say, for professionals, the processes of observation and measurement were no longer naively unproblematic but were to be handled in the same, statistical manner regardless of the spatial phenomenon being observed (Woodward 1992, 52; Goodchild 2015a; Buttenfield 2015; see, e.g., Miller and Schaetzl 2014). Once again, cartography appears as a universal and singular endeavor.

New Mapping Professions

The professionalization of map making in the nineteenth century—with the formation of specialized communities and institutions governing credentials and career advancement—was a function both of the growing bureaucracies of nineteenth-century governments and of the socialization of knowledge. In Britain, for example, the Land Surveyors' Club was formed in the early nineteenth century as a means to bring structure and order to a diverse community of land surveyors and estate stewards; this society eventually developed into the Royal Institute of Chartered Surveyors (Thompson 1968, 94–100, 129–30). In particular, two trends in professionalization supported the continuing development of the ideal of cartography.

CREATION OF CARTOGRAPHIC ARCHIVES
ontogeny / materiality / discipline / singularity and universality.   Before the second half of the eighteenth century, European institutions of state paid relatively little attention to collecting and arranging maps as a special category of archival material. For the most part, maps were kept with the original letters and memoranda with which they were created and communicated; only those maps too large to be easily bound into volumes were isolated for separate storage (Skelton 1972, 26–52; Harley 1987, 8). Individual politicians, soldiers, and administrators assembled their own map collections as much for personal and intellectual reasons as for professional ones. They ranged in size from the 48 maps that the English imperial administrator William Blathwayt bound into a guard book in about 1683 (Black 1968, 1970–75) to the almost 60,000 maps, charts, and plans, plus accompanying texts, that George III had acquired by 1811 (Harley 1987, 9; Barber 2003, 2005a). Some large map libraries were formed relatively early. In France under Louis XIV's minister, Jean-Baptiste Colbert, for example, a separate map collection was in existence within the Marine by 1682 and was formally established in 1720 as the Dépôt des cartes et plans de la Marine (Chapuis 2019), while the British Museum apparently

had a separate map room at the time of its opening in 1759 (Wallis 1973). Generally, territorial bureaucracies, especially within the French government, began to establish institutional structures for collecting, storing, and arranging coherent map collections only in the later eighteenth century (Kingston 2011, 2014; Heffernan 2014; Fulton 2017).

The prosecution of general cadastral, topographical, and hydrographical surveys in the nineteenth century, as well as the origins of resource and sociological surveys by government agencies and their academic surrogates, introduced the series map, composed of multiple sheets. At first, these large assemblages of maps were commonly bound into volumes, but as the individual sheets began to be separately updated, they were increasingly stored as separates. Official map collections—and new, dedicated map libraries that served the public—accordingly grew to permit this huge array of material to be readily retrieved and used. Map librarianship was born to manage the literal map archive.

The easiest means of arranging large collections, other than series maps, was by region covered. This was the primary organizing principle employed by the Bern lawyer and politician Johann Friedrich von Ryhiner for the collection of some 16,000 works that he had assembled by 1800; Ryhiner even wrote a detailed, but unpublished, monograph explaining how best to arrange such large collections (Klöti 1994, 221). George III similarly, if loosely, organized his huge collection by region, mixing topographical, geographical, and marine works; he kept separate only his collection of detailed government survey maps of Hanover. But after 1811, his librarians imposed a higher level of organization by dividing his collection into its geographical and topographical, military, and marine components.

The new map libraries and archives of the nineteenth century further emphasized the organization of maps by region and by type (general, marine, analytic). Maps were removed from their original contexts, whether an archival letter, a book, or a broken atlas, and stored with other maps showing the same area. Within each region and broad type, the maps were then generally arranged in chronological order. Within the new libraries, maps were evaluated according to their contribution to the grand archive of spatial knowledge and to the historical narrative of the growth of that archive as a surrogate for the rise of modern civilization. The new libraries supported the common usage of "*the* cartography" of a region, usage that supposes the existence of a comprehensive archive of spatial knowledge, from which each new map is derived. The Edinburgh geographer Patrick Geddes even sought, in about 1900, to physically manifest such a comprehensive archive in a grand, national (Scottish) geographical institute in which all spatial information would be collected, reconciled, and calculated (fig. 4.12). With hindsight, we should not be surprised that his ambitious plans did not prove feasible (Withers 2001, 225–32).

The literal archive of spatial knowledge—the grand assemblage of spatial

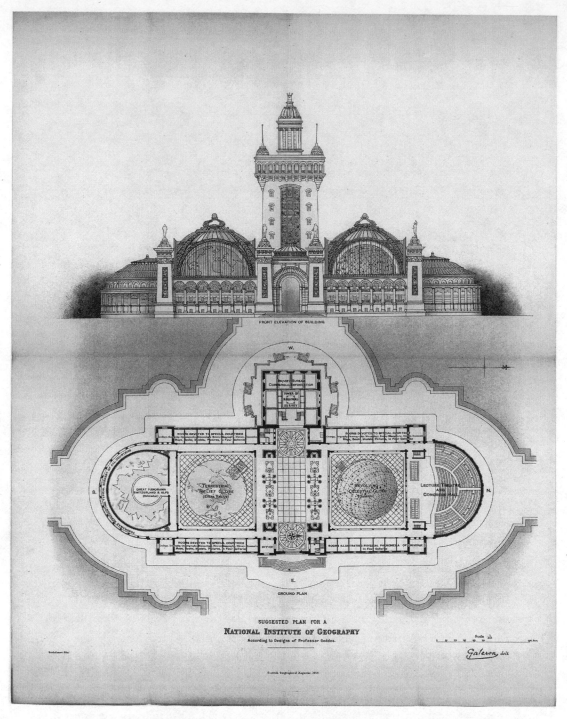

FIGURE 4.12. Paul Galeron, design for Patrick Geddes's "temple of geography," influenced by Elisée Reclus (Geddes 1902). Color lithograph, 53 × 45 cm. Courtesy of the Bartholomew Archive, National Library of Scotland (Acc.10222/PR/33a, fol. 53b).

knowledge across archives and libraries—appeared to make real the figurative archive engendered by the ideal. The apparent unification of the distinctive geometries of mapping by the projective geometry of the ideal sustained the conviction that cartography, as an endeavor, seeks to extend the archive of knowledge so that it is coincident with the world, and that maps are simply reductions of this world/archive. The culmination of this trend was, in many respects, the post-1880 attempt to create a comprehensive collection of consistently scaled spatial knowledge within the ordered sheetlines of the International Map of the World (Pearson et al. 2006; Pearson and Heffernan 2015; Rankin 2016, 23–64; Rankin 2017). The virtual world offered by digital technology in the late twentieth century has offered a new lease of life to the dream of a comprehensive archive (Schuiten and Peeters 2002; see Fall 2006), even one updateable in real time, blurring the figurative with the literal archive.

ACADEMIC CARTOGRAPHY

pictorialness / individuality / observation / morality / singularity and universality. Cartography developed as a distinct subject of academic study in central Europe after 1880, although it remained relatively weak institutionally. Map design was taught as part of geography in technical schools and universities, and only a few scholars sought to identify and study its principles. In his last, posthumously published book, Max Eckert tried to secure the institutional standing of academic cartography by arguing for its value to the Nazi authorities in Germany (Eckert 1939).* Only after 1950 did the intellectual and practical demands of nuclear-powered industrial societies support the formation of academic programs and institutions in cartography, in North America as well as in Europe.

Central European academics initially posited a variety of words for the study and codification of map design and production, including *Kartenkunde* ("map study"), *Kartenwissenschaft* ("map science"), and *Kartologie* ("systematic map study"). *Cartographie / Kartographie / cartography* was increasingly accepted as the standard term after World War II but did not completely displace the older forms, notably "cartology," until about 1970. This further meaning for "cartography" is fully in line with the neologism's various potential meanings, but it nonetheless conflates the academic study of maps with the scientific endeavor of map making. In particular, academic cartographers sought to codify the practices of drawing maps according to a rigorous logic and, eventually, to a "science" of map design. They did so in large part through historical examples (Edney 2008b, 2014a).

Academic cartographers worked hard to ensure that map design was indeed an objective science in line with the ideal. In the first half of the twentieth century,

---

*Historians are divided over whether Eckert was indeed a Nazi sympathizer (Kretschmer 2015a) or if his last book was rather a cynical attempt to secure support (Pápay 2017).

Central European academics sought to define and codify the compromises inherent in the making of any map: the mathematical impossibility of representing the curved surface of the earth on a flat piece of paper without altering the shapes and areas of regions; the necessary partiality and selectivity in deciding which features to show, and how to show them; and the complexity of depicting the shape of the landscape in planimetric form. In other words, maps cannot achieve the perfection promised under the ideal. An Austrian-trained agricultural economist and analytical map maker could thus state as an axiom that maps "have their limitations as perfect representations of the earth's surface" (Marschner 1944, 1). The goal of academic cartography was the proper codification and regulation of all these compromises, especially as mediated by the craft and aesthetics of mapping. Academic cartographers sought, therefore, as Eckert's (1908, 347) early formulation put it, to study the "logic" of cartography such that the "dictates of science" would restrain cartographers' potentially "erratic flight[s] of the imagination."

During World War II, U.S. scholars were disgusted by the willful corruption of cartographic logic by German propagandists for political ends (fig. 4.13; see Monmonier 1991, 87–112; Herb 1997; Murphy 1997). In response, they highlighted the problems of propaganda mapping and reminded geographers of best practices in order to reestablish cartography's necessary compromises on an objective footing (Speier 1941; Quam 1943; Stewart 1943; Chamberlin 1947); they also began to call for the more rigorous disciplining of map makers as professionals. In particular, J. K. Wright (1942) argued that the practices of map making have no inherent protections against political biases, personal idiosyncrasies, and other subjective elements, so that map makers henceforth would have to be thoroughly trained and disciplined in cartography's scientific foundations, and the practice would need to be regularized as a profession (also Harrison 1958). U.S. scholars thereafter spearheaded the scientific and objective study of map design (Pickles 1992; Crampton 1994; Crampton 2010, 49–61; Krygier 1995; Cosgrove 2007a, 205–6).

After the war—as the academic institutions of cartography flourished on both sides of the Atlantic—Anglophone academics came to define cartography as the "art and science of map making" (Anon. 1964; see Maling 1991). In this formulation, "art" embraced not only the aesthetics of map design but also the craft of map production, which is to say, the fundamental logic that disciplined and kept in check an individual's subjective proclivities. While the modern definition of cartography as both "art and science" appears to balance the observation and measurement of the world ("science") with individual creativity ("art"), it should more properly be read as the combination of observation and measurement with logical, systematic, and disciplined presentation. Academic cartography has thus insisted on the scientific and objective nature of the profession (also undefined) and of all professional mapping activities.

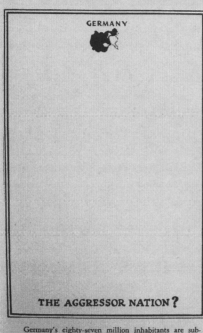

FIGURE 4.13. A classic propaganda map: "A Study in Empires," *Facts in Review* 2, no. 5 (5 February 1940): 33. One of a series of cleverly designed persuasive or propaganda maps included in a journal published by the German embassy in Washington, DC, that sought to confuse U.S. support for Britain. Other maps justified German territorial aggression. Publication ceased once Germany declared war on the United States on 11 December 1941. Lithograph, 24 × 18 cm.

Mass Mapping Literacy

The final set of factors contributing to the development and elaboration of the ideal of cartography stems from the rise of mass literacy in the nineteenth century. The restricted literacies of the early modern "public sphere" gave way, at least within industrial democracies, to compulsory primary education and a diverse mass market for printed goods, from "penny dreadfuls" to maps. That market was, in turn, supplied by the proliferation of new and cheaper printing technologies (Brückner 2017). Maps of all sorts entered the lives of more people than ever before, as tools of knowledge, navigation, education, and entertainment. By the end of the nineteenth century, maps seemed almost ubiquitous in Western society.

Writing in 1880, a French geographer argued that the renewal of geographical studies after the Franco-Prussian War (1870–71) rested on the intellectual foundation that French geographical societies had provided since 1821. Among the "prodigious" results of the explorations sponsored by the societies was a marked increase in the numbers of maps that disseminated the new knowledge to all levels of society, with demonstrably positive results:

> And, at the moment, one can say that Europe is congested: general maps and special maps; maps physical, geological, archaeological, historical, political, military, marine, hydrographic, agricultural, industrial; topographic maps, popular maps. They are of all prices, at all scales, and for all needs; they are everywhere, in offices, hotels, and even in cabarets; on many walls the map of the country has replaced the portraits of kings and emperors. The effect of this sudden increase in the means of knowing is the happiest: the concepts become positive; they spread incessantly in the crowd, whose ideas correct themselves and whose errors disappear; they facilitate corrections and allow one to keep oneself informed. The earth has its official report, prepared from year to year [i.e., the proceedings of the geographical societies]; nothing is neglected so that we can more certainly ignore nothing. (Desdevises du Dézert 1880, 17)*

Implicit in this commentary is the sense of a giant body of knowledge about the world, its features, and its inhabitants. It is a vast corpus of information that is progressively prepared and corrected by exploration and investigation, and it is expressed in graphic form by "the map." This sentiment only grew with the late nineteenth-century intensification of imperial activities and geographical pursuits (Hudson 1977; Wu 2014; see Livingstone 1992, 216–59; Driver 2001; Butlin 2009).

But the proliferation of maps of all sorts stemmed less from the wealth of information available, as Théophile-Alphonse Desdevises du Dézert (1880) would have

---

*"… et, à l'heure qu'il est, on peut dire que l'Europe en est encombrée: cartes générales et cartes spéciales, cartes physiques, géologiques, archéologiques, historiques, politiques, militaires, nautiques, hydrographiques, agricoles, industrielles, cartes topographiques, cartes populaires. Il y en a de tous les prix, à toutes les échelles, et pour tous les besoins; il y en a partout, dans les bureaux, dans les hôtels, et jusque dans les cabarets; sur bien des murs la carte de la patrie a remplacé les portraits des rois et des empereurs. L'effet produit par cette subite multiplication des moyens de connaitre est des plus heureux: les notions deviennent positives; elles se répandent incessament dans la foule, dont les idées se rectifient, et dont les erreurs disparaissent; elles facilitent les corrections, et permettent de se tenir perpétuellement au courant. La terre a son procès-verbal, rédigé d'année en année; on ne néglige rien, pour arriver plus sûrement à n'ignorer rien."

had it—a conviction of currency stemming from the disciplinary preconception—and more from the mass mobilizations of society in aid of nationalism and the myth of the nation-state, imperialism and decolonization, internationalism (communism, fascism, neoliberalism), industry, and global war, both hot and cold (Dym and Offen 2011; Bryars and Harper 2014; Barney 2015; Monmonier 2015; Harper 2016; Rankin 2016; Akerman 2017). Map imagery proliferated in public spaces: in advertisements and on walls (Barber and Harper 2010, 160–69), in classrooms and in newspapers, on stage and on screen (e.g., Houston 2005; Conley 2007). Many contemporary artists in the twentieth and twenty-first centuries have fixated on mapping motifs, using them to reflect on and to challenge popular concepts of representation, space, and place (Wood and Krygier 2006; Watson 2009). The profusion of maps and the variety of reasons for their deployment are overwhelming. Nonetheless, three particular trends led general populaces to engage actively with particular kinds of maps and mapping.

PRIMARY EDUCATION AND GEOGRAPHICAL INSTRUCTION
pictorialness / individuality / materiality / observation / discipline / singularity and universality.   Over the course of the nineteenth century, primary education in geography across Europe featured two approaches, the "analytical" and the "synthetic" (Potter 1891, 419–20). The analytical approach to geography began with the idea of the globe before introducing coarse-resolution world and other geographical maps as abstractions of the actual globe; only then would each part of the world be considered in turn. Some implementations of this "analytical method" required children to copy, or construct, coarse-resolution maps of the world and its major parts, whether drawn on paper or embroidered on cloth (fig. 4.14; Tyner 2015b; Schulten 2017).

The synthetic method originated in Jean-Jacques Rousseau's (1762) proposal to teach geography by working progressively outward from the home to the ends of the earth. The radical, anticlerical pedagogue Edme Mentelle, who had originally adhered to the analytical method of working from the general to the particular, implemented Rousseau's ideas with great effect in his *La géographie enseigné par une méthode nouvelle* of 1795 (Heffernan 2005, 279, 293). Pedagogues in the young American republic, keenly aware of their hard-won independence, followed suit. Emma Willard and William Woodbridge, for example, developed this method in the 1820s and 1830s, explaining how geographical education should start with students mapping their own classroom at a very fine resolution and then working outward to the immediate environs of the schoolhouse, then to the town (fig. 4.15), and so on to the entire country, and then, by means of voyages, to mapping the whole world. In this pedagogic methodology, the process of map making is clearly elucidated as the process of observing and measuring the landscape (Schulten 2007, 2017).

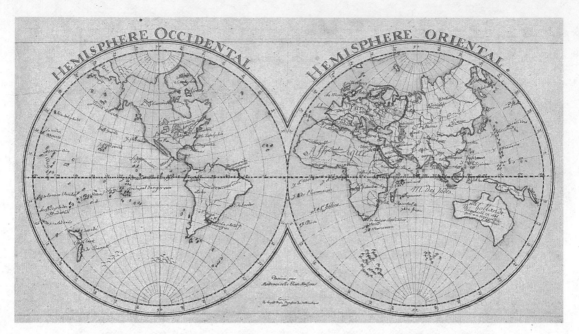

FIGURE 4.14. An example of a child's neatly drawn geographical map, the product of the analytical approach to teaching geography. Elise Massieu, double-hemisphere world map, in "Etrennes dediées a Mlle. C. Fromaget par sa nièce" (St. Quentin, Aisne, France, 1806), fol. 2r. Manuscript, 38 × 26 cm. Courtesy of the Osher Map Library and Smith Center for Cartographic Education, University of Southern Maine (Osher Collection); www.oshermaps.org/map/46798.0009.

In Germany, in about 1800, Christian Gotthilf Salzmann developed the synthetic approach into what he called *Heimatskunde*, which would become a central plank of primary education under Prussia's reforms in primary education fifty years later. Children were again first introduced to their home community (*Heimat*) and surrounding environment through topographical maps and relief models, before the scope of their geographical studies expanded to cover the province, *Reich*, Europe, and world (Keltie 1886, 477–81). Prussia's stunning defeat of France in 1870–71 and the subsequent formation of the new German empire led other European countries to rapidly adopt the evidently superior Prussian curriculum for primary education, including geography. For example, in the United Kingdom, an editor completely revamped Robert Sullivan's highly successful instructional manual for teaching geography. With some sixty editions of this manual having begun according to the analytical method, with instructions for teaching the nature of the globe, after 1874 new editions of the manual began with an exercise to have students make maps of their playground or schoolroom through a process of observation and measurement. The result, cribbed directly from Woodbridge, would image the room or playground "as it would appear to a person looking down from the ceiling." Only after establishing the basic act of surveying and mapping the smallest details of the world did the new edition of the manual move onto the more general

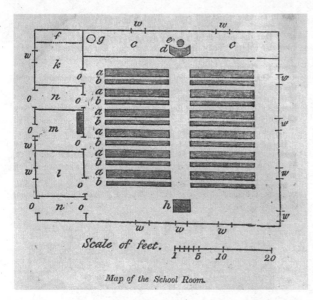
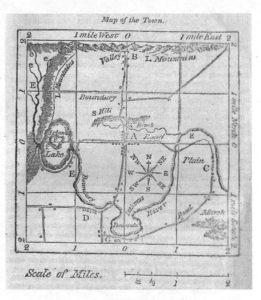

FIGURE 4.15. "Map of the School Room" and "Map of the Town" (Woodbridge 1831, 3 and 8), reproduced here from the work, included with its own pagination, in Woodbridge (1838). These exemplify the synthetic approach to teaching geography. Woodcut diagrams, each ca. 8 cm high. Images courtesy of the Harvard Map Collection, Harvard University; reproduced courtesy of Harvard College Libraries, Harvard University (Widener KC 10337).

mapping of the world. The exercise concluded with the statement, "geography is commenced, *as it should be*, with TOPOGRAPHY" (Sullivan 1883, 8; italics added). Or, as a Thai schoolbook explained in 1902, as part of a state initiative to inculcate a "modern" attitude toward mapping and science, maps were the same as plans and similarly imaged the world as if seen from above (Winichakul 1994, 51–52).

In presenting detailed topographical knowledge as the foundation of all spatial knowledge, *Heimatskunde* expressed the ideal's assumption that all mapping is based on the creation of a detailed and fine-resolution archive of knowledge, from which all maps are derived. This conviction is the basis of academic cartography's fixation on the processes of "generalization," by which the topographical archive is turned into coarser resolution maps. For example, in the first of three lectures in 1909 to the Royal Geographical Society, Edward Reeves stated as axiomatic that "all maps are, or at least are supposed to be, based upon some kind of survey, and for any surveying, instruments for exact measurement are essential." In considering the history of cartography, he asserted that

> it is reasonable to imagine that the earliest of cartographical representations would consist of maps and plans of comparatively small areas, produced to meet some demand of the times, and it would only be later on, in more advanced conditions, that any attempt would be made at geographical generalization. (Reeves 1910, 5, 4)

THE IDEAL OF CARTOGRAPHY EMERGES

It is, of course, only reasonable to imagine so if one believes that the fundamental and sole function of mapping is to generate and store an archive of spatial information.

MAPPING FOR PERSONAL MOBILITY

pictorialness / individuality / materiality / efficacy / discipline / singularity and universality. The ubiquity of maps in modern culture was furthered, through the nineteenth century and then explosively in the twentieth, by the rise of personal mobility (Akerman and Nekola 2016). First came the popular pastime of recreational hiking, aided by the romantic obsession with mountains and Napoleon's road-building through the Alps, and then the practice of orienteering (Zentai 2015). The new transportation technologies of railroads, bicycles, and finally automobiles facilitated individual travel and mobility. Taken together, this meant that large numbers of people, including women (Dando 2007, 2017), increasingly used maps for the first time, and they did so in an instrumental manner. By 1893, the general public, at least in the form of the well-off tourist, had joined specialized professionals in the ranks of map *users*:

> Everyone has at hand maps on which are traced the ground to the smallest depression; each consults them, the military officer for his defense plans, the engineer for his projects, the tourist for his pleasure.

And, the same commentator confessed, even though all the new users of territorial maps knew little about how they were made, the maps had become "so familiar and appealing" by frequent study that everyone accepted that their "tangle of lines" did indeed rest on a "stable" "backbone" (Dallet 1893, 11).*

Traditional muscle- and wind-powered travel had depended on the assistance of professional guides, mariners, and, when travelers were uncertain or lost, local informants (Delano Smith 2006). Mechanized forms of travel progressively replaced human assistance with specialized technologies created and maintained by modern states, from lighthouses to printed schedules to road signs to special-purpose maps (fig. 4.16). Within these complex technological systems, such maps permitted individuals to plan and conduct their own travels without further human assistance or control. The use of these maps, which seem to grant individuals the ability to control their own movements, effaces the complex technological systems and gives the impression of autonomy (Hornsey 2016).

---

*"Tout le monde a sous la main des cartes sur lesquelles sont tracées jusqu'aux moindres dépressions du sol; chacun les consulte, le militaire pour ses plans de défense, l'ingénieur pour ses projets, le touriste pour son plaisir. | On sait peu, en général, comment se font ces cartes dont l'étude est si familière et si attrayante. On sent bien que ces linéaments tracés sur la feuille reposent sur un fond stable, on découvre au travers de ce fouillis de lignes, le squelette qui leur sert d'appui."

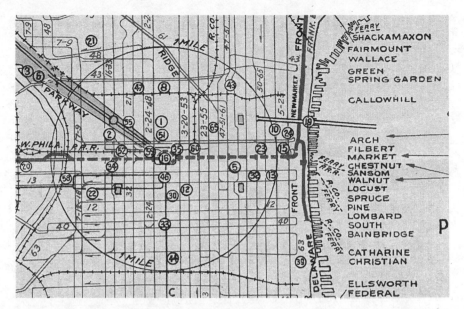

FIGURE 4.16. Detail of Philadelphia Rapid Transit Company, *Trolley and Bus Routes and Sight Seeing Places* (Philadelphia, 20 June 1926). An example of a map produced within a complex technological system. Color lithograph, 71.5 × 43 cm (entire), 9 × 14 cm (detail). Courtesy of the Osher Map Library and Smith Center for Cartographic Education, University of Southern Maine (OML Collection); www.oshermaps .org/map/45505.0001.

The spread of map use for travel across modern society has led to the routine disciplining of maps. The newly ubiquitous route and road maps are tested against the depicted world, repeatedly proving or disproving each map's accuracy (fig. 4.17). If found wanting, the offending maps might be annotated with the necessary information (Akerman 2000) or discarded. An early instance reveals the pattern. Henry David Thoreau, on the first night north of Bangor on his first trip into the Maine Woods, in 1846, stayed in a tavern in Madawamkeag, where he found a late state of Moses Greenleaf's *Map of the State of Maine with the Province of New Brunswick* (fig. 4.18). Because he "had no pocket map," Thoreau

> resolved to trace [Greenleaf's work as] a map of the lake country: so dipping a wad of tow into the lamp, [he] oiled a sheet of paper on the oiled table cloth, and, in good faith, traced what [he] afterwards ascertained to be a labyrinth of errors, carefully following the outlines of the imaginary lakes which the map contains.

Thoreau's testing of Greenleaf's map against the world itself exemplifies the modern ritual in which travelers of all sorts test maps—whether oral, paper, or digital—against the world to prove or disprove their quality. Indeed, Thoreau went on to imply that works that fail the evaluation process should not even be called

157

FIGURE 4.17. The cover of an early twentieth-century road map, indicating both the reliance of personal mobility on a technological complex, including the production of specialized maps, and the manner in which the road map is tested against the world. Standard Oil's 1929 map of New Jersey, distributed by Standard Oil's Service Station no. 126, Maplewood, NJ. Color lithograph, 20 × 29 cm (image). Courtesy of the Osher Map Library and Smith Center for Cartographic Education, University of Southern Maine (French Collection); www.oshermaps.org/map/1003656.0001.

"maps," stating that George W. Coffin's 1835 "Map of the Public Lands of Maine and Massachusetts is the only one I have seen that at all *deserves the name*" (Thoreau 1848, 73, emphasis added; see Ryden 2001, 96–134).

Moreover, the sporadic nineteenth-century practice of sticking pins into maps, especially to follow distant (fig. 4.19) or even local conflicts,* was intensified and standardized through the 1900s, complete with specialized commercial products, such as Rand McNally's "Map Tack System." The practice became so ubiquitous that the pushpin would be digitally reconfigured as the default location marker

---

*In an imagined dialogue, in his "Prolegomena to an Apology for Pragmaticism" of 1906, Peirce (1931–58, 4: 411–12, ¶530 [1906]) imagined a general explaining one use of maps: "had he replied that he found details in the maps that were so far from being 'right there,' that they were within the enemy's lines, I ought to have pressed the question, 'Am I right, then, in understanding that, if you were thoroughly and perfectly familiar with the country, as, for example, if it lay just about the scenes of your childhood, no map of it would then be of the smallest use to you in laying out your detailed plans?' To that he could only have rejoined, 'No, I do not say that, since I might probably desire the maps to stick pins into, so as to mark each anticipated day's change in the situations of the two armies.'"

FIGURE 4.18. Detail of the interior "lake country" of Maine from Moses Greenleaf's *Map of the State of Maine with the Province of New Brunswick* ([Portland], 1844). This map was originally published in 1829. Lithograph in 4 sheets, dissected into 16 segments; entire map, 136 × 106 cm. Courtesy of the Osher Map Library and Smith Center for Cartographic Education, University of Southern Maine (Osher Collection); www.oshermaps.org/map/1027.0001.

in online map applications such as Google Maps, which first went online in 2005 (Ehrenberg 2006, 231; Wallace 2015; Wallace 2016, 25–63). Rand McNally's system was one of many technologies developed in the nineteenth and twentieth centuries by ingenious U.S. inventors, who reacted to the increasing integration of maps into daily life by creating new addressing systems, new ways to fold maps, technologies for automotive navigation, and so on (Wyckoff 2016; Monmonier 2017; Monmonier et al. 2018).

There have been three particular results of the increasing prevalence of map use among the publics of the industrialized world. First, the presumed tie of the map to the territory was intensified, especially through cycling and hiking maps that are overtly concerned with representing relief in detail. Second, cartography's status as a science was reinforced, in that its products were constantly tested and refined by their users. And third, because automobile maps became the predominant popular map form in the twentieth century, a new, functional conviction developed within the ideal that *all* maps are instruments of navigation. Overall, the repeated use of maps for travel and for keeping track of business activities enforced the belief that all maps mirror and model the physical world.

159

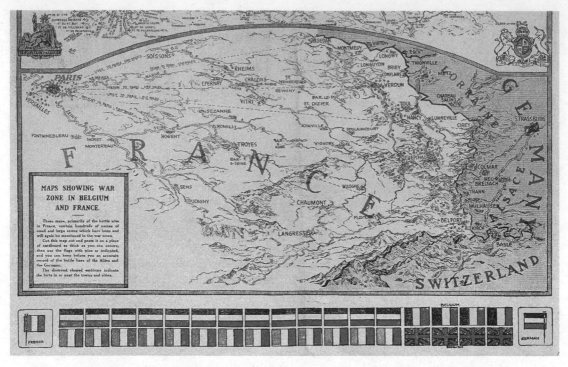

FIGURE 4.19. Detail of "Maps Showing War Zone in Belgium and France," *Boston Sunday Post* (20 September 1914): front cover. The text invites the reader to add the flags at bottom to pins, to be used "to keep an accurate record of the battle lines of the Allies and the Germans." Courtesy of the Osher Map Library and Smith Center for Cartographic Education, University of Southern Maine (Yensen Collection); www.oshermaps.org/map/42761.

POPULAR AND PICTORIAL MAPPING

pictorialness / observation / efficacy / singularity and universality. Popular culture reacted to the increasing dominance of the ideal by playing with the idea of the map. Early modern commentators had used mapping strategies for satirical and allegorical purposes, to show particular relationships and concepts; geographical writing also supported some transgressive works that cast women as continents, with pornographic intent (e.g., Reitinger 1999; Lewes 2000). The nineteenth century was a golden age of moralistic map imagery, in which strategies of both regional and place mapping were used to allegorize human relationships and life journeys, with a heavy emphasis on the many, broad paths to Hell and the single, narrow path to Heaven (Reitinger 2008). Moreover, the ideal's emphasis on direct observation gave rise to a new aspect of the carto-pornographic tradition that equated women's bodies to landscapes, to be subjected to the male gaze and traveled over in search of "treasure" (McClintock 1995, 1–4; Edney 2007a). The inverse of this practice is Liz Gutowski's 1980 artwork, "*Coitus topographicus*," which commented on the excessively clinical nature of sex manuals (reproduced by Holmes

FIGURE 4.20. The location of hidden treasure: "X marks the spot," not on the map but in the landscape itself. Detail of Kathleen Voute, "Ye Mappe of Happie Girlhood wherein is Shewn ye Camp Fire Girls—How They Disporte Themselves and Grow in Health and Service into a Blythe and Useful Citizenry," *Everygirl's: The Magazine of the Camp Fire Girls* (March 1926): cover image. Courtesy of P. J. Mode Collection of Persuasive Cartography, Cornell University (2234); digital.library.cornell.edu/catalog/ss:19343633.

1991, 186; Chwast, Heller, and Venezky 2004; and Besse and Tiberghien 2017, 84; see DeRogatis 2005).

Such strategies combined with the anxieties of modern capitalism to give rise to the completely fictitious category of "pirate treasure maps," common in twentieth-century juvenile adventure literature, which guide the fortunate to riches on this earth (Zähringer 2017, especially 14). Responsibility for this common literary trope lies with the famous map in Robert Louis Stevenson's *Treasure Island* (1883); the map was prominently displayed in the many twentieth-century film and television adaptations of the novel and also in the advertisements for them. Stevenson's conceit stemmed in large part from the manner in which he had conceived, planned, and written the novel as a creative outgrowth of a topographical map of an island that he had idly drawn (Stevenson 1894). In a further, satirical permutation, the X on the map that "marks the spot" has been construed as a feature of the landscape itself (fig. 4.20), which, like any other small feature, can be manipulated (as in the Warner Brothers "Bugs Bunny" and "Wile E. Coyote" cartoons).

In the twentieth century, and especially from the 1920s through the 1960s, a genre of popular and sometimes surreal mapping flourished in Europe and North

THE IDEAL OF CARTOGRAPHY EMERGES

FIGURE 4.21. Pierre Lissac, "Demandez le Nouveau Plan de Paris!" *Vie Parisienne* (14 February 1920). A humorous map of the craze for jazz dance following the 1918 Armistice. Color lithograph, 33 × 52 cm. Courtesy of the P. J. Mode Collection of Persuasive Cartography, Cornell University (1209); digital.library .cornell.edu/catalog/ss:3293871.

America (fig. 4.21). Popular mapping cut across several modes to express and construct a variety of national, regional, urban, and place identities. Such mapping employed a variety of design strategies borrowed from the graphic arts—and especially poster design and animated films—to characterize and caricature social, cultural, economic, and commercial differences. Some were staid, some humorous; some were intended to make education fun; some were for tourists, others for advertisements (Griffin 2013, 2017; Hornsby 2017). But each genre of popular mapping relied for effect on consciously referring back to the ideal. The significance and humor of each pictorial or metaphorical map derived from its self-evident difference from the normative map.

The awareness of producers and consumers alike that such maps were unorthodox served only to reinforce the validity of the ideal of cartography. Indeed, popular mapping did not challenge the ideal. Those cartographically abnormal occasions when people have shaped continents or countries as people or animals or have produced maps of imaginary places were construed simply as playful fun unrelated to actual, proper practice; such allegorical and satirical maps were the cartographic equivalent of a "busman's holiday" (Tooley 1963, 3). The manifest idiosyncrasy of cartographic "oddities" and "curiosities" required each to be considered on its own terms, as a unique work, and as such they have been unsuited for use in mounting any sustained critique of the nature of the normative map (Hill 1978).

Popular maps could have enormous reach. When the stars of the hugely successful U.S. radio show *Amos 'n' Andy* sought in 1930 to demonstrate the size of their audience, as part of their negotiations for a new contract, they offered their listeners a free map of the fictitious Weber City, the subject of a long-running story line about a real estate development. One million people—almost one percent of the U.S. population at the time—wrote in for a copy of the map (MacDonald 1979, 33). While it does seem improbable that so many of these maps were actually printed and distributed, I have seen impressions dated as late as 1935, so perhaps it did reach such a large audience. Saul Steinberg's parodic "View of the World from 9th Avenue," which graced the cover of *the New Yorker* in March 1976, reached the half-million or so people who then subscribed to the magazine, only a fifth of whom resided in New York itself (Travis 2000, 255–56; Downs 2015, fig. 655, reproduced the original artwork). Many more millions of people in the United States and the rest of the world have undoubtedly seen the map in its frequent reproductions; its many imitations rely for their effect on the fame of Steinberg's original.

Popular, pictorial maps have become a prominent element in recent map scholarship (Hanna and Del Casino 2003; Bryars and Harper 2014; Monmonier 2015; Harper 2016; Hornsby 2017), and have served as the basis of several works with a popular bent that have promoted sociocultural interpretations of maps (e.g., Holmes 1991; Harmon 2004, 2016; Harzinski 2010). This new trend has effectively operationalized Brian Harley's (1989, 3) observation that the sociocultural critique "demands a search for metaphor and rhetoric in maps where previously scholars had found only measurement and topography." Yet, while an appreciation of metaphor and rhetoric in maps has helped to reveal the ideal's limitations, it does nothing to challenge the ideal's core convictions. Overtly rhetorical and popular maps are compelling precisely because of their impropriety and the manner in which they diverge from the ideal's tenets; they seem exotic and illicit, and there is something kinky about their quirkiness. Their existence only emphasizes that the proper, default state of mapping is the maintenance of a geometrically rigorous relationship of image to world.

Forging the Web

The ideal of cartography developed and grew over the course of the nineteenth and twentieth centuries. In 1800, mapping continued to be understood as several different practices, variously producing and consuming maps (specific), charts, and plans. But already by the 1820s, the projective geometry of systematic, territorial mapping seemed, to some specialists and experts at least, to unify all mapping activities within a single endeavor. Further factors continued to reinforce and publicize this idealized understanding of mapping, from the spread of the neologism "cartography," to pedagogic techniques, to the increasing frequency with which the public used maps instrumentally, to the unreflective application of numerical ratios to all maps. This review can, however, barely summarize the ideal's emergence within modern culture, its rise to hegemonic status, and the naturalization of the generic, normative map.

In large part, the manner in which the ideal pervades modern culture stems from its history. Each new factor reinforced already developed convictions. Academic cartographers, in particular, inherited an existing ideal and worked in accordance with its precepts. Each new factor added further nuance and complexity. The deeply entrenched preconception that maps are necessarily functional, and more specifically are intended to guide movement, grew out of the proliferation of route-finding maps for public consumption, not out of any logical corollary to the convictions established before the late nineteenth century. If the ideal was a logically structured whole, it would be easy to overcome it: contradict one conviction, and the entire ideal would become untenable; the whole would unravel if we were just to pull on one thread. But the ideal is not a logical construct. It is a multifaceted, reflexive, and frequently contradictory body of beliefs and convictions that together possess a remarkably resilient internal flexibility. Narrowly targeted critiques can be easily sidestepped. Different elements can simply be brought into play as need dictates and without any concern for contradictions.

Cartography thus includes concepts that have meant different things to different people at different times. It comprises an extensive archive of fine-resolution spatial data from which all coarser resolution maps are derived, according to the logical procedures of cartographic generalization. "The map" is a factual record of the world, but it also simplifies the world in line with objective logic to promote the visualization and understanding of the world. Cartography is a mathematical process, a transformation from data sets in three and two dimensions. The map is a functional tool, intended to aid human navigation across land and sea. The map is the product of vision, and more especially of vision from above, even as it is a tool of visualization. The map is the model for metaphors of representation and knowledge as well as for satires and allegories. The map is a statement of fact. And if something looks like a map, but is self-evidently wrong, inadequate, or even playful, then it is not really a map but just a *cartograph* or picture-map or maplike object, set outside the fence that completely encircles and bounds the true endeavor of cartography. Cartography is a science; cartography is an art; cartography is a technology. Cartography is moral.

These multiple factors have manifested within the ideal as a series of preconceptions about what maps and cartography should be. Despite the significant and substantial advances made since 1980 by the sociocultural critique of maps, the ideal's preconceptions continue to have power over academic and popular approaches to maps, not the least of which is that the singular endeavor of cartography exists. The following chapter explicates these preconceptions in detail, to establish the myriad ways in which the ideal of cartography sustains inadequate, misleading, and fundamentally wrong understandings of maps and their history.

5

Map Scale and Cartography's Idealized Geometry

Cartography is paradoxical

Yet another paradox endemic to the ideal of cartography stems from the apparently universal concept of map scale. This particular paradox is not veiled and obscured, to be revealed only through lighthearted satire. Instead, academic textbooks have highlighted it, even as map scholars have refused to face up to its implications. Consider a representative statement from the last, multiauthored edition of Arthur Robinson's *Elements of Cartography*, the textbook that trained several postwar generations of cartography students in the United States, including myself:

> Maps, to be useful, are necessarily smaller than the areas mapped. Consequently every map must state the ratio or proportion between measurements on the map to those on the earth. This ratio is called the map scale and should be the first thing the map user notices. Map scale is an elusive thing because . . . transformation from globe to map means that the map's scale will vary from place to place. It can even vary in different directions at one place. (Robinson et al. 1995, 92)

The simple numerical ratio of map scale is thus foundational. It is *the* defining characteristic of "every map" and it is the "first thing" that any map user should notice. Yet, at the same time, map scale is something else. Far from being foundational, it is an "elusive thing." Rather than being simple in nature, it is an inherently variable and complex measure. So, which is it: a fundamental measure of the nature of the map or a phenomenon that, were it given physical expression, would tear the map apart?

Unlike the paradoxes discussed in previous chapters, which derived from the ideal's internal inconsistencies, this one stems from the dichotomy between the idealization of the normative map and actual mapping practices and processes. The ideal holds that *all* maps are made in proportion to the world, that the nature of any map is determined *solely* by the degree of its proportionality, and that this degree of proportionality is quantifiable and measurable. The ideal promotes the same term—"scale" or, more properly, "map scale"—for both the quality and the quantity of proportionality. This conceptual duality conflates the map-to-world relationship with the metric used to express and to explain the nature of that relationship. Implicit is the expectation that map scale is consistent across any and every map.

In practice, however, the quality of proportionality depends upon a particular geometry that many maps do not possess, so that it is nonsensical to quantify their map scale. The ideal's postulates are, in fact, wrong. Map scale is a core property of the normative map, but it is not a property of all actual maps. The design of persuasive or propaganda maps relies in large part on the general public's misapprehension that map scale is a constant on any map and does not vary (Tyner 2017, 443). Nonetheless, mapping professionals abide by the concept of map scale, they routinely make comments such as "the smaller the scale of the map, the more generalized it has to be" (Darkes 2017, 292), and they willingly accept the logical incongruences that ensue when one quantifies the proportionality of *every* map.

The fundamental issue is that map scale is understood as a generic numerical ratio in the form 1:*x*. The numerical ratio works regardless of the units involved. For example, on early U.S. Geological Survey topographical maps constructed at 1:62,500, 1 inch on the map equates to a terrestrial distance of 62,500 inches (5,208.33 feet), 1 centimeter to 62,500 cm (625 m), and 1 *cùn* 寸 (a Chinese "inch," standardized in 1930 as 3.33 cm and one-tenth of a *chǐ* 尺) to 62,500 *cùn* (or 3.472 *lǐ* 里). The numerical ratio is so commonly used that historians and librarians routinely calculate it even when the maps themselves do not mention it.

The ratio is explained verbally in many textbooks and dictionaries as a ratio of distances, as in the quotation above from *Elements of Cartography*. A fuller example is the explanation provided by the International Cartographic Association in its multilingual dictionary of technical terms, which defined map scale as "the ratio of distances on a map, globe, relief model or (vertical) section to the actual distances they represent" on the earth's surface (Meynen 1973, 59). The rhetoric of such definitions is telling. The ratio is singular, but the distances are plural: *any* pair of map/ground distances will produce the same ratio, it is implied, even though this is manifestly not always the case, as Robinson and colleagues (1995) noted. Furthermore, only like things are related—distance to distance, both measured in the same units—although people conceptualize *lengths* on a map differently from *distances* on the ground and use different measures to express them. As commonly defined, map scale is utterly idealized.

The ideal of cartography construes the correspondence of map to world in simple and direct terms: the map is a measured reduction of the world; therefore "the quantity of features and the way they are shown should be proportional to the scale of the map" (Darkes 2017, 293). Indeed, the numerical ratio provides a "single representative parameter" for what is otherwise the "complex process" of map generalization (Goodchild 2015b, 1383). The numerical ratio apparently governs the nature of any and every map, from the kind and density of its content, through its readability, to its predictive or instrumental value (Freitag 1962). The numerical ratio is therefore "commonly employed as a primary means of classification for different categories of map" (Maling 1989, 15). In other words, under the ideal, the nature of any and every map is determined *solely* by the degree to which it reduces the world. The role of the numerical ratio as a universal metric for the nature and character of any map is enshrined in its common Anglophone label as the *representative fraction*, which is to say the fraction that stands in for—that defines—a map's nature. The manifest diversity of maps and the intricacies of mapping are both reduced to a single variable.

The largely unquestioning acceptance of map scale extends to the work of map historians, who have accepted the ideal's conviction that the quality of proportionality is a universal and fundamental characteristic of all true maps, whenever made. The one aspect of map scale that map historians have addressed is the technical ability to achieve a consistent proportionality. Paul Harvey, in particular, held "the *observance* of a fixed proportion between distances on the map and distances on the ground" to be a hallmark of the recognizably modern survey "plat" or "plan" (Harvey 1987, 466, emphasis added; also Harvey 1980, 1993; Imhof 1964; Wallis and Robinson 1987, 45–52). But the quality itself has seemed universal to historians, regardless of how well it is observed. In their summary of cartographic concepts and techniques, Helen Wallis and Arthur Robinson (1987, 163–204) accordingly failed to include map scale in the sizeable section on "reference systems/geodetic concepts," even as they traced the historical development of such geometrical structures as cardinal direction, latitude, longitude, the ellipsoid, and grids. At best, map historians have considered either the various ways employed to express map scale (Wagner 1914, 3–10), and especially the graphic methods deployed (Kretschmer 1986b), or the development of statistical and graphical techniques for modeling the variability of map scale on coarser resolution regional and world maps (Wagner 1914, 15–24; Kretschmer 1986a; Snyder 1993, 76–91, 147–49; Goodchild 2015b, 1383–85). Yet such work nonetheless treats map scale as an innate attribute of maps. Indeed, no map historian has wondered when, let alone why, map makers began to quantify map proportionality with the abstract, unitless numerical ratio, beyond a vague sense that the numerical ratio somehow developed in conjunction with the rise of the metric system after 1790 (Wagner 1914, 10, repeated without attribution by Skelton 1958, 600; Robinson 1960, 8).

This chapter therefore explores why one particular kind of geometrical relationship, that of proportionality, should have become so privileged. It explores the different geometries employed in different modes of mapping, even in the modern era. Of necessity, I omit two important concepts from consideration. The geometries of marine mapping are too complex to be easily integrated into this chapter. This is not to say that marine mapping was insignificant. Far from it. Rather, the explanations required to avoid the complexities engendered by the misunderstood nature of marine mapping—especially the confusions of plane and cosmographical geometries and the mythic status of the Mercator projection—do little to advance my argument and would only distract the reader (see chapter 2). Nor do I consider the several systems of perspective as deployed since the Renaissance in the imaging of landscapes and urban places; perspective is core to the observational preconception, and underpins the conviction that maps are mimetic, but it is largely irrelevant to the comparative histories of the geometries of mapping and the development of map scale.

With the different geometries established, I then trace how the concept of map scale, the use of the numerical ratio, and the conviction of maps' proportionality combined to create the idealized, normative map. The fulcrum in this narrative, and in the transition from the early modern to the modern era, is the invention, in 1802–3, of the numerical ratio by the French military engineer Pierre Alexandre Joseph Allent. Allent did not think that it was meaningful to apply the numerical ratio to all maps, but the numerical ratio was progressively applied to more kinds of maps over the course of the nineteenth century, marking the intensification of the ideal's conceptual hegemony. As the ideal gained dominance, the semiotic strategies for expressing proportionality lost their connotative aspects (and the *scale* was renamed and degraded to a strictly denotative status as the *scale bar*). I conclude with some reflections on how to refer to the character of maps without reference to the flawed concept of map scale.

I have discussed the substance of this chapter with a range of map scholars, and many seem to find it hard to break away from the idealized concept of map scale. The conceptual duality of map scale gets in the way: as discussion proceeds, comprehension slips back and forth between the quality and the quantity of proportionality. Several scholars have thus misunderstood my argument because they think I argue that no one before 1800 possessed an understanding of the quality of proportionality. That would indeed be a perverse and troubling argument, given that consistently proportioned plans go back at least to about 2120 BCE, when a scaled plan of the enclosing wall of a temple, complete with a graduated scale, was carved on a tablet on the lap of a statue of Gudea, prince of the Sumerian state of Lagash (Harvey 1980, 122–25; Millard 1987, 109; Rochberg 2012, 18–20). A few scholars appear to have concluded that my poststructuralist tendencies have irreparably damaged my capacity for reason.

Let me therefore be clear about my terminology and concepts. The remainder of this chapter provides fuller explanations and examples for each:

"Correspondence": all maps, regardless of form (graphic, verbal [oral, written], physical, gestural, performative, numeric) correspond in some way to the world (real, allegorical, imaginary, virtual).

The correspondence of some maps, and especially those produced in the Western tradition since the fifteenth century, has been construed in geometrical terms. But different kinds of geometry construe correspondence differently. In particular:

"Proportionality": the quality of a consistent correspondence adopted within those modes of mapping specifically grounded in plane geometry, such that the map is uniformly proportional to the world depicted. Proportionality is not a feature of other modes grounded in other geometries.

Before 1800, map makers expressed proportionality through three semiotic strategies, each of which bore specific connotations:

"Direct expression": the length of a line or the area of a region as annotated directly on a map.

"Scale": a consistently graduated line that mimics the steps of a ladder (*scala* in Latin), whether a physical ruler or the similarly divided line on a map. It is the graduated ruler that has been referenced, since the Renaissance, in the common phrase "*to scale*," which refers to the use of a ruler to ensure a consistent correspondence.

"Verbal expression": the statement in words of a map's correspondence to the world, comparing the customary units used on rulers with those used to measure the world.

Shortly after 1800, a fourth strategy was adopted, initially among engineers and specifically for expressing proportionality:

"Numerical ratio": the ratio, in the form $1:x$, such that 1 unit on the map equates to x units on the ground, which establishes an abstract metric of the degree (quantity) of proportionality. Anglophone professionals commonly use the label *representative fraction*, which is to say, the fraction that stands in for—that defines—a map's nature, while in most other European languages the ratio is simply called "numerical scale" (*numerischer Maßstab, échelle numérique,* or *escala numérica*; Meynen 1973, 61–62). Confusion between terms can arise because the numerical ratio is often simply referred to as "scale" or "map scale."

The adoption of the numerical ratio and its application to any and all maps was integral to the idealization that proportionality is a necessary and universal attribute of *all* maps, regardless of their underlying geometry, so that *all* maps are therefore defined *solely* by the degree to which they reduce the world. Therefore:

> *"Map scale"*: the *idealized* quality of proportionality with the world deemed to be possessed by any map, regardless of its underlying geometry. Although a complete fiction, map scale is held to be a real quality under the ideal of cartography.

In order to break apart the conceptual duality of "map scale," I distinguish between the idealized quality of universal proportionality (map scale) and the metric that quantifies that proportionality (numerical ratio).

Technical Points Concerning the Numerical Ratio

Before getting into the history of the geometries of mapping and the rise of the numerical ratio, it is necessary to clarify two technical issues that derive from the numerical ratio. The first addresses the classification of maps as being variously of large, medium, or small scale. This classification has permeated discussion of maps and mapping since the mid-nineteenth century, but it has never been firmly defined. Moreover, since the sociocultural critique opened up map studies to scholars across the humanities and social sciences, the terminology of relative scale categories has been increasingly confused by colloquial usage. The second issue is the reason why the numerical ratio varies across maps. I cannot presume that readers already know the reasons for the variation, and until they do, they will not appreciate the paradoxical nature of the idealization of map scale.

THE RELATIVE CLASSIFICATION OF LARGE SCALE VS. SMALL SCALE

The adoption of the numerical ratio after 1800 gave rise to a more general system of classifying maps by the relative degree to which they generalize the world. These categories depend on expressing numerical ratios as rational numbers. Consider the example of two ratios:

1:1,000 is the same as $^1/_{1,000}$ and resolves to 0.001
1:1,000,000 is the same as $^1/_{1,000,000}$ and resolves to 0.000001

As a number, 1:1,000 is small in absolute terms but is nonetheless three orders of magnitude larger than the number to which 1:1,000,000 resolves. The map scale of 1:1,000 is therefore "larger" than one of 1:1,000,000. The smaller the ratio's denominator, the larger the number to which the map scale equates. Such comparisons gave rise to the broad categories of "large scale" and "small scale." Use of these categories eventually spawned an intermediate category of "medium scale" as well as the extreme categories of "very large" and "very small" scales.

The relative classification of maps as variously small, medium, or large scale has served in the twentieth century as a universal means for comparing maps, regardless of their nature or their historical and cultural origins. Roughly contemporary maps of the same place will be readily understood to be quite different in their form, content, and context according to the degree of reduction indicated by their map scale. The truth of this deterministic relationship is demonstrated in many texts by the comparison of details, placed side by side, of the same area from maps at different map scales. For example, Daniel Dorling and David Fairbairn (1997, 39) compared two maps of Bern, at 1:25,000 and 1:100,000; Michael Goodchild (2015b, 1384, fig. 883) compared three of Madison, Wisconsin, at 1:24,000, 1:100,000, and 1:250,000; and Bill Rankin (2016, 31) compared three centered on Washington Island, Wisconsin, from 1:250,000 to 1:2,000,000. Such comparisons are effective because the maps from which details are taken are specifically selected to be largely similar in style, if not in their degree of detail. In each of these examples, the numerical ratios of the selected maps all fall within one order of magnitude of each other.* But even with like compared to like within a limited range of scalar difference, there is sufficient variation to imply that such comparisons are extensible to other maps with quite different numerical ratios. Map scale thus appears to be a legitimately universal measure of reduction and generalization from the world to the map.

All map scholars rely on the relative categories of large, medium, and small scale to quickly and efficiently summarize the nature and character of any given map. Thus, large-scale maps are widely understood to be the product of surveying by engineers, while small-scale maps are understood to be based on data visualization by social scientists. In this respect, cartography seems to comprise at least two major dialects, of "surveying and mapping." Yet such institutional distinctions have never been allowed to undermine the coherence of cartography as a single endeavor (see chapter 3). Regardless of their institutional origins, both large- and small-scale maps are still conceptualized as being shaped and defined by their respective map scale. Both kinds of mapping seem to adhere to cartography's idealized ontology and the conviction that the character of any and every map is determined by the degree to which it reduces the world. As Darkes (2017, 291) asserted, "accuracy is . . . a relative rather than an absolute concept, and is affected by the available information and by the scale of the map. What is deemed 'accurate' at a small scale may be hopelessly inaccurate at a large map scale."

Given how map scholars have routinely relied on these relative categories to

---

*I'm rounding down for Goodchild's set. His range of numerical ratios, 1:24,000 to 1:250,000, represents a difference only fractionally over one order of magnitude (1.042).

TABLE 2. Two different classifications of relative categories of map scale, by ranges of numerical ratio, reflecting different intellectual contexts: Derek Maling (1989, 15) based his system on the work of the Ordnance Survey in the U.K.; Bill Rankin (2016, vii) aimed for a more generic and inclusive approach. Each classification is compared to Allent's *triades* of numerical ratios (see table 3). Allent himself distinguished the application of numeric ratios to maps as being valid (I–IV), approximately valid (V–VI), and invalid (VII–VIII). For graphic comparison, note that the map reproduced in figure 5.3 is at 1:1,980; fig. 4.2, at 1:86,400; fig. 4.18, at 1:580,000; fig. 5.21, at 1:1,641,836; and fig. 5.23, at 1:45,000,000.

Scale Category	Range of Numerical Ratios	Kinds of Maps	Allent's *Triades*
MALING (1989)			
Large	– 1:12,500	detailed plans	I–V
Medium	1:13,000 – 1:126,720	detailed territorial maps	V–VI
Small	1:130,000 – 1:1,000,000	territorial maps	VI
Very small	1:1,000,000 –	geographical maps	VII–VIII
RANKIN (2016)			
Very large	– 1:10,000	detailed plans	I–IV
Large	1:10,000 – 1:100,000	detailed territorial maps	V
Medium	1:100,000 – 1:1,000,000	territorial maps	VI
Small	1:1,000,000 – 1:10,000,000	regional maps	VII
Very small	1:10,000,000 –	world maps	VIII

provide an easy shorthand for the nature of maps, and given how they have sought for more than a century to codify and standardize map design, one might expect them to have advanced clear-cut boundaries between the categories. But they have not. The difficulty is that the categories of large scale and small scale are only relative; the nature of their relation is defined within each thread of spatial discourse, which inevitably means that definitions of the two categories vary widely. Mary Lynette Larsgaard (1984, 3–9) reviewed the professional literature on territorial mapping and found no agreement among forty-two definitions of "large scale" by government surveyors. For some, "large scale" is larger than 1:10,000 and "very large scale" is larger than 1:1,000; for others, "large scale" is larger than 1:100,000 (also Steward 1974, 25–26). Bring regional and world maps into the equation, and certainty is impossible: the "boundary between large- and small-scale maps is subject to enormous subjective individual variation" and is therefore inherently imprecise (Dorling and Fairbairn 1997, 25). Map scholars have been left to establish their own practical guidelines (table 2). They have done so by accepting certain maps as prototypes for each category; each newly encountered map is quickly compared against the array of prototypes held in memory to identify its category. Any attempt to actually codify scale categories in precise terms, and to justify that codification, would however require a careful scrutiny of the actual nature of the numerical ratio.

The failure to codify the relative categories of map scale has only encouraged a terminological confusion caused by the contradictory meanings of "large scale" and "small scale" in technical as opposed to colloquial usage. Colloquially, "large scale" means extensive and widespread, while "small scale" means tight and constrained, so commentators have frequently miscategorized small-scale regional maps as being large-scale because they cover extensive areas, and vice versa. The U.S. Geological Survey (2002) accordingly began an online fact sheet about map scales with the heading "Large Is Small." Confused? Too many people are, especially those scholars who in recent decades have come to map studies from other fields. Most of the time the confusion is inconsequential, but on occasion it has detracted significantly from the analysis (e.g., Silbernagel 1997). Rather than seeing this as a reason to define precisely the relative categories, or even to reconsider the concept of map scale altogether, map scholars have instead simply allowed correct usage of the technical terminology to become a shibboleth.

EXPLAINING THE PARADOX OF MAP SCALE

The ambiguity and paradox of map scale indicated by Robinson et al. (1995) stems from the basic fact that maps are flat, the earth's surface is curved, and it is impossible to flatten a curved surface without tearing and forcefully shrinking or stretching it. The usual analogy is to think about what happens when one tries to flatten an orange peel. In large-scale mapping, the plane of the map is almost coincident with the curved surface of the earth, so the latter can be flattened with so little contraction or spreading that the alterations are largely unappreciable and undetectable to the human eye. But this is not the case for small-scale mapping. For such mapping, the alterations are readily apparent, and they can be measured as variations in map scale, as defined by the numerical ratio, across the surface of a map.

Such variation is famously apparent on world maps constructed on the Mercator projection (fig. 5.1). This projection converts meridians of longitude, which on the globe are lines of equal length that converge toward the poles, into parallel lines of increasing length; parallels of latitude, which on the globe are lines of variant length that do not converge, remain parallel but are now all of uniform length. The poles, which in reality are points of zero length, appear as lines the same length as the equator. Or, at least, they would so appear were it actually possible to show them: the geometry of the Mercator projection is such that the poles are both infinitely distant from the equator!

The effect of these transformations on the numerical ratio is pronounced. Because each parallel is stretched out uniformly, we can easily calculate the numerical ratio for each; in doing so, we find incredible variation. If the world were taken to be a sphere, its radius would be about 6371.1 km, its circumference about 40,031 km (Maling 1973, 42). On the world map reproduced in figure 5.1, the straight line of the equator is 34.5 cm, which equates to 4,003,100,000 cm on the earth. Therefore:

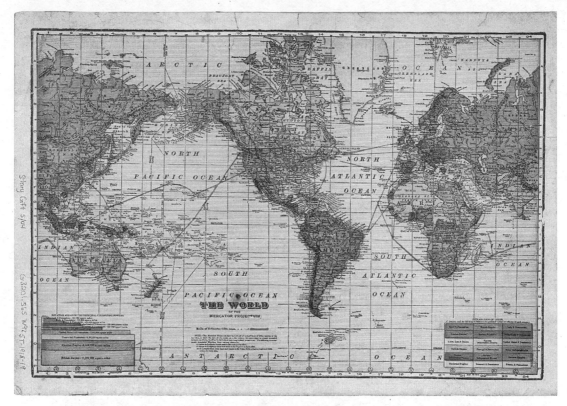

FIGURE 5.1. *The World on the Mercator Projection* (Anon. 1918, rear cover). Colored lithograph, 17. 24 × 35.5 cm. Courtesy of Osher Map Library and Smith Center for Cartographic Education, University of Southern Maine (Story Collection); www.oshermaps.org/map/2314.0009.

34.5 cm on the map along the equator = 4,003,100,000 cm on the earth

1 cm on the map along the equator = $^{4,003,100,000\ cm}/_{34.5\ cm}$ on the earth

1 cm along the equator = 116,031,884.058 cm on the earth

The numerical ratio is therefore about 1:116,000,000. On the earth, the parallels at 60° north and south latitude are each half the length of the equator, but are also drawn 34.5 cm long on this map:

34.5 cm on the map along the parallel of 60° = 2,001,550,000 cm on the earth

1 cm along in the map along the parallel of 60° = $^{2,001,550,000\ cm}/_{34.5\ cm}$ on the earth

1 cm along the equator = 58,015,942.029 cm on the earth

The numerical ratio along those parallels is therefore twice that along the equator, about 1:58,000,000. The denominator of the numerical ratio for the parallels continues to decline toward the poles: very, very close to either pole are parallels of latitude that are 34.5 cm in circumference, so they would have a numerical ratio

175

of 1:1 were they able to be shown on figure 5.1. The denominator reduces further so that, at each pole, if they could be shown, the numerical ratio along the parallel would be an infinite 1:0.

Similar variations in numerical ratio are evident along the meridians on the Mercator projection. Although the ratio constantly changes along the meridians, we can calculate approximate numerical ratios for each small portion of the meridian. On the map reproduced in figure 5.1, a one-degree portion of a meridian straddling the equator would have a numerical ratio of about 1:116,000,000, but the ratio again increases to 1:0, or infinity, at either pole.

Such variations in map scale, and accordingly alterations in the size or shape of the world's features, are evident in every regional or world map, albeit usually to a less pronounced degree than on the Mercator projection. They are the origin of the paradox that map scale is simultaneously uncomplicated and tortuous. Rather than accept that the universal concept of map scale is flawed and should be abandoned—or, at least, that it is inappropriate in its ideological burden and should be reconsidered—adherents of the idealized endeavor of cartography and of the normative map have clung to the simple quality of proportionality offered by map scale, paradox be damned. As Robinson and his colleagues (1995) indicated, according to the ideal, the variation of numerical ratio on coarser resolution maps is simply a mathematical quirk that does not detract from the absolute conviction that map scale is a universal property of all maps, that map scale defines the nature of every map, and that map scale is meaningfully determined by the simple metric offered by the numerical ratio.

Map scholars generally refer to the geometrical alterations inherent in map projections—the contraction and stretching of the shapes and sizes of the world's features—as "distortions" or "deformations." These terms are quite correct, strictly speaking; a map projection does indeed deform the earth's spherical surface to make it fit the plane of the map. However, both words entail a sense of pejorative complaint: a projection prevents the map from possessing the ideal quality of proportionality (deformation); a projection alters the proportions of the world's features and does so for the worse (distortion). So as not to inadvertently perpetuate the ideal's ontological preconception, I therefore use the nonpejorative phrase "geometrical alterations" instead.

The Geometries of Western Mapping

The common conviction that the origin of modern cartography lay in the rationalization and geometricization of all aspects of mapping and vision during the Renaissance (see chapter 1) rests upon the further conviction that cartography features a single geometry. The ideal has not encouraged any reflection about the nature of this geometry, which is routinely identified as being both Euclidean or

Cartesian in nature, even though these two kinds of geometry represent markedly different operational schemas when it comes to mapping practices. Euclid and Descartes both conceptualized the geometrical plane as an abstract and indefinitely extensible surface, but there the similarities end. Euclidean plane geometry is implemented as a series of lines that are set at angles and have length relative to each other; Cartesian plane geometry is implemented as a series of points defined by their perpendicular distances (the abscissa [x] and ordinate [y]) from perpendicular axes. Euclidean geometry emphasizes the graphic construction of straight lines and curves, such that their angles and lengths possess certain properties, in a relative space; Cartesian geometry emphasizes the analytical description of lines and curves on a plane anchored by an absolute point (the "origin," intersection of the two axes). In practice, however, different modes of mapping have employed a variety of geometries, even in the present day. Moreover, we cannot really say that any of these disparate mapping practices before the digital revolution have adhered to either Euclidean or Cartesian geometry, in that none actually construes the surfaces it constructs on a plane to be indefinitely extensible. The discursive conditions of mapping always entail a sense of closure and limitedness to the part of the world being mapped. To construe mapped surfaces to be infinitely extensible requires either the forceful contravention of those conditions—as, for example, by Charles Sanders Peirce's treatment of the Mercator projection (chapter 4)—or an unquestioning adherence to the ideal's preconceptions.

Significant differences must also be recognized among those who sought to geometricize mapping. Early modern "mathematical practitioners" might all have espoused the application of mathematics and mathematical instruments to all walks of life, but their community was riven by a fundamental divide between practitioners *per se* (i.e., those who did the work and who often used craft practices) and their patrons and the academics who advocated the use of higher forms of mathematics. The historical record contains many instances in which the practitioners and their patrons were directly at odds over the implementation of new techniques, even as implementation of higher mathematics served as an efficient means for men of skill but humble origins to climb the social ladder (see especially Bennett 1991; Edney 1994b; Higton 2001; Silverberg 2015). The apparent conceptual unity offered by applied geometry did not extend to an actual unity of map geometry. Tracing the geometrical systems of early modern and modern mapping thus requires some care, to keep in mind the distinctions between proposed and actual practices.

Each of the different geometries employed for mapping expressed a different correspondence of map to world. Until the widespread adoption of the numerical ratio, there was no consistent or uniform sense of map scale across all spatial discourses. Some maps used direct expressions, scales, and verbal expressions, or some combination thereof, to denote the correspondence for instrumental ends or to connote the quality and range of their source materials. Other maps eschewed such devices completely. In other words, expressions of geometry and correspondence were as conventional as any other signs used on maps before and after 1800.

This is not to say that early modern Europeans did not differentiate between or classify types of maps. They did. The absolute category of "universal" maps comprised maps of the entire earth with, generally, cosmographical elements. All other maps were defined according to the relative and rather impressionistic distinction between the "general" versus the "particular" or "special." General maps covered larger areas than particular ones, with a further implication that particular maps might potentially be assembled to create new general maps. This comparative usage was also deployed metaphorically, as when the English jurist William Blackstone (1758, lxvi) suggested that a course of legal studies should be like "a general map of the law, marking out the shape of the country, it's [*sic*] connexions and boundaries, it's [*sic*] greater divisions and principal cities," and should not be like "particular" maps that "describe minutely the subordinate limits."* Eighteenth-century French practice also distinguished in a relative manner between maps and charts as being either *à grand point* (i.e., in great detail, in fine resolution, or limited extent) or *à petit point* (i.e., in small detail, in coarse resolution, or extensive extent) (Chapuis 1999, 310, 729). *Grand point* has been rendered as "large scale" in modern English, and *petit point* as "small scale," but the concepts are neither cognate nor equivalent, and the use of the modern terms is anachronistic and misleading.

PLANE GEOMETRY: MAPPINGS OF PROPERTY, PLACE, AND REGION

It is something of a truism that the independent development of plane geometry in ancient cultures likely stemmed from architecture and the delineation and representation of property boundaries. In the fifth century BCE, Herodotus attributed the origins of geometry to the work of Egyptian "rope stretchers" as they redefined field boundaries after each annual flooding of the Nile, although his further comments that such surveys were primarily to regulate property taxes have generally been overlooked (Herodotus 2008, 135–36 [bk. 2, ¶109]; see also Proclus 1970, 51–52). It has similarly been argued that the Vedic cultures of South Asia developed plane geometry in conjunction with property measurement and mapping rather than, as orientalist scholars had previously asserted, solely through the religious practices of altar construction (Rajanikant 1974).

In the early modern West, a variety of social, cultural, and economic factors led, by the sixteenth century, to the application of plane geometry to the representation of property, places, and even regions; the relevant modes are those listed in the first section of table 1. As some of my examples indicate, these plane geometries have continued to be sustained by multiple spatial discourses within the modern period.

---

*My thanks to Isabella Alexander for this reference.

It is beyond the scope of the present work to explain the factors giving rise to these kinds of mapping, but we must nonetheless keep in mind that their effects varied widely between districts and even within institutions. The adoption of geometrically consistent property maps, in particular, was neither universal nor consistent in the early modern and modern eras (Fletcher 1995; De Keyzer, Jongepier, and Soens 2014; see also Harvey 1980).

The variety of factors contributing to the rise of plane surveying, and the different social levels at which they operated, can be seen in the several terms adopted in early modern Europe for the specific practice of reducing the earth's surface to a series of lines whose lengths and angles of intersection might be measured in some way. The craft-based nature of such measurement was reflected in terms such as the German *Feldmessen* ("field measurement") or the French *arpenter* ("to survey") or *arpentage* ("surveying"), an *arpent* being an agricultural unit of area varying between 3,420 and 5,110 m^2 (Zupko 1978, 5). More learned expressions were "geodesy" and "mensuration." "Geodesy" derived from the Greek γεωδαισία (from *gē*, "earth," and *daíō*, "to divide"). Its original meaning as any kind of mathematical-based survey that divides up the earth is common across Europe even today, although the term has also come to bear the more specific meaning of the scientific practice of determining the size and shape of the earth. "Mensuration" derived from the Latin *mensurare* ("to measure") and in English usage specifically meant the measurement of lengths, areas, and volumes. In addition, the English *surveying* stemmed from the authoritative act of "looking over" a situation (*surveier* in Anglo-Norman French) and has been applied generally to managerial and supervisory practices; the respective practices of *surveying* and *mensuration* have thus been at odds for most of the early modern and modern periods. Finally, as some practitioners and scholars sought to codify and regularize the practices of measurement, they promoted the term *géométrie pratique / praktische Geometrie / practical geometry* as well. There is an implicit social conflict in these terms, between the elites' definition and regulation of property and places, and the preservation of local and customary rights (see, e.g., Sullivan 1998).

It has been tempting to lump these otherwise disparate practices under the catchall term of "surveying." Certainly, early modern Europeans depicted a Platonic ideal of measurement and of the graphic replication of its results through the playful depiction of angelic putti engaged in surveying and measurement (fig. 5.2). The putti signified the almost divine nature of the process of re-creating the world *in parvo* by maintaining its proportions on paper (Heilbron 2000, 5–9; also Dempsey 2001).* The plane surveying in which people engaged when the curvature of the earth's surface was simply irrelevant—whether for a property survey of a small parcel of land or a chorographical survey over a county or province—

---

*My special thanks to Mary Pedley and Emily Brill for this point.

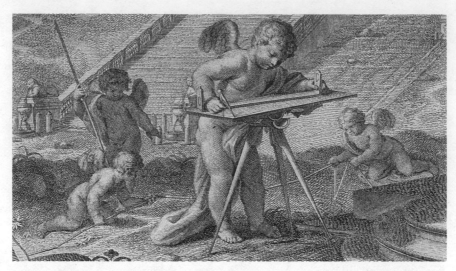

FIGURE 5.2. Putti surveying a place with chain and plane table. Detail of Giovanni Battista Nolli, *Nuova pianta di Roma* (Rome, 1748), sheet 12. This map bore a scale of *palmi Romani d'architettura*. Courtesy of the Osher Map Library and Smith Center for Cartographic Education, University of Southern Maine (Osher Collection); www.oshermaps.org/map/13798.0011.

appeared to be the same because it possessed a pragmatic unity of techniques and instrumentation and because, before the proliferation and professionalization of mapping specialists in the nineteenth century, many individuals were employed across the different mapping modes. The property mapper who made chorographical maps in the sixteenth and seventeenth centuries and the military engineer who mapped fortifications and regions in the eighteenth deployed an essentially uniform set of skills and techniques. Moreover, surveying practices originally featured relatively simple techniques for measuring angles and lengths that permitted the lines traced by the surveyor to be easily and directly recombined on paper; the drafting of a plan actively recapitulated the acts of measurement. As a result, numerical calculations were limited to just a few tasks, notably the determination of areas and the solution of similar triangles (as in the ancient problem of determining the distance from an artillery piece to a target; Cavelti Hammer 2019).

Such proportional mapping on the plane might therefore be said to manifest "Euclidean" geometry (as Edney 2017b), in that the multifarious practices all appear to implement the kind of geometry codified by Euclid in his *Elements*, early in the third century BCE. However, the great majority of plane surveying work does not actually deserve such a prestigious label. No matter the extent to which authors prefaced their textbooks on "practical geometry" with long introductions to "theoretical geometry"—complete with dry diagrams of different curves, triangles, quadrilaterals, and other shapes—property and place mapping remained craft practices of the observation and measurement of well-delimited or at least restricted areas. Moreover, the ends to which plane geometry was directed were sufficiently diverse that we can see significant differences in the application of geometry to the mappings of property, place, and region.

In property mapping, local customs and regulatory regimes compete within various spatial discourses focused on demarcating, creating, and regulating real property for legal, administrative, financial, commercial, or political reasons. Semiotically, the mode is broadly concerned with physically marking property boundaries in the land, with describing them in words and perhaps images, and with determining areas. The physical, verbal, and geometrical practices followed by George Washington in Virginia in about 1750 (fig. 5.3) were basically the same as those of Henry David Thoreau when he mapped woodlots in eastern Massachusetts a century later (fig. 5.4; Chura 2010), and as those of the many surveyors across the United States who work today to relocate and remonument lost property markers.

By contrast, the various civilian, military, and engineering mappings of place have been strongly aligned with visual and poetic practices of depicting the lay of the land. A wonderful example is a 1932 plan of George Washington's estate at Mount Vernon, drawn by the landscape architect and artist, B. Ashburton Tripp (fig. 5.5). The graphic style and its thematically paired marginal vignettes emphasize that this is a work of art as much as of measurement. Indeed, Tripp exhibited the work at the 1932 May Show at the Cleveland Museum of Art. This combination of measurement and artistry will be familiar to the historian of maps of fortresses and their environs made by military engineers, starting in the sixteenth century. It is a combination facilitated by the plane table (see fig. 5.2). Initially designed to permit the direct graphic construction of a plan's geometry, and used as such by property mappers, the plane table also permitted the place mapper to sketch the landscape directly. Fast and easy to use, plane tables were a favored instrument for landscape mapping well into the twentieth century. With enough manpower, they could be deployed over truly extensive areas: from 1763 to 1799, the Austrian monarchy's *Josephinische Landesaufnahme* covered some 570,000 km^2 with little or no geometrical control (see most recently Veres 2015; Tebel 2019); only the particular survey of the Austrian Netherlands was based on triangulation (Vervust 2016a, 2016b).

The plane surveying of property and place mapping could also be deployed to observe and measure larger regions in chorographical surveys (see table 1) that might later be compiled into maps of still greater extent. This is likely the method employed by Christopher Saxton for his famous maps of the English counties, surveyed and printed in 1573–79 (fig. 5.6). While map historians have long taken Saxton's work to have been a "national survey" that could only have been based on a countrywide geometrical framework of triangulation (see Ravenhill 1983), Peter Barber (2007, 1628–29) demonstrated that the survey proceeded county by county and that each county map was a compilation of existing materials. Saxton likely used existing itineraries and perhaps some new road surveys to provide a rough geometrical structure within which to assemble other materials (Andrews 2013).

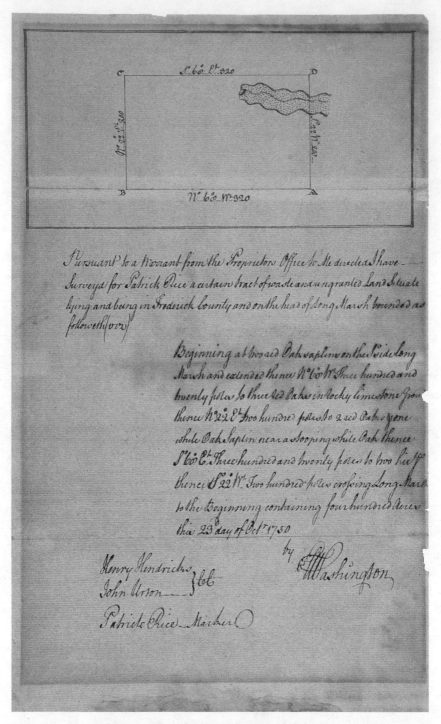

FIGURE 5.3. George Washington, "I have survey[e]d for Patrick Rice a certain tract of waste and un-granted land situate lying and being in Frederick County" (23 October 1750). The graphic plan recapitu-lated the legally binding metes-and-bounds description, below, which verbally explained the length and bearing of each side of the property between marks set physically in the landscape: "Beginning at two red oak saplin[g]s on the S° [south] side Long Marsh and extended thence N° [north] 60° W [west] three hundred and twenty poles to three red oaks in rocky limestone ground," leading ultimately back "to the Beginning containing four hundred acres." Manuscript, 31 × 18 cm. Courtesy of the Osher Map Library and Smith Center for Cartographic Education, University of Southern Maine (Osher Collection); www .oshermaps.org/map/665.

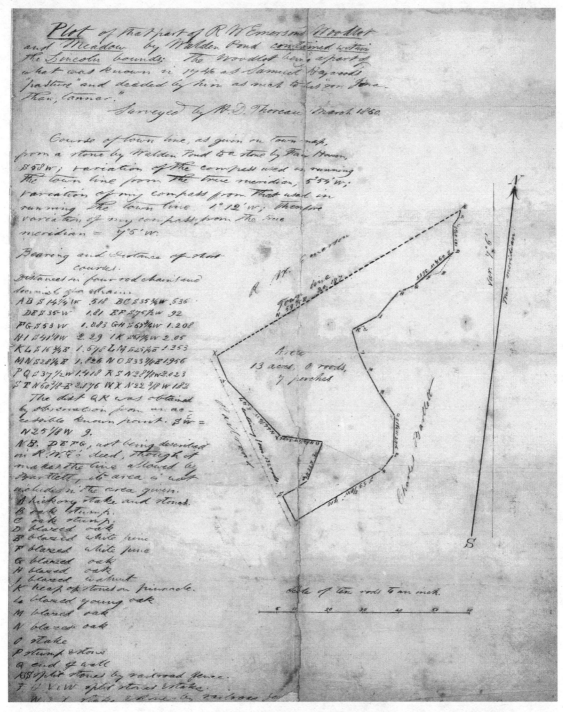

FIGURE 5.4. Henry David Thoreau, "Plot of that part of R W Emerson's Woodlot and Meadow by Walden Pond contained within the Lincoln bounds; the woodlot being a part of what was known in 1746 as Samuel Haywood's 'pasture' and deeded by him as such to his 'son Jonathan, tanner'" (March 1850). The table to the left of the plan lists the bearings and distances of each side, the table at lower left the nature of the witness trees marking the corners of the property. Manuscript, 51 × 40.5 cm. Courtesy of the American Antiquarian Society, Worcester, Massachusetts.

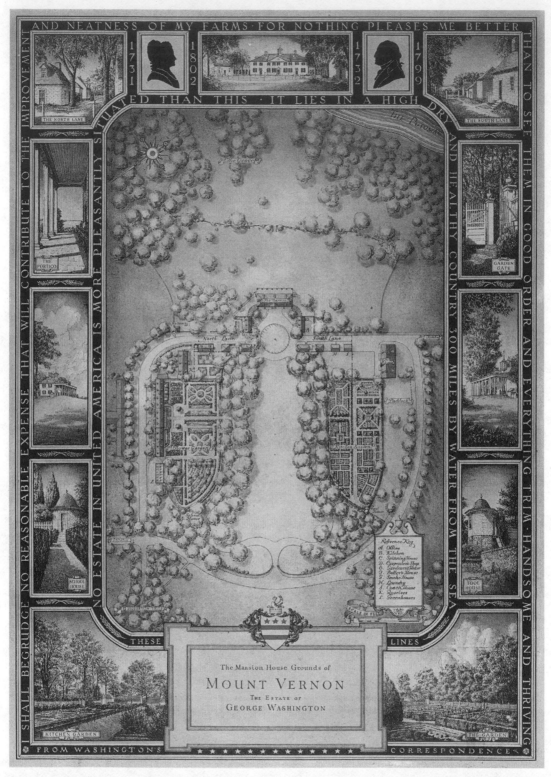

FIGURE 5.5. A manuscript map of place, being a planimetric map accomplished in the style of landscape architecture, with a border of views. B. Ashburton Tripp, "The Mansion House Grounds of Mount Vernon, the Estate of George Washington" (1932). This was one of fourteen works that Tripp submitted to the May Show at the Cleveland Museum of Art, between 1925 and 1936, which the museum's "May Show Database" (http://library.clevelandart.org/search_mayshow) variously described as "architectural rendering," "illustration," "decorative painting," "miscellaneous handicraft," and "decorative design for process reproduction." Manuscript, 78 × 55 cm. Courtesy of the David Rumsey Collection (8643001); www.davidrumsey.com.

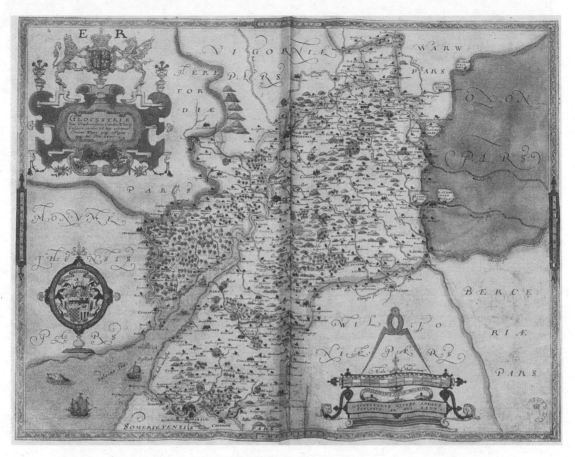

FIGURE 5.6. An example of chorography: mapping regions with plane geometry. Christopher Saxton, *Glocestriæ [Gloucestershire] . . . verus Tipus atq. Effigies Ano. Dni. 1577*, from his 1579 atlas of England and Wales. The royal coat of arms at upper left (above the large title cartouche) is matched in size by that at center-left of Thomas Seckford, the midlevel official who had initiated the survey and had hired Saxton to undertake it. Hand-colored copper engraving, 41 × 54.5 cm. Courtesy of the Bibliothèque nationale de France (Département des cartes et plans, GE DD-1440 [RES], map 12); gallica.bnf.fr.

This same methodology of compilation was used throughout early modern Europe. John Ogilby (1675, unpaginated preface), for example, referred to the "Itinerary Way as the most Regular and Absolute" means of constructing chorographical maps. Ogilby and his surveyors were aided in this work by attaching mechanical counters to wheels to measure the distance traveled (Holwell 1678, 191–225).* In some instances, such as the many English county maps derived from Saxton's, marginal indications of latitude and longitude, and perhaps also meridians and parallels, were added, as if the survey work had been integrated with the cosmographical geometry of world and regional maps. However, only in the

---

*My thanks to Peter Barber for the information that Holwell was one of Ogilby's surveyors (see Holwell 1678, 190, 195).

FIGURE 5.7. Direct expression of correspondence on a property plan. Each side of the property bore statements of both bearing (as "No 68° Wt") and length (as "320" poles). A "pole," the measure specified in the accompanying metes-and-bounds description, is an alternate name for a rod or perch (5.03 m). See figure 5.3 for full image. Courtesy of Osher Map Library and Smith Center for Cartographic Education, University of Southern Maine (Osher Collection); www.oshermaps.org/map/665.

nineteenth century were field observations for latitude and longitude sufficiently common to permit the actual realignment of regional surveys to cosmographical geometry (Edney 2019b).

The common grounding of these modes of property, place, and chorographical mapping on plane geometry is manifested in their deployment of one or more of the same three strategies to indicate the nature of the maps' proportionality to the world. They also often included a north arrow of some sort. What they lacked, except for the few chorographical maps that bore indications of latitude or longitude, was an indication of the location of the mapped property, place, or region within the wider world.

The first and perhaps the simplest strategy to indicate the correspondence between a map constructed on plane geometry and the world is simply to write distances directly on the map, next to a line: the line on the map equates to the specified distance. Such direct expressions can be found on early modern property maps, when surveyors annotated portions of a property's boundaries with their lengths. This practice reflects the manner in which early property maps were often created as graphic plots of traverses that replicated the verbal metes-and-bounds description of the property (fig. 5.7). Property mappers could also inscribe the calculated area of each field or lot on the map. Such annotations establish an immediate correspondence between the map and the land, but the correspondence is necessarily ungeneralized. Each line or area has its own correspondence to the world; there is no overall and general relationship determining a consistent proportionality of all lines on the map to the land.

Similarly, some regional map makers were led by their use of road itineraries to write the length of each road segment on their maps. Direct statements of distance also appear on strip maps, whether Ogilby's from the seventeenth century or the "Triptiks" prepared by the American Automobile Association in the later twentieth century. In such cases, the specified length refers to the length of the road, not the direct distance. Once again, each direct expression does not indicate a generalized and consistent correspondence of map to world.

By contrast, a generalized correspondence, and therefore proper proportionality, could be expressed by the second strategy, that of the scale. On a map, the scale is simply a line, generally subdivided, whose length corresponds to a specific ground

FIGURE 5.8. A scale. Detail of Tripp's 1932 plan of Mount Vernon; see fig. 5.5 for full image. Courtesy of the David Rumsey Collection (8643001); www.davidrumsey.com.

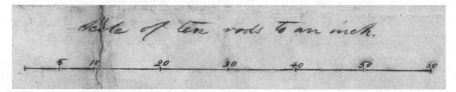

FIGURE 5.9. A scale surmounted by a verbal statement, from 1850: "Scale of ten rods to an inch." See figure 5.4 for full image. Courtesy of the American Antiquarian Society, Worcester, Masschusetts.

distance (fig. 5.8). A scale connotes a generalized correspondence between the map and the world. The presence of such a scale is an assertion that all lengths on the map, and not just those that are accompanied by a direct statement of distance, obey the same map-to-ground correspondence embedded in the graphic line (La Chapelle 1755, 248). Scales permitted map readers to readily measure distances, especially through the use of a pair of dividers; the frequent presence of dividers atop scales on early modern geographical maps, as in figure 5.6, seems to have actively invited map readers to take measurements, or at least proclaimed that because map readers could do so, the map had a consistent proportionality.

The third strategy used to indicate a map's correspondence to the world within a system of plane geometry was the verbal expression. For example, the phrase "ten rods to an inch" indicates that 1 inch (2.54 cm) on the map represents 10 rods (50.3 m) across the ground (fig. 5.9). The units in either part of the expression are necessarily different: lengths on the map are expressed in the particular measures used to precisely subdivide rulers, such as the inch or *ligne*, while distances on the ground are expressed in measures used for property mapping, such as rods or *arpents*, or for topographic or chorographic mapping, such as miles or leagues. Verbal expressions could occur by themselves, without other expressions of correspondence, or in conjunction with a scale.

Verbal expressions also indicate generalized correspondence and consistent proportionality. In this respect, they seem to have been a product of the standardization of surveying practices, whether informally (as in property mapping, when communities of surveyors and property owners settled on common practices) or through the adoption of formal protocols (as when the efforts of many military surveyors needed to be coordinated). However, like the few scales that were ungraduated, verbal expressions were not directly instrumental; after all, surveyors generally also included graduated scales on their maps. Rather, verbal expressions served the rhetorical function of emphasizing that a map was grounded in a consistent system of measurement and geometry, which is to say, in a single survey.

In some instances, the presence of a verbal expression served to claim that a map was based on a single survey when such was manifestly not the case. Consider, for example, William Douglass's *Plan of the British Dominions of New England*, published posthumously in London in 1755, which claimed to be the product of a single survey but was instead a compilation of several different property, boundary, and chorographical surveys. As published, the *Plan* bore no scale but instead the verbal expression, "5 English Mile . . . to one Inch" (Edney 2003b).

Scales and verbal expressions both embody the markedly different practices and customary measures used for measuring different kinds of things within distinct social circumstances. Official attempts to regulate measures did lead to the definition of categories of measures by reference to some physical standard, in the process giving rise to those irregular conversion factors that used to bedevil British schoolchildren, such as a statute English rod (or pole, or perch) being equivalent to 5½ statute yards, the English standard for linear measure, or 40 rods to a furlong, or 8 furlongs to a mile. But in everyday practice, customary measures were not compared one to another because there was simply no practical reason to do so. Implicit in scales, and especially verbal expressions, therefore, was the functional difference between the units, so that both strategies acknowledged the power of the map to transform one metrological regime into another.

This neat argument is complicated by something of a tradition among early modern Dutch surveyors to employ a curious verbal expression that is almost tantamount to a numerical ratio, albeit an inverted one. On one map—the environs map in the corner of the plan of Bourbourg (Flemish Broekburg), near Dunkirk in France, surveyed in 1644 by one Vaast (Vedastus) du Plouich and published by Joan Blaeu in his 1649 town atlas of the Netherlands—the 7 cm long scale represented a length of 3 miles, or 4,200 rods, and bore the statement:

> 165 miles of this scale make a Flemish Rod of 14 Feet. 231,000 of these miles make a mile of 1,400 rods. 231,000 distances taken on this map make the actual distance across the earth's surface. Brouckburch Ambacht is 53,361,000,000 times bigger on the earth than it is in this map. (Van der Krogt 2017)*

---

*"165 Mijlen van dese Scala maken een Vlaemsche Roede van 14 Voeten. 231000 van dese Mijlen maken een mijle van 1400 roed. 231000 Distantien genomen op dese Caerte maken de selve distantie op d'Aerde-Cloot. Brouckburch Ambacht is 53361000000 maels alsoo groot op Aerdrijck als het is in deze

That is: 165 times the length of one mile division on the scale (i.e., 2⅓ cm) constitutes the length of one rod of 14 feet, so that a mile of 1,400 rods must comprise
231,000 mile-equivalents; or, as the annotation further explains, 231,000 "distances"
on the map is 1 "distance" on the globe. This statement verbally expressed an inverted numerical ratio of 231,000:1. I am, however, unclear as to the utility of con-
ceptualizing geometrical correspondence in this way; it seems to be a hindrance in
calculating distances and in this respect seems once again to be a rhetorical device
to emphasize the consistency of the chorographic map's proportionality with the
world. Clearly, more work is needed to identify the different practices, and their
connotations, for expressing geometrical correspondence and proportionality in
plane mapping.

COSMOGRAPHICAL GEOMETRY: MAPPING THE WORLD AND ITS REGIONS

The image of the world—*world* being a flexible, cultural concept that is not necessarily coincident with either *earth* or *globe* (Woodward 2001, 58–60; Cosgrove
2007b, 67–69)—is a core element of early modern and modern spatial discourses
concerned with organizing knowledge about distant places and, indeed, of the
entirety of creation and the relationship of the human to the divine. Geographical and cosmographical representations of the world in early modern Europe did
on occasion reference the schematic and ageometrical world maps of medieval
Europe, such as Heinrich Bünting's 1581 map of the Old World as a cloverleaf
centered on Jerusalem (Van der Heijden 1998; Shirley 2001, no. 143), and some referenced the marine-style planisphere (see fig. 2.5), but for the most part they relied
on the cosmographical geometry of latitude and longitude. This is the geometry
of the cosmos, of the relationship between the geocentric earth and the celestial
sphere. Stars can be precisely located by their location along great circles, either as
declination and right ascension with respect to the celestial equator or as celestial
latitude and longitude with respect to the ecliptic. The same coordinates locate terrestrial features: longitude, the angular distance around the equator; and latitude,
the angular distance along a meridian between the equator and the poles (fig. 5.10).

Both terrestrial latitude and longitude were determined by reference to celestial phenomena. World maps included multiple references to cosmographical
elements, from the indication of the tropics and the ecliptic on the maps themselves, via celestial hemispheres, to the iconographic representations of the planets, seasons, and elements. The origins of modern cosmographical geometry lie in
Hellenistic astrology perpetuated by Renaissance metaphysics. The key moment
was the translation into Latin by Jacopo Angeli, in about 1410, of Claudius Ptolemy's *Geography* (Ptolemy 1991, 2000). The *Geography* had originally been part of

---

Caerte." My thanks to Peter van der Krogt for bringing this map, and a couple of other early modern
Dutch works with similar formulas, to my attention.

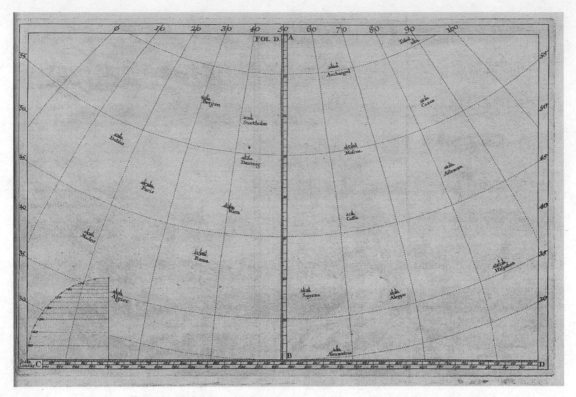

FIGURE 5.10. The framework of cosmographical geometry—meridians and parallels—within which known geographical places might be plotted. This untitled diagram (Scherer 1710, pl. D) illustrates the initial plotting of known places around which to compile a map of Europe and western Asia. Copper engraving, 22.5 × 34 cm. Courtesy of the Osher Map Library and Smith Center for Cartographic Education, University of Southern Maine (Smith Collection); www.oshermaps.org/map/13974.

an interconnected set of works by this second-century CE scholar: the practice of astrology, laid out in his *Tetrabiblos*, required both a mathematical model of the cosmos to locate planets, laid out in his *Almagest*, and a gazetteer of locations on the earth's surface, in the *Geography*.* A modern equivalent of Ptolemy's *Geography* is Eugene Dernay's (1945) detailed U.S. gazetteer specifically intended to assist in the casting of horoscopes.

Cosmographical elements were used to structure knowledge of the earth and its inhabitants. In particular, a consistent argument held that the five Aristotelian climatic zones—northern frigid, northern temperate, torrid, southern temperate, and southern frigid, delimited by the tropics and Arctic and Antarctic circles—determined the distribution of the human races as well as plants and animals. This concept proved especially obvious and persisted well into the nineteenth century. But other cosmographical concepts had steadily eroded before 1800, notably the

---

*I owe this point to discussions with Leif Isaksen.

concept of micro-macrocosmic relations and the "doctrine of the sphere," which lay at the core of the use of pairs of globes as pedagogic tools (Dekker 2002). Maps of the world gradually ceased to be part of the study of the fabric of the cosmos (see "cosmography" in table 1) and were integrated into new trends in studying the peoples of the world and their histories (geography) (Relaño 2001; Cattaneo 2009; Tessicini 2011). As early as the fifteenth century, European scholars built on earlier Islamic and Byzantine practices to apply the cosmographical geometry of latitude and longitude to organize knowledge of particular, politically and historically defined regions. This "top down" geographical mapping on cosmographical geometry contrasted with the "bottom up" chorographical surveying of regions on plane geometry.

Construction *de novo* of geographical maps of the world and its regions required the construction of a network of meridians and parallels (fig. 5.11) and then the plotting of locations of known latitude and longitude (see fig. 5.10). Early modern scholars had inherited three ways to represent the celestial sphere on a plane from ancient astronomers: the gnomonic, stereographic, and orthographic "projections." All three used geometrical perspective to translate the heavens onto a map centered on the north celestial pole, as if seen respectively from the center of the cosmos, the south celestial pole, and from infinity. While other early modern arrangements of meridians and parallels have similarly been called "projections," they did not actually rely on the same rules of perspective. Rather, other networks of latitude and longitude were designed to be easy to draw with straight lines and arcs of circles, to capture on paper some sense of the earth's curvature, and to correspond to the arrangement of the earth's meridians and parallels (Snyder 1993; Morrison and Wintle 2019).

Once the framework of meridians and parallels was constructed, the geographer could graphically interpolate other information drawn from a variety of sources, including itineraries and chorographic maps. The cosmographical geometry might be indicated on the final image by the lines of selected meridians and parallels, and also the other cosmographical circles, or by marginal gradations of latitude and longitude (Edney 2019b). These indications of cosmographical geometry explicitly tied the mapped region to a specific portion on the earth's surface and further related it to neighboring regions, obviating the need for any other expression of correspondence.

Nonetheless, early modern geographers did often add one or more scales to their regional maps. In large part, they did so specifically to permit the instrumental use of the maps in determining distances between towns. For example, J. B. B. d'Anville (1777, 25–27, 32, 54, 62) only mentioned scales in such instrumental terms and more specifically with respect to their lack on some maps and to the care needed in their construction. As the *Encyclopédie* explained in 1755:

FIGURE 5.11. Three projections for a world map: (upper) world in two hemispheres, each on an equatorial aspect stereographic projection; (middle) a trapezoidal projection, replicating the manner in which meridians converge toward the poles; (lower) a *plate carrée*, or an equal-spaced grid of meridians and parallels. From an anonymous and untitled French manuscript, ca. 1600, fol. 1r. Hand-colored manuscript, 30.5 × 22.5 cm (paper). Courtesy of the Osher Map Library and Smith Center for Cartographic Education, University of Southern Maine (Osher Collection); www.oshermaps.org/map/42502.0001.

FIGURE 5.12. Putti measuring scales in Spanish leagues (*Leucæ Hispanicæ*), common German miles (*Milliaria Germanica communia*), and common French miles (*Milliaria Gallica communia*). Detail of John Ogilby's 1671 version of Arnoldus Montanus's map of Chile. Courtesy of Osher Map Library and Smith Center for Cartographic Education, University of Southern Maine (Smith Collection); www.oshermaps .org/map/1759.0001.

> To find on a map the distance between two towns, one takes the interval between
> them with a compass [i.e., pair of dividers], and applying this interval to the scale
> of the map, one judges by the number of divisions which it encompasses, the
> distance between the two towns. (La Chapelle 1755, 248)*

Scales were only added to maps that were of sufficiently small extent that the projection's geometrical alterations were not pronounced. The presence on a regional map of a single scale connoted that it had originated—or was claimed to have originated—in itineraries and chorographical surveys. As some geographers published works in multiple countries, with different common units of itinerary measure, they provided multiple scales to make their maps relevant to as many readers as possible (fig. 5.12). Significantly, no early modern geographical maps that bore scales also bore verbal expressions, nor were they ever described as being drawn to scale. The lack of such elements indicates that scales on regional maps were solely devices to determine distances; they were *not* expressions of proportionality.

Scales bore further connotations on the new maps compiled by eighteenth-century critical geographers. The compilers provided scales for each of the linear measures used in their source materials (fig. 5.13). The measure for each scale was carefully defined as so many units of the basic measures used in the geographer's

---

*"Pour trouver sur une carte la distance entre deux villes, on en prend l'intervalle avec un compass; & appliquant cet intervalle sur l'*échelle* de la carte, on jugera par le nombre de divisions qu'il renferme, de la distance des deux villes."

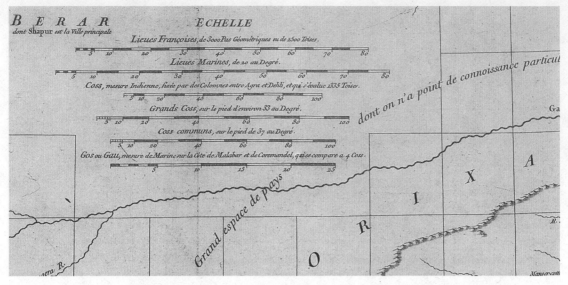

FIGURE 5.13. Detail of scales on J. B. B. d'Anville's *Carte de l'Inde* (Paris, 1751). The scales relate to d'Anville's sources: two scales for French leagues (land and marine) and for four different incarnations of the South Asian *coss*. This image also exemplifies the placement of a set of scales to fill empty space, labeled as "a great extent of country of which we have no particular knowledge." Courtesy of Osher Map Library and Smith Center for Cartographic Education, University of Southern Maine (Smith Collection); www.oshermaps.org/map/1915.0050.

own country (*toises* in the case of fig. 5.13) or as so many measures to one degree of latitude. The proliferation of scales on such maps would of course be helpful to a reader trying to relate some other source material to the map, but, more important, they served as graphic testimony to the multitude of sources from which the geographer had carefully and laboriously compiled his new map and therefore to the geographer's skill and expertise.

Scales were not applied to all geographical maps; there was a limit to their use. For maps covering extensive regions, such as Russia, a continent, or an entire hemisphere, the projection's geometrical alterations were so pronounced that there could be no pretense that one scale could be even approximately valid over the map's entire surface. While the map's meridians and parallels established a correspondence with the world, they could not establish proportionality, and in the early modern era no one expected them to do so. We can look—almost*—in vain

---

*Excluding the printed versions of marine-style world maps intended for geographical consumption (such as fig. 2.5), and world maps that geographers made for mariners and that bore scales in emulation of marine practice (see chapter 2), I have encountered just two early world maps structured on a cosmographical framework of latitude and longitude that also bore scales: the 1558 and 1570 derivatives of Caspar Vopel's now lost six-sheet wall map, each known in one impression (Shirley 2001, nos. 102 and 123), which I cannot explain, and Nicola van Sype's ca. 1583 one-sheet map celebrating Sir Francis Drake's circumnavigation (Shirley 2001, nos. 149 and 151), where the scale relates once again to the maritime navigation theme.

for scales on geographical maps of the entire earth even in the nineteenth century. The problem was most acute for maps on the Mercator projection. As a note below the title of the 1918 world map reproduced in figure 5.1 stated:

> The Mercator Projection does not permit of a fixed Scale of Miles because of the fact that, to show the face of the Globe on a flat surface, the [map] scale must be greatly extended towards the Poles, both as to latitudinal and longitudinal proportions.

Thus, until well into the twentieth century, geographical maps of the whole world used their lines of latitude and longitude to establish correspondence and made no claim to proportionality.

Some geographers have compromised when it comes to mapping regions so extensive that scales are inappropriate. Recognizing that their readers might still try to determine distances across the map, geographers have on occasion provided an alternate, indirect form of verbal expression. These comments simply specified the length of a degree (of latitude or along the equator); the map user could then relate the interval between two towns, say, not to a scale but to a nearby meridian. In 1592, for example, Theodore de Bry's *Americae pars magis cognita* bore a large pair of instrumental-looking compasses set above a statement explaining that 3 degrees comprise either 80 French leagues or 240 Italian miles, and that 1 degree comprises precisely 27 French leagues or 80 Italian miles. Didier Robert de Vaugondy added a note to his 1750 maps of eastern and western Russia—*Partie orientale de l'empire de Russie and Partie occidentale . . .*—that "one degree of the earth contains 104 versts or Russian miles, or 250 Chinese *li*" (Robert de Vaugondy 1755, 151, 280–81; see Pedley 1992, 53 and nos. 404–5). And an early twentieth-century pair of terrestrial hemispheres that advertised Admiral Peary's exploits at the poles bore the note, "Distance Between Each 10 Degrees of Latitude Approximately 690 Miles" (Peary 1941 [1919]). Not every such extensive map bore such a comment, but many did, specifically in support of the desire of map readers to determine distances, and not to claim any quality of proportionality.

On occasion, eighteenth-century geographers used "scale" more in a conceptual manner than as a reference to the instrumental device situated on their regional maps. For example, Robert de Vaugondy, in the context of explaining how he had compiled his maps, repeatedly used *échelle* in noting how particular maps of "equal/equivalent scale" (*égalité d'échelle*, *même échelle*, *uniformité d'échelle*, or *conformité d'échelle*) could be directly combined into general maps. Most of his usage seems to have simply been shorthand for having two scales on different maps representing the same unit with the same size gradations (Robert de Vaugondy 1755, 154, 207, 223, 232–33, 235, 290, 295, 308, 316). Yet he also used *échelle* metonymically, reconfiguring the graphic line or ruler as a concept. Specifically, he referred to having to reduce extensive objects "to a scale so small" (*à une échelle si petite*) that they could be mapped; to a map "double of scale" (*une carte double d'échelle*); and to having maps "on the same footing of scale" (*sur le même pied d'échelle*) (Robert de Vaugondy 1755, 266, 311, 352). Guillaume Delisle (1728) had previously made similar comments that shaded on the metonymic when comparing maps of Paris and London. Such usages would seem to prefigure the nineteenth-century development of an idealized understanding of map scale.

TRIGONOMETRICAL GEOMETRY: GEODESY

Something of a transitional arena between plane and cosmographical geometries was formed by an extension of plane geometry to cover the kinds of extensive areas otherwise mapped by cosmographical geometry. This extension took the form of a new technique, specifically that of triangulation, which, in its more advanced applications featuring trigonometry, sought to take into account the curvature of the earth's surface. The new technique contrasted markedly with other techniques of property and place mapping, which rarely involved mathematics more complicated than multiplication, division, and the relationships of similar triangles.

The Dutch mathematician and cosmographer Gemma Frisius described triangulation in 1533 as a method for constructing geometrical frameworks for regional surveys (Pogo 1935, 474–79, 486–506 [facsimile]; Haasbroek 1968, 11–14; Lindgren 2007, 483–84). Triangulation comprises a network of interconnecting triangles whose interior angles are measured; the triangles are sized by the known length of at least one side, however determined. The result, as in the triangulation that underpinned P. P. Burdett's survey of Cheshire in the 1770s (fig. 5.14), is a rigid structure that can readily control the more detailed survey of topographical features. Yet there was significant variation in just how early modern surveyors implemented this process.

As Frisius described the technique, the re-creation of the triangles on paper was a graphic process. The one side whose length is known is drawn on the paper; the surveyor uses a protractor to construct the other two sides of the triangle from measured angles, and those sides can then be used as the basis for constructing the adjacent triangles; the process is repeated until the triangulation is plotted out. Most local surveys based on triangulation, such as Burdett's of Cheshire and the other eighteenth-century county surveys in England, seem to have used this process, which was little different from other plane-geometry practices.

Frisius also observed that, while it would be possible to calculate the lengths of all the sides of the triangles with trigonometry, doing so would be "too difficult for the common man" (Haasbroek 1968, 14). Moreover, knowing the lengths of each side of the triangle had little effect on how the surveyor would graphically reconstruct the triangulation on paper. But the use of trigonometrical functions was relevant to another kind of triangulation survey, specifically, that intended

196

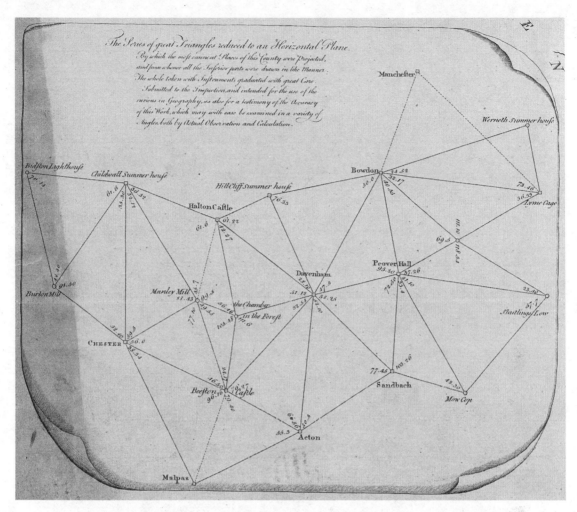

FIGURE 5.14. The triangulation diagram, "The Series of great Triangles reduced to an Horizontal Plane," on sheet 3 of P. P. Burdett, *This survey of the county palatine of Chester*, 2nd ed. (London: William Faden, 1794; originally published 1777). This triangulation was constructed graphically, without trigonometric cal culations, and Burdett's claim to have reduced the triangles to the plane was pure exaggeration. Courtesy of the Biblioteca virtual del Patrimonio bibliográfico (Colección Mendoza, Biblioteca nacional de España, Madrid); www.europeana.eu.

to determine the size and shape of the earth. After a hesitant beginning with Willebrord Snel van Royen's survey in the Netherlands in 1615–17 to measure the earth's size, Jean Picard refined the technique in his survey of a chain of triangles along the meridian from the Paris Observatory to Amiens in 1668–70. The linear style of this triangulation is evident in figure 5.15. Picard carefully measured a long baseline directly on the earth with measuring rods, along a long, straight, and flat road; from this baseline he trigonometrically calculated the lengths of all the sides of the triangles, and in particular those that lay close to the meridian. Reducing these particular lengths to the length of the meridian itself, and observing the

197

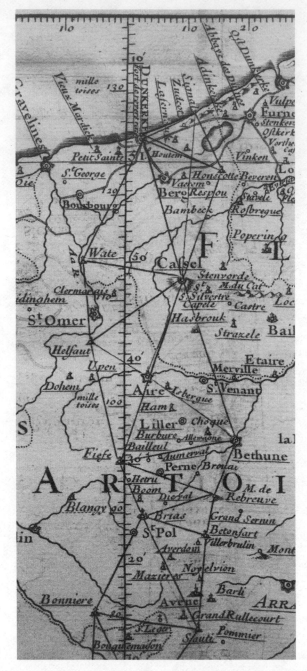

FIGURE 5.15. Jacques Cassini's indication of the calculated length of the Paris meridian and latitude. The lefthand (west) side of the meridian line is calibrated directly in thousands of *toises* north or south of the Paris Observatory; the righthand (east) side, in degrees and minutes of latitude. Cassini had specifically undertaken the extension of Picard's triangulation to investigate the earth's shape. Detail of the first sheet (of three) of *Carte des provinces de France traversées par la méridie.ne de Paris* (appended to Cassini 1720, vol. 2). Courtesy of the Department of Special Collections, Memorial Library, University of Wisconsin-Madison.

difference in latitude between the extremes of the triangulation, he could then calculate a value for the circumference of the earth, then still considered a sphere (Bendall 2019; Edney 2019a).

The hallmark of Picard's and similar trigonometrical surveys was their goal of relating the plane geometry of the survey and of the trigonometrical calculations to the earth's curved surface.* Once the length of a meridian had been defined, the meridian could be graduated with both degrees of latitude and linear measures (see fig. 5.15). Geodesists were also able to calculate the latitude and longitude of every survey station that formed a vertex of a triangle. Conversely, true to their origins in plane geometry, surveyors' maps of triangulation networks largely remained abstract geometrical diagrams with little reference either to specific features or to how the networks corresponded to the world.

PROJECTIVE GEOMETRY: SYSTEMATIC MAPPING

Members of the Académie des sciences in Paris were the first, in the eighteenth century, to blend plane and cosmographical geometries, via trigonometrical surveys, within a system of projective geometry. Doing so was not, however, the original goal of their efforts to make a consistent map of France. We can trace the process whereby plane, cosmographical, and trigonometrical geometries were integrated through the history of the efforts to map the French state before 1750 (Pelletier 2013, 2019; Laboulais 2019).

The first plan developed by the academicians in the 1660s, at the request of Louis XIV's minister Jean-Baptiste Colbert, was to perfect the cosmographical geometrical framework within which the provincial maps could be fitted together into a general map of the entire country. This plan featured two main elements. First, the academicians would use precise astronomical observations to determine latitudes and longitudes, the latter using the new technique of timing the eclipses of Jupiter's moons (see fig. 4.4; Sandman 2019). Second, a careful measurement of the earth's size, in the form of Picard's geodetic survey, would permit itinerary distances to be properly converted to differences in latitude and longitude, and so be readily incorporated into a cosmographical framework.

The existing provincial maps were so geometrically inconsistent, however, that the academicians quickly realized that it would be insufficient to improve only the cosmographical technologies. The provincial maps would also have to be recompiled, perhaps even resurveyed. And for that, the academicians argued, Picard's geodetic survey of part of the meridian of Paris had to be expanded into a grid

---

*There is no paradox here. Before 1800, instrumentation was still too imprecise to measure the spherical excess in triangles (the surplus, above 180°, in the sum of the interior angles of a spherical triangle) and, although William Roy and Adrien-Marie Legendre anticipated the need to account for the phenomenon in the 1780s, they nonetheless continued to use plane trigonometry to calculate triangle sides (Edney 2019a).

of triangulation covering all of France. The work proceeded rather fitfully, until it was finally completed by César-François Cassini de Thury in 1744. To guide the preparation of new provincial maps, Cassini de Thury and his colleagues prepared:

- A large, eighteen-sheet map of the triangulation of France, *Carte qui comprend touts les lieux de la France qui ont étés déterminés par les opérations géométriques*, that measured 1.4 m × 1.25 m when assembled (Konvitz 1987, 15–16; Pelletier 2019)
- A summary, single-sheet map of the triangulation, *Nouvelle carte qui comprend les principaux triangles qui servent de fondement à la description géométrique de la France* (fig. 5.16), complete with a long table of the latitudes and longitudes of 442 key locations across the country, together with their distance from the Paris Observatory (fig. 5.17)
- A long, detailed memoir that explained the survey and its results to map makers but was not, in the end, published (Cassini de Thury [1744]; see Cassini de Thury 1749; Konvitz 1987, 18–19)

Even so, the plane and cosmographical geometries were not yet completely integrated. Other map makers were still expected to make provincial maps on the frameworks of cosmographical geometry, guided by the dense network of places that the triangulations had fixed in latitude and longitude.

But even as Cassini de Thury presented the triangulation as being of service to the geographical mapping of France's regions, in November 1745, he also suggested that it would be possible to use the triangulation as the basis for a single map of the entire country that would go into much greater detail than could be achieved by traditional, cosmographical/geographical techniques (Cassini de Thury 1749). Cassini de Thury perhaps built upon an earlier proposal, made in 1735–36 by the geographer Philippe Buache, to use the triangulation of France as the basis of a 116-sheet map of all France (Konvitz 1987, 15).

Cassini de Thury's proposal required a striking innovation that present-day scholars tend to take for granted: the integration of plane and cosmographical geometry within a new, projective geometry. Specifically, he described a simple, transverse cylindrical projection with every location in France located by an abscissa and ordinate defined by the meridian through the Paris Observatory and its perpendicular. He used this projection for his maps of the triangulation. Both the one-sheet *Nouvelle carte* and the eighteen-sheet *Carte qui comprend touts les lieux de la France* featured a straight meridian through the observatory; curved, concave meridians to either side, as indicated by the increasingly angled marginal gradations for longitudes east and west of the observatory (apparent in fig. 5.18); and the perpendiculars to the meridian at increments of 60,000 *toises* north and south of the observatory. Note that neither map showed the vertical grid lines parallel to the observatory's meridian.

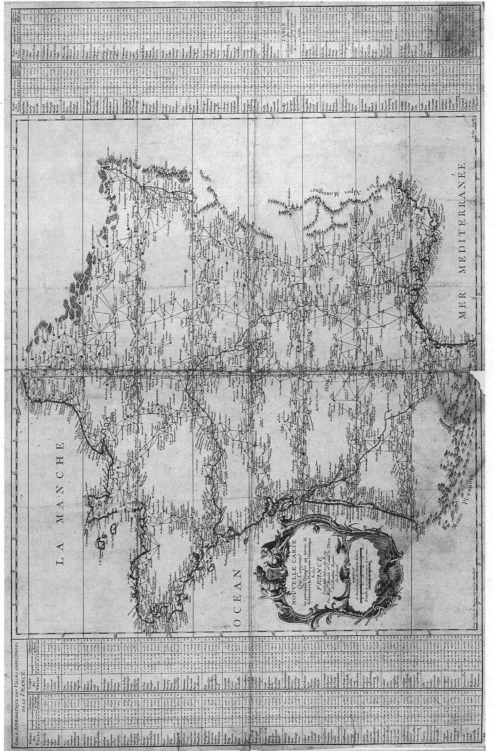

FIGURE 5.16. The primary triangulation of France, completed 1668-1744. Giovanni Domenico Maraldi and César-François Cassini de Thury, *Nouvelle carte qui comprend les principaux triangles qui servent de fondement à la description géométrique de la France levée par ordre du Roy* (Paris, [1745]). This first variant of this map was drawn in 1744 and presented to the Académie des sciences in 1745, when it was likely published. Copper engraving on one sheet, 59 × 91 cm (the map is 63 cm wide without the columns of text on either side); first state. Courtesy of the Geography and Map Division, Library of Congress (G5831.B3 1744 .M3); www.loc.gov/item/2004629162.

TABLE ALPHABETIQUE DES VILLES PRINCIPALES DE LA FRANCE.

NOMS des VILLES	Longitude D. M. S.	Latitude D. M. S.	Distances à l'Observatoire de PARIS. en Toises	Lieues	NOMS des VILLES	Longitude D. M. S.	Latitude D. M. S.	Distances à l'Observatoire de PARIS. en Toises	Lieues
Abbeville	0. 30. 2	50. 7. 2	75443.	57.	Crauzon	6. 50. 34	48. 14. 52	260636.	130.
Agde	1. 8. 11	43. 18. 57	318301.	159.	Creil	0. 8. 11	49. 15. 38.	24744.	12.
Agen	1. 44. 11	44. 12. 7.	273096.	136.	Cremieu	2. 55. 13	45. 43. 20.	210586.	105.
Aire	0. 3. 18	50. 38. 18.	101927.	51.	Croisic	4. 51. 42	47. 17. 40.	205174.	102.
Airvaut	2. 35. 29	46. 49. 55.	151396.	75.	Dammartin	0. 20. 42	49. 3. 14.	17942.	8.
Aix	3. 6. 34	43. 31. 35.	327142.	163.	Dax	3. 23. 55.	43. 42. 23	321824.	160.
Albi	0. 11. 16.	43. 55. 44.	280103.	140.	Decise	1. 6. 18	46. 50. 24.	121507.	60.
Amance	3. 57. 9.	48. 43. 8	148630.	74.	Desvre	0. 30. 8.	50. 40. 4.	106142.	53.
Ambleteuse	0. 44. 14.	50. 48. 13.	115570.	57.	Dieppe	1. 15. 48.	49. 55. 17.	77713.	38.
Amboise	1. 20. 53.	47. 24. 54.	95984.	47.	Digouin	1. 38. 40.	46. 28. 55.	148507.	74.

FIGURE 5.17. Detail of the beginning of the table of latitudes and longitudes, with distances from Paris, from the upper-left corner of fig. 5.16.

Cassini de Thury completed the new projective geometry after 1747, when Louis XV gave permission for a statewide territorial survey. The marker of the full implementation of the new geometry is the third variant of the *Nouvelle carte*, which not only delineated all the secondary triangulations undertaken in 1748–55 between the main chains but also the 182 sheets of what Cassini de Thury originally called the *Carte générale et particulière de la France* (fig. 5.18). Cassini de Thury's title indicated the manner in which the survey presented an utterly new kind of map, one that was not general or particular, but that was both general and particular at the same time. Both names that the work later acquired—*Carte de France* and *Carte de Cassini*—accepted the ideal's ontological claim to create a single cartographic archive in *the* map (unqualified and normative); under the ideal, there has been no need to draw attention to just how innovative the French survey was in its combination of general and particular mapping.

Cassini de Thury defined each sheet of the *Carte générale et particulière* by constructing lines parallel to the meridian and perpendicular, and set at constant intervals of 25,000 *toises* (north to south) and 40,000 *toises* (east to west). The values of each sheetline's boundaries were offset by half of these values because the origin of the coordinate system, the Paris Observatory, was placed at the center of the first sheet. Other academicians had suggested alternative projections, but they were rejected for requiring several extra steps in calculating the coordinates; Cassini de Thury's projection was adopted because it was easy to construct without actual reference to mathematical formulas (Bret 2019).

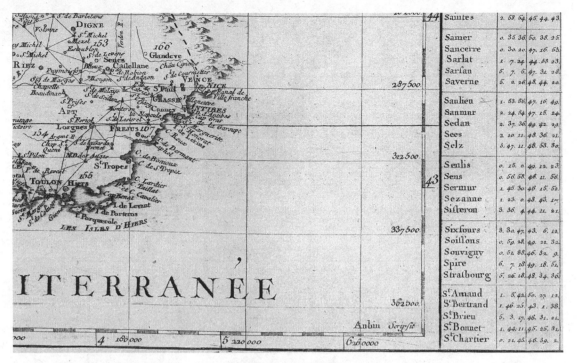

Saintes	2.	53. 54.	45. 44. 43.
Samer	o.	35. 36.	50. 38. 25.
Sancerre	o.	30. 20.	47. 16. 53.
Sarlat	1.	7. 24.	44. 53. 23.
Sarſau	5.	7. 6.	47. 31. 28.
Saverne	5.	2. 26.	48. 44. 20.
Saulieu	1.	53. 56.	47. 16. 49.
Sanmur	2.	24. 54.	47. 25. 24.
Sedan	2.	37. 36.	49. 42. 20.
Sees	2.	10. 11.	48. 36. 21.
Selz	3.	47. 11.	48. 53. 30.
Senlis	o.	16. o.	49. 12. 23.
Sens	o.	56. 58.	48. 11. 56.
Sermur	1.	45. 30.	46. 15. 52.
Sezanne	1.	23. o.	48. 43. 1.
Siſteron	3.	36. 4.	44. 11. 21.
Sixfours	3.	30. 47.	43. 6. 12.
Soiſſons	o.	59. 28.	49. 22. 32.
Souvigny	o.	51. 88.	46. 32. 9.
Spire	6.	7. 18.	49. 18. 51.
Straſbourg	5.	26. 18.	48. 34. 36.
St Amand	1.	5. 42.	50. 27. 12.
St Bertrand	1.	46. 25.	43. 1. 38.
St Brieu	5.	3. 17.	48. 31. 01.
St Bonnet	1.	44. 11.	45. 25. 81.
St Chartier	o.	21. 45.	46. 39. 2.

FIGURE 5.18. The realization of projective geometry. Detail of the lower-right corner of the third variant of the *Nouvelle carte qui comprend les principaux triangles qui servent de fondement à la description géométrique de la France* (see fig. 5.16 for first variant). The second variant of this map was marked by the addition of a date (1744), the secondary triangulation in northern France, a long textual note, and the proposed sheet lines. This third variant further added the remainder of the secondary triangulation and the coordinates of the sheet lines; it was probably first issued in 1755 or 1756, between the completion of the secondary triangulation and the issue of the first sheet (Paris) of the *Carte générale et particulière de la France* (see fig. 4.2). In addition to the new marginal indication of projected coordinates, this detail shows how the numbers for each sheet had to be crammed in to fit the map's details: here 153, 154, 155, 166, and 167. Courtesy of the David Rumsey Collection (5694200); www.davidrumsey.com.

Cassini de Thury constructed the sheets of the *Carte générale et particulière* at a correspondence of 1 *ligne* (2.25 mm) to 100 *toises* (194.904 m), as expressed in a single scale at the bottom of the sheet (see fig. 4.2). The lone scale suggested that the entire map was a work of plane geometry carried out over the entire surface of France. Of course, the map represented a projected, pseudo-plane, but at the contemporary level of precision of measurement, it could be treated and used as if it were actually a plane. The *Carte générale et particulière* thus offered only an approximate proportionality.

The individual sheets of the *Carte générale et particulière* included no expression of their correspondence to the earth, other than the scale and the indication of the distance of each corner from the meridian of the Paris Observatory (abscissa) and the perpendicular thereto (ordinate). Instead, Cassini de Thury issued each sheet of the map along with a table of places falling within the sheet, listing each place's precise abscissa and ordinate. In this respect, the *Carte générale et particulière* only held out and did not fulfill the intellectual potential of projective geometry; users could not directly take measurements or define locations by those coordinates (Rankin 2016, 128). This situation would change. As other European states adopted the same system after 1790, their surveyors extended Johann Heinrich Lambert's mathematization of geographical map projections (see chapter 4) to topographical map projections, to establish systems of projected $(x[\phi, \lambda], y[\phi, \lambda])$ coordinates. And, as detailed hydrographical surveys were increasingly undertaken from the rigorous foundation of onshore triangulations, so these projective geographies came to be understood as unifying the geometries of all mapping modes, including marine mapping, to establish the single geometry of cartography.

The spread in the nineteenth century of territorial surveys as an integral element of the modern state—as a basis for taxation and censuses, geological and botanical studies, economic assessments, military and civil planning, and eventually personal mobility—promoted the use of projective geometries to the point, in the mid-nineteenth century, where British professionals needed to distinguish between the two meshes of lines. They began to use "graticule," previously used only for a copyist's grid, for the network (*réseau/Netz/net*) of meridians and parallels, whether on the earth's surface or projected onto the plane of the map; they retained "grid" for the rectilinear grid for plotting projected $(x[\phi, \lambda], y[\phi, \lambda])$ coordinates and delimiting sheet lines in an ordered manner. The earliest instance I have found of "graticule" in its cartographic sense is in a work from the Survey of India (Strachey 1848, 537; see also Wallis and Robinson 1987, 172–74).

The fulfillment of this trend and of the interchangeability of coordinate systems came with the imposition of rectilinear grids to theater maps during the Franco-Prussian War (1870–71) and World War I (1914–18). Called "artillery squares," the grids assisted in the targeting of long-range artillery pieces. In 1917, the Germans and Austrians agreed to a common coordinate system for their topographical mapping. By the 1930s, grids were being added even to the civil incarnations of European territorial maps, and they proliferated after 1945. Examples include the United Kingdom's "national grid," the U.S. "state plane coordinate system," and NATO's "universal transverse Mercator" system, although the latter was anticipated by German military map production during World War II (1939–45). The result has been the rise of the practice of "coordinate surveying," in which plane geometry is displaced by projective geometry (Morrison 2015; Rankin 2016, 119–201; Buchroithner and Pfahlbusch 2017). Finally, the technological development of radionavigation and global positioning systems have further reified what Bill Rankin (2016, 205–99) called the "stitching" of the map to the territory, the wedding of the projective geometry of detailed territorial mapping to the cosmographical geometry of the curved surface of the earth.

If cosmographical geometry is about the cosmos transferred to the earth as great and small circles, expressed numerically by coordinate pairs of latitude and longitude (ϕ, λ), then projective geometry is ultimately concerned with coordinate pairs of the form $(x[\phi, \lambda], y[\phi, \lambda])$. Moreover, the geometries are reversible: not only can one determine $(x[\phi, \lambda], y[\phi, \lambda])$, one can also, in theory, determine cosmographical from projected coordinates $(\phi[x, y], \lambda[x, y])$. This projective geometry should not be mistaken for Cartesian (as Edney 2017b), if only because the earth is not indefinite in extent.

From this utterly modern perspective, mapping has but a single geometry. All the distinctions between plane, cosmographical, and trigonometrical geometries collapse in a single grand system: no matter how a map is made, its particular geometry is interchangeable with the others; even when particular mapping practices continue to be undertaken on other geometries, they are *really* part and parcel of a singular cartography. It is within this system that cartography acquired the idealized singularity of an indefinitely precise archive that can incorporate, and evaluate, all spatial knowledge, however generated and expressed.

Projective Geometry, Numerical Ratios, and Map Scale

The widespread adoption of projective geometry for systematic territorial and hydrographic mapping brought about both the usage of numerical ratios and the concept of map scale. By collapsing the distinctions between plane, trigonometrical, and cosmographical geometries, projective geometry made meaningful the idea that one could use a single form of expression to represent any map's correspondence to the world. That a numerical ratio could have been derived from certain verbal expressions of proportionality is evident from the case of the 1649 map of Bourbourg, discussed above. Such early instances of the numerical ratio—should any further be found in the historical record without the curious inversion of expression on the Bourbourg example—were undoubtedly restricted to plane mapping. Only after 1800 did the quality of map scale, engendered by the entirely new expression of the numerical ratio, emerge as the universally applicable relationship of maps to the world. Map scale thus became a key element in the formation of the ideal of cartography.

PIERRE ALEXANDRE JOSEPH ALLENT'S REDUCTION FACTORS

The immediate stimulus to the conceptualization of map scale was the promulgation in the 1790s of the metric system. The revolutionary government in France sought to achieve what natural philosophers had advocated for several centuries: to replace customary practices of measurement, governed by antiquated social and technological relationships, with a strictly rational and universal system of measurement. When First Consul Napoleon Bonaparte convened a *Commission topo-*

graphique in 1802 to establish a universal system of cartographic conventions—to rationalize the work of the many mapping agencies then active within the French state and to promote the exchange of cartographic materials between them (Anon. 1803; see Bret 2008)—the stage was set for an attempt to rationalize the manner in which maps correspond to the world.

One member of the *Commission topographique*, the leading military engineer and later politician Pierre Alexandre Joseph Allent, proposed two mathematical series that he called *échelles décimales* or *échelles métriques*. This particular meaning of *échelle* stemmed from a variant ladder metaphor that describes mathematical and musical series with consistent intervals. Allent's two series together defined specific degrees of graphic enlargement or reduction, either as a series of abstract terms in the form $x/1$ to express enlargement (ascending terms) or $1/x$ to express reduction (descending terms). For Allent, x was a factor of ten (1, 2, 5) progressively multiplied by powers of ten. His reduction series was thus:

1 [i.e., ¹⁄₁], ½, ⅕, ¹⁄₁₀, ¹⁄₂₀, ¹⁄₅₀, ¹⁄₁₀₀, ¹⁄₂₀₀, ¹⁄₅₀₀, ¹⁄₁,₀₀₀, ¹⁄₂,₀₀₀, ¹⁄₅,₀₀₀, ¹⁄₁₀,₀₀₀, . . .

The inverse constituted the enlargement series. Allent argued that these mathematical series could be used to define preset design parameters, including the consistent sizing of lettering and signs, in order to permit consistency across all kinds of engineering imagery (Anon. 1803, 11–16; Allent 1803).

Within the year, Allent had reconfigured his reduction series into a series of predetermined reduction factors that engineers could use to ensure consistency across all their imagery; this revision evidently circulated in manuscript before being finally published almost three decades later (Allent 1831). In the process, Allent applied the term *échelle métrique* not to the entire reduction and enlargement series but to the individual terms within each series. Here, in about 1803, is the birth of the numerical ratio.

Allent's "metric scales" were applicable, in principle, to all kinds of images that reduced the world to paper. As Allent explained:

> The *descending series* contains, beyond the first term [i.e., ¹⁄₁], all metric scales in which a millimeter [in the image] represents a larger extent [in the world]. This series gives all scales that measure, on their projections [i.e., perspective views], the body or the ground when the smallest details one must notice are larger than a millimeter. It descends down to world maps, and even to projections of the solar system. In the same series are found the scales most used in topography, and in general in the various public services. (Allent 1831, 45–46)*

---

*"La *série descendante* contient, au delà du premier terme, toutes les échelles métriques dans lesquelles le millimètre représente une grandeur plus considérable. Cette série donne toutes les échelles qui servent à mesurer, sur leurs projections, les corps ou le terrain, lorsque les plus petits détails, qu'il im-

TABLE 3. Summary of Pierre-Alexandre-Joseph Allent's revised "Tableau des premières triades de la série descendante des échelles métriques, indiquant leurs rapports et l'emploi qu'on en fait dans les services publics" (published in Allent 1831, 47–51), including Allent's own commentary in the final column. This table substantially revised Allent's "Tableau de la série générale des échelles métriques, indiquant leurs applications au service du génie militaire" (Allent 1803, table 1). Originally, Allent had started *triade* I with ²⁄₁, so that the *triades* were misaligned (e.g., *triade* II contained ⅕, ⅒, and ¹⁄₂₀). He had also specified *nombres abstraits* as both ordinary (⅕) and decimal (0.2) fractions, specified the *rapport* for metric and premetric units, and explained the application of each *triade* only to military engineering. My thanks to Salim Mohammed at Stanford University for supplying images of Allent's original 1803 table.

TRIADE	*NOMBRES ABSTRAITS*	*RAPPORT DES ÉCHELLES* (1 MM TO ? ON THE GROUND)	USE
I	1 ½ ⅕	1, 2, 5 mm	instruments, machines, etc.
II	⅒ ¹⁄₂₀ ¹⁄₅₀	1, 2, 5 cm	architectural details, whether civil, military, or marine
III	¹⁄₁₀₀ ¹⁄₂₀₀ ¹⁄₅₀₀	10, 20, 50 cm	topographical and property maps requiring high precision, especially in the interior of towns where property is valuable and sold by the square meter; the Comité des fortifications had already adopted ¹⁄₅₀₀ as a replacement for the older standard scale of 1 *pied* to 100 *toises* (1/600)
IV	¹⁄₁,₀₀₀ ¹⁄₂,₀₀₀ ¹⁄₅,₀₀₀	1, 2, 5 m	topographical and property maps of less precision; all 3 scales had been adopted for different kinds of military mapping, ¹⁄₂,₀₀₀ replacing the standard scale specified in 1776 for many fortification plans of 4 *pouces* to 100 *toises* (¹⁄₁,₈₀₀)
V	¹⁄₁₀,₀₀₀ ¹⁄₂₀,₀₀₀ ¹⁄₅₀,₀₀₀	10, 20, 50 m	most topographical plans of a country, as respectively adopted by the Dépôt de la Guerre for the surveying, reduction, and engraving of a new map of France; a law of 1810 required all maps of mining rights to be at ¹⁄₂₀,₀₀₀
VI	¹⁄₁₀₀,₀₀₀ ¹⁄₂₀₀,₀₀₀ ¹⁄₅₀₀,₀₀₀	100, 200, 500 m	general chorographical maps
VII	¹⁄₁,₀₀₀,₀₀₀ ¹⁄₂,₀₀₀,₀₀₀ ¹⁄₅,₀₀₀,₀₀₀	1, 2, 5 km	geographical maps of countries and parts of the world; Maxime Auguste Denaix made his beautiful physical map of Europe at ¹⁄₅,₀₀₀,₀₀₀
VIII	¹⁄₁₀,₀₀₀,₀₀₀ ¹⁄₂₀,₀₀₀,₀₀₀ ¹⁄₅₀,₀₀₀,₀₀₀	10, 20, 50 km	geographical maps *à petit point*, including world maps and terrestrial globes

Allent grouped the terms in sets of three (*triades*, reinforcing the parallel with musical scales) and related each *triade* to different kinds of engineering practice (table 3).

But we should not overemphasize Allent's provision of an abstract, numerical ratio (*nombre abstrait*) for each degree of reduction. He intended his overall

porte d'apprécier, sont plus grands que le millimètre. Elle descend jusqu'aux mappemondes, et jusqu'aux projections du système solaire. C'est dans la même série que se trouvent les échelles les plus usitées en topographie, et en général dans les divers services publics."

scheme to guide engineers in selecting from a series of standardized reduction factors appropriate for imaging a particular subject matter in plan, profile or elevation, and oblique view. The engineer would first identify the size of the smallest thing that had to be represented—whether the thread of a screw or the width of a road, with the assumption that this dimension would be drawn on the paper as 1 mm in extent—and then select the appropriate standardized reduction factor for the overall image. Allent therefore used his table to highlight the correspondence of one millimeter on the image to world distance, effectively recapitulating the existing practice of providing verbal expressions. His scheme was not an explanation of a universal concept of map scale but the deduction from first principles of a series of generic reduction factors for engineers who, like him, were well versed in the rules of perspective.

The problem, as Allent (1831, 58–59) admitted, was that those engineering rules applied only to small phenomena that could be properly delineated by linear perspective and plane geometry, which is to say plans and elevations of instruments, architectural features, and perhaps even entire fortifications (*triades* I through IV in table 3). For larger phenomena, such as the areas covered by territorial and chorographical maps (*triades* V and VI), linear perspective proved inadequate, and Allent's reduction factors could only ever be approximate. And linear perspective was quite irrelevant to the delineation of still larger areas of regions and the entire world, which nominally fell within *triades* VII and VIII, so that such maps could not feature a single, constant reduction factor. A natural element of fine-resolution engineering practice, Allent's reduction factors could thus be no more than an approximate or a vague descriptor for other kinds of mapping.

ADOPTION OF NUMERICAL RATIOS BY ENGINEERS

As secretary to the *Comité des fortifications*, Allent had a certain degree of influence. Anne Godlewska (1999, 162–63, 171–72) and Nicolas Verdier (2015, 308) have both noted the wide adoption of his prescriptions for the methods to be deployed in military reconnaissance surveying. He could thus ensure that his two particular innovations—predefined and standardized ratios and their expression as a number (the numerical ratio) rather than as a scale or verbal expression—were adopted by French military engineers. After 1815, the *Dépôt de la guerre* set out to resurvey the territory of France using the ratios defined in *triade* V: original fieldwork at 1:10,000, reduced to neat copies at 1:20,000, with publication at 1:50,000. The cost of such detailed work quickly proved too great, yet it was not possible to reorganize the work around the ratios of *triade* VI; in that case, the final maps would be published at 1:500,000, which would be insufficiently detailed for military use. The eventual compromise abandoned Allent's neat scheme: the final sheets of what became known as the *Carte de l'état-major* were published at 1:80,000, a rounding down to the nearest, pseudo-metric denominator of the numerical ratio calcu-

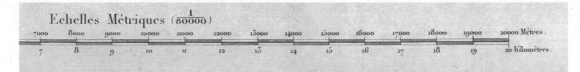

FIGURE 5.19. Detail of part of the *échelles métriques* and numerical ratio on *Ferrette*, sheet 115 of the *Carte de l'état-major* (Paris: Dépôt de la guerre, 1835). Courtesy of the Bibliothèque nationale et universitaire de Strasbourg (MCARTE299); http://gallica.bnf.fr/ark:/12148/btv1b102342131.r.

lated for the eighteenth-century *Carte générale et particulière de la France*. When the sheets of the new map series began to be published in 1832, each carried three scales, one for *toises* (*échelle en toises*), one for leagues (*échelle en lieues*), and one for both meters and kilometers that was surmounted by the ratio ⅟₈₀,₀₀₀ (fig. 5.19). The label for the last, "*échelles métriques*," is in the plural, indicating that *échelles* still referred to the two graduated lines, the scales *per se*, rather than to the sole numerical ratio. In addition, the margins of each sheet were graduated in latitude and longitude, both in degrees and in the new metric grade (or gradian) of 100^G to $90°$ (Berthaut 1898–99; Huguenin 1948).

Allent's concept of using a numerical ratio to express the proportionality of certain maps to the world was slowly accepted as a means to express *any* map's correspondence to the world in terms of an absolute ratio, 1:x, rather than as a means to predefine a degree of reduction in preparing an engineering drawing. This new practice first took root within mapping modes concerned with the detailed mapping of the earth's surface, especially as undertaken by engineers.

Some official and commercial German map makers were using numerical ratios for territorial and geographical mapping as early as the 1820s (Wagner 1914, 10–12). Numerical ratios had certainly entered the consciousness of British engineers and policy makers by the 1850s. A point of contention during the parliamentary "Battle of the Scales" (1851–59) over the future of the Ordnance Survey was whether the survey's maps should be constructed at *natural scales* that used verbal expressions involving conventional measures, such as "6 inches to the mile,"[*] or at *rational scales* that used neatly rounded denominators inspired by the metric system, e.g., 1:2,500 (Oliver 2014, 211–12).

Numerical ratios were accepted by Britain's Royal Engineers by the 1850s. One of the first instances was the Ordnance Survey's 1850 survey of the city of York, part of a series of city surveys intended to aid the installation of sewage systems (Oliver 2014, 183–93; Kain and Oliver 2015, 212–20). The map's 21 sheets, published in 1852, were constructed at 5 feet to 1 mile and each bore the numerical ratio of

---

[*]Anglophone usage for *natural scale* subsequently shifted, with its application to any numerical ratio as opposed to the rounded denominators of specifically rational scales. See, e.g., Ravenstein, Close, and Clarke (1911, 629) and Boggs and Lewis (1945, 31). Such usage of *natural scale* by Ravenstein (in Anon. 1901, 100–101) led Wagner (1914, especially 9) to adopt the term.

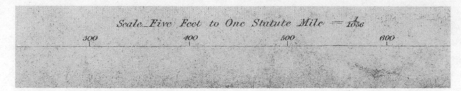

FIGURE 5.20. Detail of the scale and numerical ratio on sheet 9 of the Ordnance Survey plan of York in 21 sheets (1852), surveyed by Henry Tucker. Courtesy of David Rumsey Collection, David Rumsey Map Center, Stanford Libraries (6740009); purl.stanford.edu/cf177xz3720.

1:1,056 (fig. 5.20). Such practice would soon be codified in the manuals written by instructors at the Royal Engineers academy in Woolwich. In one essay, chief instructor Joseph Ellison Portlock used *scale* for both graduated rulers and the line on a map to indicate proportionality, thereby emphasizing how "to scale" meant that the work was prepared using scaled rulers. He also used *representative fraction* for the degree of reduction and correspondence:

> The fraction showing the ratio between 1 inch and the length of 60 inches, which it here represents, viz. 1/60, is called the *representative fraction of the scale*, a line in the drawing 1 inch long representing one in the original 60 inches long. Such a fraction should always be written over a scale, and may even be of use if placed alone on the drawing, as it gives, at once, the proportion between the length of any horizontal line on the drawing and that of the corresponding one on the original, whether such line is measured by miles, feet, yards, or other units; and it enables the person using the drawing to find the length of any of those lines, or to make a scale for himself, in terms of that unit that best meets his requirements, without regard to that [at] which it was at first constructed, which may often be advantageous when using plans of foreign works or countries, if their scales are only indicated by fractions. (Portlock 1858, xv, original emphasis)

This passage perpetuated the older practical distinctions between different strategies for expressing proportionality and did not yet reveal a fully modern conception of "map scale." But it did employ a new term, *representative fraction*, today the common term for numerical ratio among Anglophone academics and professionals, which conveys succinctly how the numerical ratio gives an immediate sense of the degree to which the world has been reduced within engineering plans.

That the concept of "map scale" was just then taking root in the practices of the Royal Engineers is further indicated by a slightly earlier essay, in which Portlock had referred generally to *scale* as the degree of a map's reduction of the world that could be specified by verbal expressions, by scales, or by the numerical ratio (which Portlock then called the *proportional scale*, a term as evocative as *representative frac-*

tion). For Portlock, the numerical ratio's particular benefit was that it allowed the map reader to use familiar linear units when working with maps constructed with unfamiliar ones. Moreover, in order to distinguish different kinds of engineering and territorial maps, and so ensure that only similar kinds of maps were compared and combined, Portlock used what seems to have been an already established concept: the relative classification of "large scale" versus "small scale" maps. Portlock's engineering perspective gave him a fairly restricted definition of these categories:

> Finally it may be said, that if a proportional scale be desirable for very large scale maps, which can rarely have any but a very local application, it is far more so for small scale maps; in the one case there can be rarely any necessity for the comparison of distant localities, in the other such comparisons may be frequently necessary. In the small scale map the unit of comparison will be large, as the mile, the kilometre, the werst [Russian verst], &c.; whilst in the large scale map, the unit will be some familiar measure of less extent, such as the chain or the perch. (Portlock 1855, xiv)

Portlock distinguished between "large scale" maps of estates and properties that were used "for agricultural and other purposes," which would fall into Allent's *triades* III and IV, and "small scale" territorial maps like the Ordnance Survey's one-inch series (1:63,360) or the maps of the *Carte de l'état-major* (1:80,000), which would fall into *triade* V. Like Allent, he did not extend a concept that was natural to plane engineering work to nonengineering and nonterritorial mapping modes dependent on cosmographical geometry.

ADOPTION OF NUMERICAL RATIOS AND MAP SCALE BY NON-ENGINEERS

Nonetheless, a few non-engineers did begin, quite early, to apply the presumption of proportionality implicit in the numerical ratio even to regional and world maps that fall within Allent's *triades* VI–VIII. This significant enhancement of the concept seems to have originated in programs to map extensive areas in a consistent manner. The eventual result was the foundation of map scale as a universal concept.

Perhaps the first non-engineer to use the numerical ratio was Pierre Grégoire Chanlaire. In 1792, Chanlaire had published an *Atlas national de la France en départemens conformément aux nouvelles divisions de territoire*, in which each map delineated one of France's newly formed *départements*. The new *départements* had, in a fit of revolutionary zeal and rationality, been created to each have approximately the same areal extent (Ozouf-Marignier 1992). Chanlaire accordingly mapped each *département* at a consistent correspondence so that they could all be fitted together into a single, large map of the entire country, but he did not then reference a numerical ratio. Later, however, as Allent's *échelles métriques* took hold, Chanlaire added a note to the title page of the revised, 1810 edition of his atlas, giving a verbal expression of the numerical ratio:

> Note: These maps, designed on the same scale of 1 on the paper to 259,000 on the terrain (one *ligne* to 300 *toises*), can be reunited and reassembled perfectly. This is what we did, a few years ago, to make the great map that was displayed in one of the rooms of the Imperial Palace of the Tuileries.*

This numerical ratio placed the atlas within Allent's *triade* VI, where its use remained a largely valid approximation.

The Belgian geographer Philippe Vandermaelen soon pushed the concept of a consistent, multisheet map into the inappropriate region of *triade* VII, with his six-volume *Atlas universel de géographie physique, politique, statistique et minéralogique* (published in installments from 1825 to 1827). This atlas's 380 sheets were projected so that they could, in theory, be assembled across the surface of a large ball, about 7.75 m in diameter, to make a huge terrestrial globe (fig. 5.21; Delaney 2011; Silvestre 2016). The title page of each volume bore a similar statement to Chanlaire's, now making the numerical ratio explicit: "on the scale of 1/1,641,836 or of 1 *ligne* to 1,900 *toises*."

Most atlases, of course, depicted all regions at one size, that of the paper, regardless of the regions' actual extent; each map's correspondence to the world therefore differed from its atlas mates. By mid-century, some atlas makers were providing numerical ratios on regional maps in order to describe and characterize each map and to permit readers to understand the differences between them (Wagner 1914, 12–14). For example, the table of contents to Auguste Henri Dufour's *Atlas universel, physique, historique et politique de géographie ancienne et moderne* (1863) listed each of the forty maps in the atlas, together with the numerical ratio of each. Many of the maps themselves also bore the ratio; the smallest ratio specified was for a map of the world as known to the ancients, at 1:20,000,000 (fig. 5.22).

But Dufour did not give a numerical ratio to three of the maps in his atlas: the map of the entire modern world and two regional maps, one of North and Central America and one of Oceania. A note on the table of contents explained that these three maps covered "increasing latitudes" and "therefore had variable [map]

---

*The title page to volume 1 of the Library of Congress's copy ("revu et augmenté en 1810") bears this statement: "*Nota.* Ces Cartes, dressées sur la même Échelle de *un* sur le papier, à 259,000 sur le terrain (une ligne pour trois cent toises), peuvent se réunir et se rassemblent parfaitement. C'est ce qu'on a exécuté, il y a quelques années, pour obtenir la grande Carte qui a été exposée dans une des salles du Palais Impérial des Tuileries." My thanks to Ed Redmond for information about the Library of Congress's volume. The statement did not appear in the 1792 edition. Wagner (1914, 10) noted that the statement first appeared in 1806, but while some impressions of individual maps are dated to 1806, I have found no catalog entry for an 1806 edition of the entire atlas.

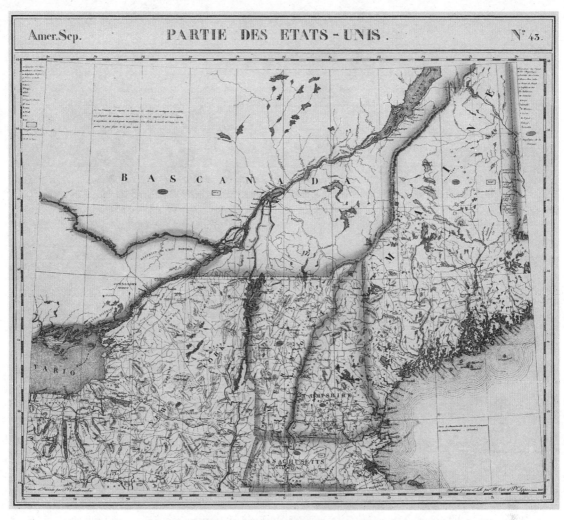

FIGURE 5.21. One sheet from Philippe Vandermaelen's six-volume *Atlas universel de géographie physique, politique, statistique et minéralogique* (Brussels, 1827), 5: sheet 45. Hand-colored lithograph, 47 × 51 cm. Courtesy of the David Rumsey Collection (2212248); www.davidrumsey.com.

scale."* Yet other regional maps in the same atlas covered much larger latitudinal extents and had been assigned numerical ratios. Dufour's misleading statement suggests a lingering uncertainty about the meaningfulness of applying numerical ratios to maps within Allent's *triades* VII and VIII.

The issue to which Dufour's *Atlas* alluded so misleadingly is the same that Allent had realized: that the numerical ratio is not consistent across regional and world maps, which undermines the validity of the concept of map scale for such

---

*"Les cartes pour lesquelles l'échelle est marquée » sont en latitudes croissants, et par conséquent à échelles variables."

MAP SCALE AND CARTOGRAPHY'S IDEALIZED GEOMETRY

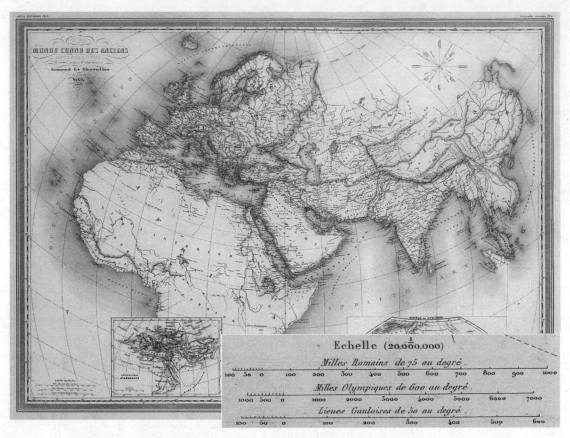

FIGURE 5.22. A map of the ancient world with (inset) a detail of its scales and numerical ratio: *Monde connu des anciens*, in Dufour ([1864], pl. 2). Courtesy of the David Rumsey Collection, David Rumsey Map Center, Stanford Libraries (G1046.C1 D8 [1864] FF); purl.stanford.edu/jp980bk7033.

maps. This issue made makers of geographical maps understandably reluctant to embrace the concept of numerical ratio, especially for world maps, not just those on the Mercator projection, and indeed for regional maps, too.

By 1899, however, the cause of international geographical science was seen as being hindered by the lack of a common "cartographic language" that was in large part constituted by the numerical ratio. At the International Geographical Congress held that year in Berlin, in a session on the development of international standards for measurement, especially in light of the British and Russian refusal to adopt the metric system, Alexander Supan, editor of *Petermanns geographische Mitteilungen*, proclaimed:

It is important that we ask cartographers to specify the scale of maps not only graphically, but also with figures. This remark is not superfluous; there are even atlases where there is no numerical expression for the scale on any single map, as for example with the otherwise very appealing Citizen's Atlas by Bartholomew that appeared only a year ago. It is easy to see how much the comparability of maps suffers. But it suffers just as much when one uses different numerical expressions, one of which is the known fractional form 1:x, the other the ratio of an inch on the map to miles or to wersts in reality, or vice versa, the relation of a statute mile or a nautical mile to an inch on the map. (Anon. 1901, 98–99)*

In other words, so long as some people insisted on using nonmetric units of measure, numerical ratios would be needed to allow readers to compare and evaluate maps against each other.

In his address to the congress, Supan referred mostly to territorial maps, for which the numerical ratio was, of course, largely valid, but by calling out "atlases" generally, and specifically Bartholomew's *Citizen's Atlas of the World* (1898)—a high-end but not overly large atlas of world and regional maps covering the entire earth—Supan suggested that he accepted the principle that regional and even world maps are necessarily based on more detailed maps, so that even those geographical maps should bear an abstract numerical ratio. Supan thus effectively asserted the principle that every map is a proportional image of the world. After some debate over the proper wording of Supan's motion, the delegates finally passed it unanimously: *all* maps should bear both a scale and a numerical ratio (Anon. 1901, 110).

Supan's integration of small- with large-scale maps as a single category of imagery defined by numerical ratio was paralleled by Max Eckert's (1907, 545) distinction between "concrete" and "abstract" maps. Previously, German geographers had prepared manuals of map design by focusing on two sets of problems faced by geographers in their research: when mapping regions, they needed to address the design and construction of map projections; when mapping landscapes, they needed effective methods for depicting relief (Ormeling 2007, 184–85). Eckert now argued that all such maps, both small- and large-scale, should be considered as "geographically *concrete* maps" because they all "reproduce facts as they exist in nature, such as the distribution of land and water and of heights and depressions." By contrast,

---

*"Vorausgeschickt muss werden, dass wir an die Kartographen die unerlässliche Forderung erheben, den Maassstab der Karten nicht blos graphisch, sondern auch durch Zahlen anzugeben. Diese Bemerkung ist nicht überflüssig; es giebt sogar Atlanten, wo auf keiner einzigen Karte ein ziffermässiger Ausdruck für den Maassstab sich findet, wie z.B. in dem erst vor einem Jahr erschienenen, sonst sehr ansprechenden Citizen's Atlas von Bartholomew. Es ist ohne grosse Überlegung ersichtlich, wie sehr die Vergleichbarkeit der Karten darunter leidet. Aber sie leidet ebenso sehr, wenn man verschiedene ziffermässige Ausdrucksweisen anwendet, der eine die bekannte Bruchform 1:x, der andere das Verhältniss von einem Zoll auf der Karte zu Meilen oder zu Wersten in der Wirklichkeit, oder umgekehrt das Verhältniss von einer Statute-Mile oder einer Seemeile zu Zollen auf der Karte."

geographically *abstract* maps ... present, in cartographic form, the results of scientific induction and deduction and, in most cases, can be traced back to the study of the scientist. To this class belong all general economic, commercial, statistical, ethnographic, population, and physical maps. (Eckert 1908, 346; emphasis added)

Eckert argued that the core of his desired *Kartenwissenschaft* (map science) would be the codification and regularization of the design of abstract (i.e., analytic) maps, although he remained unable to break away from the institutional reality that such maps were produced across the social and natural sciences and not just by geographers (Scharfe 1986).

The salient point is that both Supan and Eckert shared an understanding that there was no difference in principle between the cosmographical geometries of geographical mapping and the plane geometries of place mapping, mediated as they were by the projective geometry of systematic mapping. With such sentiments prevalent among influential geographers, it is not surprising that official and high-end map publishers would soon be adding numerical ratios not only to regional maps but to world maps as well (fig. 5.23).

What Allent had created in 1802–3 as a mechanism to standardize the preparation of all sorts of engineering images, including plans, so as to permit their easy combination and comparison had by 1900 become the means to describe *all* maps, whether in absolute (1:x) or relative (large scale / small scale) terms. And with numerical ratios being applied to *all* maps, even world maps, map scale was accepted as a universal attribute of all maps. The map was now completely normalized.

And herein lies the paradox with which this chapter began, in which "the map" stands as a single category when we should, like Allent, distinguish three categories of maps according to the validity of treating them as inherently proportional. First, there are those maps for which the numerical ratio is a valid and appropriate descriptor. These are the maps and plans that fall within Allent's *triades* I–IV and depict areas that are so small in extent that they can be measured and drawn with plane geometry, without reference to the curvature of the earth, and as such can be truly proportional. Second, there are those maps for which the numerical ratio is an approximately valid descriptor. These are the territorial maps that fall within *triades* V–VI, such as those of the Ordnance Survey or the *Carte de l'état-major*, that cover large areas and were therefore projected to take into account the curvature of the earth in such a manner (often as a series of zones) that variability in the numerical ratio is inappreciable to the user Finally, there are those regional and world maps, falling into *triades* VII–VIII, for which the numerical ratio is quite invalid. Such maps are projected so as to cover extensive areas in an uninterrupted fashion, so that variation in the numerical ratio is appreciable, and can be modeled, making the single ratio effectively meaningless.

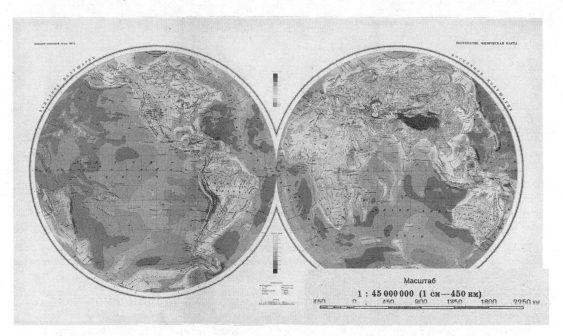

FIGURE 5.23. A twentieth-century world map with (inset) a detail of scale and numerical ratio: physical map of the world (Anon. 1937, pl. 2). Courtesy of Osher Map Library and Smith Center for Cartographic Education, University of Southern Maine (Gibb Collection); www.oshermaps.org/map/39508.0005.

The paradox stems from the insistence that even maps in the third category are properly proportional and that they are meaningfully described by the numerical ratio or, at least, by the relative designation of "small scale." This insistence sustains the claims that map scale is universal. It requires that otherwise discrete mapping practices are construed both as merely aspects of the single endeavor of cartography and as practices fundamentally of observation and measurement. The widespread application of numerical ratios and scales to geographical maps in the nineteenth and early twentieth centuries marks the triumph of the ideal of cartography; only with the ratio's general acceptance could actual differences in mapping practices be denied and ignored.

Numerical Ratios and Map Scale in the Twentieth Century

The acceptance of numerical ratios as a requirement for *all* maps, even world maps, completed the idealization that map scale and proportionality are indeed a universal attribute of maps. Map scale idealizes the map-to-world correspondence as a relationship between conceptually the same phenomena expressed in the same units. It ignores the messy realities of customary measures and the translation between metrological regimes that were integral to scales and verbal expressions. The ubiquitous classification of any map as variously small-, medium-, or

217

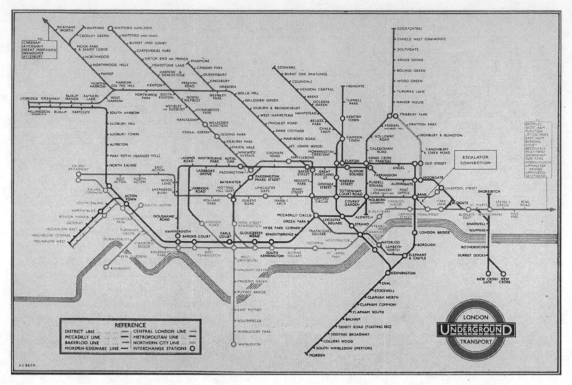

FIGURE 5.24. The epitome of topologically structured maps. In his famous design for the London Underground, first published in 1933, Henry Beck eschewed the actual distances and directions between stations. He instead spaced them at constant intervals along lines drawn only vertically, horizontally, or at 45° diagonals so as to emphasize the lines and their interconnections. Henry C. Beck, *Railway Map, No.1* (London: London Underground Transport, 1937). Size of the original (paper): 16 × 23 cm. Image courtesy of Osher Map Library and Smith Center for Cartographic Education, University of Southern Maine (Auletta Collection).

large-scale requires the conviction that all maps are shaped and determined by the degree to which they reduce the world, as measured by the metric of the numerical ratio. The ideal's ontological preconception is sustained.

This argument depends on a useful sleight of hand: spatial images for which no numerical ratio can be calculated, such as Henry Beck's famous 1933 map of the London Underground (fig. 5.24), are simply and emphatically discounted as maps. They have been variously called "diagrams," "sketches," "maplike objects," "cartograms," or "cartifacts," but not "maps" (Delano Smith 2001, 284–85). This conceptual constraint has long been challenged, initially by humanistic geographers who have sought to represent personal experiences (e.g., Wood 1978) and more recently by ethnographic analyses of mapping practices by indigenous peoples (Woodward and Lewis 1998). Still, common terminology continues to suggest an only grudging acceptance of nonscalar works as maps. Tim Bryars and Tom Harper (2014, 8) might have acknowledged that Beck's map has now attained the status of "cartographic superstar," but they nonetheless insisted on calling the work a "diagram" or "schematic diagram." Another recent popular work was similarly conflicted, beginning by calling Beck's image a "renowned schematic diagram" before relenting and calling it a "map" (Anon. 2015a, 298), while a graphic design guru has stated that the reason Beck's map works is "because it's not really a map at all" (Bierut 2018, 0m00–30s). Studies of the map and its significance have proliferated, embracing quite distinct intellectual viewpoints (Garland 1994; Hane 1996; Hadlaw 2003; Roberts 2005; Vertesi 2008; Dobbin 2012; Hornsey 2012; Merrill 2013; Cartwright 2015; Long 2015). Yet this extensive attention and the still greater proliferation of parodic images—in which the stations of London's Underground system are variously renamed, or the design is applied to other kinds of phenomena—have been driven less by the map's iconic status and more by its difference from *proper* maps (Field and Cartwright 2014).

The back formation *cartograph* was coined for the then emergent genre of popular, pictorial mapping, whose products were maps, yes, but not normative maps. Dori Griffin (2013, 7–9; 2017) argued that the U.S. graphic designer Ruth Taylor Watson, who produced numerous popular maps, coined the neologism in 1929.[*] The new term has been adopted as a poetic alternative to the quotidian *map* (e.g., Rakosi 1931). More recently, *cartograph* has been similarly applied to nonscalar or other not-real maps by indigenous peoples (Beyersdorff 2007, 153, 155, 156), nineteenth-century landscape artists (Moser 2017), and modern statisticians (Friendly 2008b, 514). The term received wide exposure when Nicolas Cage's character in *National Treasure* (dir. John Turteltaub, 2004) used it to refer to the treasure map supposedly written in invisible ink on the verso of the manuscript of the U.S. Declaration of Independence; in this instance, it is unclear if *cartograph* was used because of the map's abnormal state of invisibility or its completely fantastical nature.

Statistical maps that modify topographical relationships to represent data values have, since their initial elaboration in the later nineteenth century, often been called *cartograms*. This contraction of "cartographic diagram" was originally used in a general manner to indicate an image that departed in some respect from cartographic orthodoxy (Hooper 1883, 476, 504, 509–13; see Dorling 2015; Slocum and Kessler 2015, 1503), and it still retains some of that sense of unorthodoxy. Furthermore, Martin Brückner (2015; 2017, 311–16) referred to the preparation of bird's-eye views and similar "map-related" images as constituting the "cartoral arts," once again actively distancing these kinds of images from the ideal of cartography. It is easy to argue that map scale is a universal attribute of all maps, if the only works considered as maps are those that possess a consistent map scale!

---

[*] It was perhaps inevitable that someone would apply *cartograph* to popular maps. The back formation was already being applied to several newly designed, map-related devices: a mechanism connected to a car's wheels to automatically scroll a road map (Anon. 1907, 553); a particular brand of plane-table alidade that included a form of tachymetry (Wallis 1911); and a complex German photogrammetric stereoscope (Jones and Griffiths 1925, 20, 36, 47).

The hegemonic status of map scale in the twentieth century is demonstrated by the practices of map librarians: when faced with a map lacking a numerical ratio, their response has been to calculate one. How can one gauge the nature of any map from a library catalog record unless a numerical ratio is included as part of the bibliographic metadata? To simplify the complex process of calculating numerical ratios, especially when a map lacks even a scale, map librarians have created several specialized tools, such as Samuel Boggs's 1935 "natural scale indicator," still widely used today, which consists of a ruler with different scales that permits the direct measurement of numerical ratios.* And it is common practice to conceptualize early maps in terms of their apparent numerical ratio, even when their makers did not conceive of the correspondence between their maps and the world in such terms. Danish map historians have, for example, made much about the apparent rationality of the supposedly "metric scale" of 1:20,000 deployed for the survey-ing of Denmark after 1766 (e.g., Pedersen and de Clercq 2010, x); this nice, round denominator was, however, a by-product of the chief surveyor's personal prefer-ence for using a ruler that divided each *fod* (foot) into ten *decimal-tome* (decimal inches) rather than twelve customary *tome* (Bugge 1779, 15). In order to calculate the numerical ratio for early maps, map librarians and historians have sought to codify precise modern equivalents of old linear units (e.g., Heidenreich 1975; Klein 1983; Chardon 1988; Smits 1996; Kupčík 2011, 183–96). A less fraught approach, now increasingly practiced, is to record only the map's dimensions among the catalog data.

That map scale is deeply seated within the ideal is further indicated by how hard map scholars have worked *not* to examine it in any detail. Different mapping institutions and individual scholars have developed their own classifications (as table 2), but there is no agreement as to how to precisely differentiate the cate-gories. And, without clear delineation, the categories have become linguistically confused.

There was briefly an attempt, during and after World War II, to simplify mat-ters by sidestepping the paradox. Then, the U.S. public consumed a wide variety of world maps showing information about the global conflict, and some academic and commercial cartographers sought to educate readers about the issues of the variability in map scale in projected maps, as a way to counteract common misun-derstandings of the nature of the earth and the respective sizes of countries. Some commentators suggested that it would be best to discard the established definition of map scale as the map-to-ground numerical ratio and instead adopt the *scale of projection*. This alternative is the ratio of the radius of the generating sphere, from

---

*This device was included, for example, in Boggs and Lewis (1945, 79–80), a foundational work in U.S. map cataloging. Boggs also invented what he called a *chartograph*, a device by which "it is readily possible, without any computation whatever, to determine the natural scale of any map" (80). Similar instruments, marketed by several groups, remain a key tool of U.S. map librarians.

which the map's projection was derived, to the radius of the earth itself. Scale of projection does not vary across the surface of the small-scale map.

But this argument did not catch on, for four probable reasons. First, maps made with the same generating sphere, and therefore the same scale of projection, still varied greatly in size and shape according to the projection used, as demonstrated by several commentators at the end of the war (Marschner 1944, 3; Fisher and Miller 1944, especially 127–28, who constructed all their maps with the same scale of projection of 1:500,000,000; Boggs and Lewis 1945, 31–32). Second, the use of scale of projection still ran afoul of the idealization of map scale as a universal attribute of maps. As Robert Lee Williams (1959) observed, the scale of projection served as a valid metric for any map constructed through a projection, whether territorial maps (i.e., maps in Allent's *triades* V–VI) or regional and world maps (*triades* VII–VIII), but it cannot meaningfully apply to those maps that are sufficiently detailed to be prepared on the assumption that the world is flat and therefore are not projected (*triades* I–IV). If map scale is valid for *triades* I–VI, if only approximately, scale of projection is valid only for *triades* V–VIII. Neither ratio can be the universal measure that the ideal of cartography requires. Third, arguments for the adoption of the scale of projection in lieu of map scale implicitly treated planar mappings as distinct from mappings on projective and cosmographical geometries, and as such denied the ideal's observational preconception, which would presumably have sat uncomfortably with other scholars. And, fourth, arguments on behalf of the scale of projection necessarily contradicted several of the preconceptions of the ideal of cartography—notably the ontological, pictorial, and observational—by reasserting the primacy of cosmographical geometry. In short, the scale of projection ran counter to the established idealizations of cartography and as such was irrelevant to the great majority of practitioners and academics.

One might argue that the idealization of map scale and its expression as a numerical ratio persist because they are intuitive and easy to grasp. They represent a common-sense understanding of the essential, Platonic act of measurement. This might explain the definitions of "map scale" provided in general dictionaries and the like. But why does the ideal persist even within technical cartographic texts that are intended for academics and professionals? The concept of map scale persists because of the manner in which it manifests the ideal's conviction that maps are necessarily direct reductions of the world. The numerical ratio is intuitively representative of a map's nature *only* on the presumption that *all* maps, from the smallest-scale map to the largest-scale, are the unmediated reductions of the observed and measured world.

Furthermore, map scale is intuitive only if it is assumed that *all* maps are made to be used, and especially to be measured, so as to discover distances and directions across the world's surface. Throughout the modern era, the ideal's preconception of efficacy has determined how academics and the public alike approach maps, such that it is a matter of complaint that projected maps do not actually replicate plane geometry.* More commonly there is the oft-satirized obsession with adding "not to scale" to sketch maps, which only serves to reinforce the rule that proper maps are all scalar. (The angular variant of this modern practice is the insistence on adding a compass rose or north arrow to every map, no matter the appropriateness of the indicator or the actual function of the map.) In sum, the apparent intuitiveness of map scale as the simple ratio of map distance to ground distance is not a common-sense phenomenon but a learned behavior and a function of the idealization of mapping as cartography.

Map Resolution, Not Map Scale

Map scale, the ideal of cartography, and the normative map are fundamentally interdependent. Before 1800, the several different strategies used to express a correspondence between a map and the world denoted specific geometries, but only for certain kinds of maps, and each had further connotations depending on the particular threads of spatial discourse in which they were deployed. "Scale" consistently referred to a particular graphic element of certain kinds of map. On maps grounded in plane geometry, a scale did connote the map's proportionality to the world, but scales on maps grounded in cosmographical geometry connoted only a correspondence that might be instrumentally useful. There was no understanding of "map scale" as a universal property of all maps. Nor was there the sense that a single process of mapping produced all maps (geographical, marine, topographical, and so on) so that all maps could be defined by the manner in which the putative singular process would have reduced the world to paper.

But, over the course of the nineteenth century, engineers promoted a new method for expressing the correspondence of map to world, the numerical ratio. Geographers soon accepted the numerical ratio as a valid measure of the nature of regional and world maps produced on cosmographical geometry, not only of the engineers' plans and territorial maps. The variable denotations and connotations of scales and verbal expressions were swamped, in the name of international science, by the single denotation of a direct mathematical relationship between the world and the map. The numerical ratio is also, seemingly, free of cultural connotation: map scale just *is*. And so all distinctions between geographical maps, marine charts, and detailed plans collapse into a single category of image. The strictly

---

*Hennerdal (2015, 773) revealed a corollary to this complexity in observing, "It is well established that map projections make it difficult for a map reader to correctly interpret angles, distances, and areas from a world map. A single map projection cannot ensure that all of the intuitive features of Euclidean geometry, such as angles, relative distances, and relative areas, are the same on the map and in reality." That is, map scale varies over a projected map and causes problems.

denotational and connotation-free understanding of map scale aligns neatly with common conceptions of "scale of inquiry" (macro-, meso-, micro-; Lepetit 1993, 1995) or "scale of measurement" (nominal, ordinal, interval, ratio; Doiron 1972, especially 30).

The formation of the ideal of cartography thus entailed acts of metonymic substitution. "Scale" originated as an expression for a concrete object, the ruler or graduated line, but after 1800 it increasingly came to refer to the idealized concept of "map scale." This metonymized map scale appears to be a universal attribute of all maps.

The origins of the numerical ratio lay in the schema that Pierre Alexandre Joseph Allent developed in 1802–3 to permit military engineers to standardize their work. The increasing adoption of his neatly rounded and abstracted ratios (e.g., 1:50,000) for territorial and even geographical maps manifested the adoption of the metric system and the marketing of rulers graduated in metric units, especially after the 1840s. But the real spread of the numerical ratio, and of the concept of map scale, lay in the manner in which the ratio permitted members of one nation to use and compare the maps of other nations when different units of measurement were involved. Both scales and verbal expressions mediated between separate regimes of customary measures, whose values were notoriously fluid. Numerical ratios appeared to replace the ambiguities with certainty, and so appealed to topographers and geographers alike. The result was the burgeoning use of concretely precise numerical ratios (e.g., 1:63,360) to implement a modern, generic regime of measurement divorced from older customary practices enshrined in the evocative labels used in Anglophone countries antithetical to the metric system, specifically, *natural scale* and *representative fraction*.

The metonymic concept of map scale engenders an apparently fundamental, intuitive, and common-sense understanding of the essential act of measurement, as enacted every time someone lays a ruler or tape alongside a body part, piece of furniture, or the edge of a building. Even if the measurement does not take place within an individual's physical reach—say, measurement of a field boundary or a long road—it still takes place within their visual or conceptual reach. Or, rather, map scale was accepted as an attribute of all maps because of the hegemonic status achieved by the ideal of cartography, with its complex web of preconceptions about the nature of maps and mapping. The paradox of map scale reveals the inability even of professionals and academics to abandon their fundamental belief that maps are all properly derived from the measurement of the world, pure and simple.

Projective geometry sustains the paradox. The ideal holds the projective geometry of systematic, territorial mapping to be *the* geometry of cartography. It seems to subsume all others, as it converts the earth to the plane and vice versa. The indefinite precision of projective coordinates ($x[\phi, \lambda]$, $y[\phi, \lambda]$) directly accounts for the ideal's false claims to ontological certainty, that the archive of spatial knowledge coincides with the world and is capable of perfection. The conceptual collapse of all mapping to the projected plane can be seen in the way in which map scholars, and especially those sociocultural critics of maps who continue to be profoundly misled by the apparent universality of cartography, use "Euclidean" and "Cartesian" as interchangeable descriptors for the geometry of "the map" (e.g., Casti 2018).

The projective geometry of territorial mapping (Allent's *triades* V–VI in table 3) mediates and blurs the distinctions between the mapping of small portions of the earth on plane geometry (*triades* I–IV) and the mapping of extensive portions of the earth on cosmographical geometry (*triades* VII–VIII). It means that what is valid for mapping on the plane is valid for territorial mapping, and what is valid for territorial mapping is valid for mapping on the spherical earth. The principles of planar mapping, in which it is possible and meaningful to directly redraft observed and measured angles and lengths, and for which the numerical ratio is a perfectly legitimate expression of the geometrical correspondence of map to world, are applied to territorial mapping, for which those principles and the numerical ratio are approximately valid, and then, in a core act of idealization, to geographical mapping, even though those principles and the numerical ratio are invalid for cosmographical geometry. In this manner, projective geometry underpins the ideal's observational preconception that all maps are properly grounded in the direct observation and measurement of the world. This logical extrapolation explains the ambiguity and contradictory nature of map scale as noted by Robinson and colleagues (1995) at the start of this chapter.

Like scales and verbal expressions, the numerical ratio is not a manifestation of the inherent nature of maps but is just one more sign to jostle on the map's semiotic surface. Map scale is subject to the same cultural conventions and social functions that engender all the other elements of maps and the mapping practices that have produced them. Each map is not defined by the degree to which it reduces the world, but by its discursive function and context. For example, a recent study of detailed territorial mapping observed significant differences in the style and content of territorial maps produced by wealthy as opposed to impoverished states, even when those maps possessed the same numerical ratio (Desimini and Waldheim 2016, 23).

Furthermore, projective geometries did not supplant other mapping geometries. In this respect, the paradox identified by Robinson and colleagues (1995) stemmed from a mismatch between the ideal of cartography and the pragmatic realities of mapping. Property mappers and engineers continued to produce work on an apparently flat earth without any reference to how the features they mapped related to others. Dick Dahlberg (1984, 148–50) accordingly identified a "gap" in modern practices between fine-resolution property mapping and coarser resolution mapping for natural resources (soils, vegetation, topographical, geological maps, and so on); his goal might have been to figure out a method to bridge the gap within land information systems, but his data indicated the fundamental distinction between fine-resolution property mapping, using plane geometry, and coarser resolution modes that seem to have been absorbed into the projective geometry. Geographers continued to make maps based on graphic constructions of latitude and longitude, interpolating information directly within the framework of meridians and parallels. Geodesists continued to draw out their dedicated chains of triangles without reference to either cosmographical or projective geometries. The apparent creation of a single archive of mapping was very much a function of the emergence of the ideal of cartography in the nineteenth and twentieth centuries.

There remains the issue of how, in our descriptions of maps, we might characterize the degree to which maps correspond to the world without running afoul of all the anachronisms and improprieties of numerical ratios and the relative categories of large, medium, and small scale. A new terminology is necessary: to permit the meaningful comparison of maps, especially in delineating the boundaries between specific threads of spatial discourse; to counter the ingrained categories of large and small scale and all the confusions to which they have given rise; and to permit discussion of those maps that lack any expression of their geometrical correspondence to the world.

One possible solution is to think in terms of the various ways in which humans perceive and physically relate to the world. Daniel Montello (1993, 315–16) suggested a four-tier categorization of spaces that might be readily extended to their representation:

"*Figural* space" is the realm of things perceived to be "smaller than the body" and "is the space of pictures [including maps!], small objects, distant landmarks, and the like" that can be apprehended without "appreciable movement of the entire body" (equating to Allent's *triades* I and II).

"*Vista* space" is what is perceived as being "as large or larger than the body" but is nonetheless "visually apprehended from a single place without appreciable locomotion, being the space of single rooms, town squares, small valleys, and horizons" (*triades* II and III).

"*Environmental* space" is perceived as being "larger than the body and surrounds it," requiring "considerable locomotion" to be apprehended directly, although their spatial properties can with sufficient time be "apprehended from direct experience alone," being "the space of buildings, neighborhoods, and cities" (*triades* III–V).

"*Geographical* space" is "much larger than the body and cannot be apprehended directly through locomotion" and must instead "be learned via symbolic representations [i.e., texts] such as maps or models that essentially reduce the geographical space to figural space."

These different combinations of embodiment, presence within, and direct or indirect perception of the world contribute to the primary distinctions between threads of spatial discourse, as people have variously apprehended the world and have sought to communicate their spatial experiences accordingly. Spatial discourses are thus not based on some absolute sense of map scale but derive from particular mixes of human behavior and the human sensorium.

Montello's behavioral analysis is thus a useful heuristic for approaching the different ways in which humans understand and represent spatial relationships. However, humans are also social and technological animals, and they have used their technologies in concert to blur these behavioral distinctions. Property mapping, for example, is clearly grounded in individual and communal practices of looking at and modifying vista space (see figs. 5.2 and 5.3), but over time property mapping has been carried out over environmental space, as in extensive estate maps, and, in the form of modern cadastral systems, over still more extensive geographical space. Symbolic texts are not limited to perceptions of geographical space. As useful as Montello's behavioral categories are, they are inapplicable to the description of maps.

A more comprehensive solution resurrects the eighteenth-century distinction between *grand point* and *petit point* by considering the "resolution" with which maps represent the world. At first sight, resolution seems akin to map scale: fine-resolution maps show a great deal of information for the area depicted and are also technically large-scale maps; coarse-resolution maps show a small amount of information for the area depicted and are akin to technically small-scale maps. However, the concept of resolution does not carry with it any assumptions of a necessary mathematical relationship that otherwise obscures and perverts our historical understanding of maps; resolution requires no expectation that every map is grounded in observation and survey. Indeed, Michael Goodchild (2015b, 1386–87) noted that the twentieth-century technologies of aerial photography and digital imaging—whose precision is constrained by the physical size of the photographic grain or the digital pixel—have already "decoupled" map scale from resolution.

Moreover, the concept of resolution is steadfastly imprecise. It is not defined by measurement or predefined categories. It is also applicable to maps of imaginary places and topologically based maps, works that otherwise preclude the calculation of a numerical ratio of map scale. For example, from the extent of territory covered, and the coarseness of information, no matter how effective its presentation, Beck's map of the London Underground is readily characterized as being at a fairly coarse resolution (see fig. 5.24). Assessments of resolution might help in terms of tracing coherent spatial discourses, but they are not in and of themselves explanatory.

We should further distinguish between spatial resolution and taxonomic resolution, or how finely phenomena are classified. While the two kinds of resolution are interrelated—we expect any large-scale map to be constructed with a fine taxonomic resolution (Dahlberg 1984, 149), although that is not always the case—maps with fine spatial resolution do not necessarily have fine taxonomic resolution. I have accordingly used resolution rather than the categories of map scale throughout this book to describe maps, and I have provided only historically appropriate measures of map scale independently of the ideal of cartography.

Not Cartography, But Mapping

Cartography is a map of mapping

There are many ways to characterize cartography. It is a mirage, the hazy refraction seen from afar of actual mapping processes; it is a chimera, something desired yet illusory. Self-referentially, cartography is a map of mapping. Such a metaphorical map does not, of course, refer to the normative map. After all, cartography is manifestly not a true and clear abstraction or reduction of the world of map making. Cartography, like all real maps, is human-made and is the product of cultural and social forces. It is an image constructed in a somewhat incoherent manner by a web of idealizations that are deeply rooted within modern Western culture. It depicts what mapping should be, not what mapping is.

The basic proof that cartography is a partial and distorting idealization of actual mapping practices is that the ideal has a history. Broadly speaking, we can identify four periods in that history: the ideal of cartography was in part prefigured in the eighteenth century, emergent in the nineteenth, and triumphant for most of the twentieth; in the twenty-first, it seems to be starting to degrade. This periodization is imprecise and flexible, and it has no explanatory power, yet we can nonetheless see clear patterns in terms of rhetoric and practice.

In thinking of the eighteenth century as the era of the *ideal prefigured*, I do not mean to invoke some teleological argument that the development of the ideal of cartography was inevitable. Rather, hindsight permits us to see that some attitudes that would become incorporated within the ideal were expressed before 1800. Two particular elements of the ideal had early modern origins. First, the activities of the

French state in undertaking the first systematic surveys, both topographical and hydrographic, were based on a projective geometry. Second, geographers treated the cosmographical geometry of latitude and longitude as providing the common "language of [geographical] maps." But the cosmographical and projective geometries did not combine to create the ideal's ontological preconception until well after 1800. Their parent practices remained distinct before 1800: despite their claims, geographers could not incorporate place and property mapping within the framework of cosmographical geometry and topographers, with the exception of the French, could not extend triangulation-based surveys across more than a province. Before 1800, *maps* were strictly geographical images.

The nineteenth century was the era of the *ideal emergent*. The proliferation of systematic surveys after 1790 stimulated the emergence of the concept that all mapping could be unified within a projective geometry, as codified in the coinage of "cartography" in the 1820s and its firm acceptance by 1850 to mean "the drawing of geographical and topographical maps, charts, and all the drawing of mensuration [i.e., plans]" (Lieber 1834, 98). Even so, mapping practices continued to be distinct. However they increasingly acquired the image of interconnection, interdependence, and eventually unity, as the new concept of "map scale" was adopted and as modes intersected and overlapped through the prosecution of systematic surveys across and around Europe, North America, and South Asia. But while the industrial and eventually nuclear-powered states of the West adopted the technologies of systematic surveys as a pragmatic necessity, the manner in which the desire for such surveys outstripped the ability to implement them suggests other issues were at play than just need and functionality. Of particular importance was the role of modern imperialism in promoting "scientific surveys." The spread of systematic mapping was thus a major component of the rise of the ideal, but it did not determine the ideal. The ideal was not inevitable, but was elaborated through multiple other factors.

We can track the ideal's rise to hegemonic status in several ways. Early commentators looked to the mid-nineteenth century as the point when, especially in Germany, cartography became "scientific" and "modern" (Wolkenhauer 1895; Bartholomew 1902) and when the modern "age of computation" in geography began (Wright 1955, 65). The ideal was clearly well embedded in modern culture by the 1870s, when systematic surveys were becoming permanent agencies within the bureaucracies of European states, and when Mark Twain and Lewis Carroll issued their first satires. The attention paid to cartography by those satires suggest that the ideal had yet to integrate completely into modern culture. By the early 1900s, it seems, the redirection of satirical commentaries to address the normative map suggests that the ideal had attained a cultural hegemony. Furthermore, the early twentieth century saw the adoption and application of the numerical ratio of map scale to all kinds of maps other than those made by engineers. The practical differences between small- and large-scale maps were completely elided, and the *carte / Karte / map* was thoroughly normalized. The numerical ratio became the foundation of a new "language of maps"—supplementing and blending with the equally new sense of cartographic language grounded in pictorialness—as internationally minded geographers sought to create a universal archive of spatial information in the face of national variation in measures. The numerical ratio served both to measure the degree to which the map reduced the world/archive of $(x[\phi, \lambda], y[\phi, \lambda])$ coordinates and to guarantee that, in this act of reduction, a map's pictorial character did not lose its proportionality and maintained its mimetic character even as symbolization and other graphic strategies might change across different scales.

The twentieth century was the era of the *ideal triumphant*. From the rise of academic cartography, through international mapping efforts, to the implementation of the ideal's preconceptions in aerial and digital technologies, map making has truly seemed to be a single universal endeavor, so much so that the many new spatial discourses engendered by the modern world all seem to partake of the ideal. Even property mapping appears, through the practices of coordinate surveying and national cadastres, to have been integrated into the whole. Cartography appears to have been made real. The continued identification of different flavors of "large scale" surveying and "small scale" mapping might seem to undercut the ideal, but the distinctiveness of the two flavors has long been construed as a difference in technical implementation rather than in underlying conceptions.

Given the seemingly fundamental nature of the image of a transcultural, universal endeavor of cartography, scholars automatically and unconsciously focused their attention on "the map" as the only relevant object of study. Within the ideal, maps are understood to be made as a matter of course. The intellectually important aspect of map making is therefore "how": how to make maps and how to improve cartographic technologies and techniques to make maps better. Questions of why maps are produced and consumed are simply irrelevant.

Finally, we can see the ideal starting to unravel and fray, as revealed by scholars' increasing use of the satires after 1970 to cast doubt upon the normative map and to advance the sociocultural critique of maps. The "why" questions began to be asked. Since 1990, even lay commentators have come to accept the sociocultural critique. An unintended consequence of the sociocultural critique has been to challenge parts of the ideal, and in doing so, it has inaugurated the era of the *ideal degenerate*, although the ideal still stands strong and as yet unbowed.

Maps have never been so accessible and ubiquitous as today. Map imagery proliferates across the internet, from "10 Maps that Explain the History of X," to intricate visualizations of big data (and also of much not-so-big data: Munroe 2017), to reproductions of early maps of all shapes and sizes. Digital spatial technologies of remote sensing, global positioning systems (GPS), geographical information systems (GIS), and virtual reality permeate everyday life and all levels of decision making in public and private institutions. In the process, lay interest in the minutiae of cartography has increased, and a more critical public stance has developed with regard to the normative map.

This trend should not be exaggerated. In particular, many digital facsimiles of early maps, together with the websites that post and discuss them, grab readers' attention because of the overt way in which the early maps differ from normative maps; in this respect, the ideal undoubtedly maintains its hegemonic status. Furthermore, most of the popular blogs about early maps rely heavily on the literature created under the ideal. Doing critical work online is hindered because of the necessary commitments of time and scholarly effort required to produce new, culturally and socially sensitive commentaries about maps and mapping; those commitments run counter to the immediacy and quick production cycles fostered by journalism and by the internet's huge and insatiable maw. Old and flawed concepts are thus repeated and the ideal affirmed. The instrumental use of maps in decision making and in personal navigation seems to validate the ideal's preconceptions of ontology, discipline, and efficacy, at the least.

But, in print, popular works on maps and mapping have begun to shed the trappings of the ideal. That the public appreciates this change is evident in a new flood of popular works on maps and map history that are steadily opening up an understanding of mapping as necessarily being cultural. The new popular works are intended both for the coffee table (Barber 2005b; Clark 2005; Harwood 2006; Brotton 2014; Bryars and Harper 2014; Mitchell and Janes 2014; Seed 2014; Anon. 2015a; Anon. 2015b; Baynton-Williams 2015; Hall 2016) and, in the wake of the remarkable success of Dava Sobel's *Longitude* (1995), for the library shelf (e.g., Harvey 2000; Lester 2009; Brotton 2012; Garfield 2013; Blanding 2014; Brooke-Hitching 2016). The trend is slow, but fascinating to watch in real time.

The *ideal degenerate* has already generated some obituary notices for cartography. The necrologists have adopted a historical narrative that coincides closely with that advanced in this book: they understand cartography to be a specific formation of the modern state and of modern technology. But where I have argued that cartography is an image sustained by a complex belief system, the necrologists once again hypostasize that image into an actual and particular endeavor. Denis Wood (2003, 4), for example, proclaimed:

Cartography Is Dead (Thank God!)
Let's admit it. Cartography is dead. And then let's thank our lucky stars that after the better part of a century mapmaking is freeing itself from the dead hand of academia.

Wood distinguished between map making, as what many thousands of professionals and amateurs have always done, and cartography, as what a few hundred state-supported professionals and academics have codified since 1945. The digital revolution has enabled map making to flourish beyond the confines of centralized state agencies, even as formalized cartography has stagnated (see also Dodge, Kitchin, and Perkins 2011, 116–21).

For their obituaries, both Timothy Barney and Bill Rankin equated cartography to the production of static, formal, paper maps that fixed territory, a production that is rapidly ending under the influence of digital technologies. For Barney (2015, especially 211), it was the development of inter-continental ballistic missiles (ICBMs) and computerized guidance systems that detached the warrior from the territory and that promoted the creation of politically activist maps. Rankin (2016, 65–116, especially 113–14) traced a profound shift in mapping activities, from the pre-1940 effort to create a global, truthful archive of paper maps ("representation-as-mimesis") to the technologically enhanced, instrumental use of mathematical coordinate systems to locate objects in space, from ICBMs to personal cars ("representation-as-practice"). For Rankin, the postwar era transformed the geometrical structure of territory from a mapped expanse to an ordered, multidimensional array of coordinates. Jessica Wang clarified the significance of Rankin's argument when she observed that "we live not so much 'after the map' [Rankin's title] as 'after *the* map,' . . . in the sense that with GPS, the goal of a singular ideal and universal map has given way to generalized accessibility of multiple maps in real time" (Wang 2017, 1016, original emphasis).

Rankin's arguments are of immense importance for understanding the rise of modern technological systems and their application to mapping and map-related tasks of navigation. But it must nonetheless be clear that the normative map still remains very much in evidence, and the real-time accessibility of maps still leaves much room for kinds of mapping other than the instrumentally efficacious. Modern society has not yet discarded the idea of "*the*" normative map.

Wood's arguments were similarly partial; indeed, he continues to reject any image as a map that does not clearly have locational indexicality (as per the ontological preconception; see, especially, Wood 2012), and in that respect he remains bound to the normative map. Even so, he has come to argue that the apparent death of cartographic authority has necessarily been a political move, as indigenous peoples and underrepresented groups adopt mapping practices to assert their right to exist; this process might have been enabled by technological developments, but it is not technologically determined (Wood 2010, 111–55).

A further insight by Rankin (2017) nonetheless seems applicable: cartography appears to be a "zombie project" that has been kept alive, despite repeated reconfigurations, by a "negative network" of cartographers who are no longer committed to the project. Rankin developed these concepts through a history of the "International Map of the World," the huge, unwieldy, and archly imperialistic project that steadily lost its way through several generations of map makers within an international consortium. Wood, Barney, and Rankin all understand the technological transformation of map making at the hands of modern global powers as having turned cartography itself into a zombie project. Not knowing that it is dead, it staggers on regardless.

But cartography is not an endeavor. It is an idealization. For all the self-congratulation by practitioners and academics that they have indeed engaged in a coherent, uniform endeavor, they have all participated within their own particular spatial discourses. And as many of those discourses have changed dramatically, and as new discourses arise to meet new spatial needs and conditions, the underlying idealizations of cartography persist. The *forms* of cartography may be dying off, but its fundamental concepts are not. It is, therefore, premature to write cartography's obituary.

Don't get me wrong: cartography deserves to die. But two concepts and all their implications must be widely accepted before cartography can deservedly breathe its last. First, cartography is an image produced by an extensive web of idealizations (and thus is much more than the formal, academic institution). Second, all the ideal's apparently common-sense convictions and preconceptions are flawed. Only with these points accepted will scholars and lay commentators alike actively avoid rehearsing the ideal's flaws. The normative map must be eliminated, likewise the conviction that it is meaningful to talk about "maps" or "the map" as a generic category of phenomena. It has been my goal in this book to demonstrate the many flaws and unnaturalness of the ideal of cartography. This book will not kill off cartography by itself; the belief system will likely stagger on among scholarly and lay commentators alike, but I hope to at least severely injure it.

An initial strategy to help kill off cartography is to cease using the word. From a historical perspective, "cartography" and all it implies is undoubtedly the product of a particular era in Western culture. The word should only be used in specific reference to the ideal and its idealized endeavor. It is certainly ill suited as a label for a socially and culturally inclusive field of study. However one defines "history of cartography" with respect to "historical cartography," both terms embody an approach to mapping and to map history that is fundamentally flawed. The inappropriateness of "cartography" for historical inquiry was recognized when a new scholarly society, the International Society for the History of the Map (ISHM), was established in December 2011. The new society's name paralleled "the history of the book," a field that has transcended its original constraint as the study of the hand-printed codices of the early modern era and that now encompasses the sociology of texts, together with the incorporative practices of orality and

233

performativity that circumscribe the inscriptive practices of writing. In this respect, as noted in chapter 1, the "history of the map" or "map history" becomes a viable and all-encompassing replacement for the conceptually limited and problematic "history of cartography."

Furthermore, the ideal of cartography sustains an ultimately unwarranted functional distinction between the map *maker* and the map *reader* or *user*. According to the ideal, the intellectual labor in mapping lies in the conversion of the world into its schematized, homologous image. The map's user has only to read the image to learn about the world. The map artifact at once binds producers to, and separates them from, consumers. Even as it connects its maker with a user distant in place and time, the material map seems to physically disrupt the semiotic process.

By contrast, linguists see no such disruptions in discussions of language. The users of a linguistic system are speakers and listeners, writers and readers, producers and consumers. A language remains a coherent semiotic system however it manifests, whether in a personal chat between two friends over coffee, in a personally directed piece of mail, in a printed book sold widely through the marketplace, or in an indiscriminately broadcast radio show. Linguists accept not only that expressions of language can take a multiplicity of forms (oral and written, or incorporative and inscriptive) and contexts (intimate, domestic, communal, professional, public), but also that both form and context underpin the identification of particular language communities. They accept the innate flexibility of language systems that permits meaning to be communicated, if inefficiently, even when the rules are not strictly adhered to. Similarly, we cannot think of mapping as being different from other semiotic systems. Fundamentally, participants in spatial discourses are *both* producers and consumers of maps within a wider array of representational strategies. People communicate spatial knowledge in a multitude of ways; mapping variously engages incorporative and inscriptive strategies across multiple contexts. Map scholars need to study the processes of mapping, which is to say, the dynamic ways in which maps are produced, circulated, and consumed. The key word here is "dynamic." Mapping processes are fluid, the maps they generate are mutable and volatile. Nothing about mapping is fixed and stable.

Avoiding the word "cartography" and all that it stands for is easier to do than reforming use of the word "map." The ideal of cartography has promoted an expectation of what the normative map should be: "the map" is a generic and universal concept encompassing any graphic image that presents, summarizes, or otherwise synthesizes an archive of data of spatial features, their locations, and their relationships, all according to a set mathematical scale. This understanding is unduly restrictive because it limits attention to certain, scalar kinds of images and precludes any appreciation of other, nonimage forms of text as spatial texts. And this understanding is wrong, because it posits all maps to be defined in their nature and character by the degree to which they reduce the world and not by the spatial discourses that engendered them. "The map" is as loaded and as historically contingent a concept as "cartography" and "map scale," but it remains a powerful concept (Edney 2015a). It is necessary to recognize that, in practice, graphic strategies for representing spatial relationships intertwine with other representational strategies, whether oral, written, numerical, graphic, performative, or physical. In this, we should remember that Western maps are as semantically flexible as non-Western maps. For example, Potter (2001) demonstrated the fluidity and flexibility of Japanese terminology, while the common Sinitic word for graphic image and map, 图 *tú*, has also entailed a "call to action" (Sivin and Ledyard 1994, 26–27; Brotton 2012, 124).

We can go further. Mapping is much more variegated in its semiotic forms than spoken and written language, employing verbal signs, graphic signs, signs physically installed in the landscape, and embodied signs of human gesture and performance. Contrary to the ideal's preconceptions, there is no single "cartographic sign system" or "map language" that unambiguously differentiates "the map" from other kinds of texts and that is specially and uniquely qualified to depict spatial relationships. There is *no* Platonic ideal of "the map." The map is not a stable, concrete thing like a marble sculpture that can be placed on a pedestal, where it can be praised by modern culture or, more recently, pilloried by sociocultural scholars. Certainly, an active focus on mapping permits the avoidance of that well-disciplined redirection under the ideal to focus on "the map" rather than address the processes and social and cultural conditions that gives rise to this idealized category of phenomena.

Even so, we cannot avoid "map" in the same way that we can avoid "cartography." Humans have conceptualized the world in many different ways, for different purposes, and have accordingly represented it in several different ways. Some of these differences are suggested by the distinctions drawn in everyday English between *map* (specific), *chart*, and *plan*: a *map*, in this stricter sense, is a general image of the world or of a region; a *chart* is an image of the seas and coasts for use by mariners; a *plan* is an image created through direct observation and measurement of part of the world. It is good historical practice always to keep these differences in mind and to take care to use the appropriate qualifiers to remind others of those differences. Yet, in practice, each type of spatial image has been interconnected with others. I noted in chapter 2, for example, how Renaissance chart makers occasionally expanded the spatial frame of their charts so as to cover as much of the globe as possible, for the benefit of landlubbers interested in geographical information rather than for the use of mariners. Unassociated with shipboard navigation, such "planispheres" were consumed as geographical works, and some ended up being printed for still broader consumption (see fig. 2.5). While sharing many of the same semiotic conventions as marine charts, such works were part and parcel of geographical discourses.

I therefore reject all the idealizations inherent in "the map," but I embrace the idea of "maps" as the interconnected aggregate of texts that represent spatial complexity. Maps in all their forms—graphic, textual, verbal, gestural, performative, physical—are products of the multiplicity of mapping processes. Indeed, it is tangible maps and not intangible mapping processes that so effectively demand our attention and call on us to undertake their analysis.

Yet the goal of map analysis must be to elucidate the multiple mapping processes deployed across a wide variety of spatial discourses in human societies. Only by careful analysis of the participants in the discourses and of how they produced, circulated, and consumed their maps can we hope to provide valid explanations of the particular forms maps have taken and of the varied roles they have played in human societies, whether in the past, today, or in the future.

I do not argue that by abandoning cartography, map scholars must cease creating what have hitherto been accepted as normative maps or cease studying modern cartographic practices. Nothing I have said in this book suggests that the wide variety of works that are embraced within the normative map—topographical maps, analytic maps, road maps, sea charts, interactive GIS data sets that can be queried, and so on—are in some way bad, inadequate, wrong, or unworthy of being made or studied. I do not suggest that the only "authentic" maps are those made by individuals from personal experience or within cultures that do not worship at the shrine of geometrical rigor. At the same time, modern Western maps are not "totalizing devices": even official, state-produced, geometrically rigorous topographical survey maps are produced, circulated, and consumed within discourses that limit their semiotic capacity. Neither do I argue that sociocultural critics, such as Brian Harley, Denis Wood, and the "French philosophers" on whom many rely—so distrusted by normative critics (as Andrews 2001; Clarke 2013; Monmonier 2013)—are right. But nor are the normative critics right.

What I do argue is that in pursuing their work, map scholars must discard the inadequate preconceptions and convictions that constitute the ideal of cartography. Map scholars must abandon the idea that what they study is all part of a mythic cartography, and instead seek to understand their work within the relevant modes or threads of spatial discourse. In short, map scholars must stop saying "maps are" and instead say "X mapping is" while being clear about the criteria that make X a valid label. Map scholars must pause and give serious thought to the benefits of such precision in setting up research agendas. I challenge normative map scholars to cease being prescriptivists, who argue that people should make maps in specific and limited ways, and instead become descriptivists, who seek to understand and explain how people variously produce, circulate, and consume maps according to their discursive conditions.

This is the task for future scholarship. In developing new approaches, in moving beyond the limitations of both normative and sociocultural map studies, and in establishing a processual approach to mapping, we need to accept that the ideal of cartography is fundamentally wrong. Each and every preconception and conviction it has engendered is at best applicable only to certain kinds of mapping in specific periods and social circumstances. The ideal has a history, it is riven by paradox, and it is quite simply indefensible.

Bibliography

Adams, Brian. 1994. "Parallel to the Meridian of Butteron Hill: Do I Laugh or Cry?" *Sheet-lines* 38: 15–18.

Adams, Douglas. 1978. "The Hitchhiker's Guide to the Galaxy." Radio series in 6 fits. BBC Radio 4. First broadcast, March 1978.

Adams, Randolph G. 1939. "William Hubbard's 'Narrative,' 1677: A Bibliographical Study." *Papers of the Bibliographical Society of America* 33: 25–39.

Ahmad, S. Maqbul. 1992. "Cartography of al-Sharīf al-Idrīsī." In Harley and Woodward, eds. (1992, 156–74).

Akerman, James R. 2000. "Private Journeys on Public Maps: A Look at Inscribed Road Maps." *Cartographic Perspectives* 35: 27–47.

———. 2007. "Finding Our Way." In *Maps: Finding Our Place in the World*, edited by James R. Akerman and Robert W. Karrow, Jr., 19–63. Chicago: University of Chicago Press.

———, ed. 2017. *Decolonizing the Map: Cartography from Colony to Nation*. Chicago: University of Chicago Press.

Akerman, James R., and Nekola, Peter. 2016. "Mapping Movement in American History and Culture." Hermon Dunlap Smith Center for the History of Cartography, The Newberry Library, Chicago. http://mappingmovement.newberry.org/.

Allent, Pierre Alexandre Joseph. 1803. "Échelles métriques: Instruction sur la série générale et indéfinie d'échelles décimales." *Mémorial de l'officier du génie* 1: 17–26.

———. 1831. "Essai sur les échelles graphiques." *Mémorial du dépôt général de la Guerre* 2: 43–98.

Andrews, J. H. 1975. "Motive and Method in Historical Cartometry." Paper presented to the International Conference on the History of Cartography, Greenwich, 7–11 September 1975.

———. 1996. "What Was a Map? The Lexicographers Reply." *Cartographica* 33, no. 4: 1–11.

————. 1998. "Definitions of the Word 'Map,' 1649–1996." MapHist Discussion Papers, 3 February 1998. http://www.maphist.nl/papers/199801.html.

————. 2001. "Introduction: Meaning, Knowledge, and Power in the Map Philosophy of J. B. Harley." In J. B. Harley, *The New Nature of Maps: Essays in the History of Cartography*, 1–32. Baltimore: Johns Hopkins University Press.

————. 2013. "A Saxton Miscellany." *Imago Mundi* 65, no. 1: 87–96.

Anonymous. 1802. "Des opérations géodésiques." *Mémorial topographique et militaire* 1: 50–144.

————. 1802–3. "Des opérations géodésiques de détail." *Mémorial topographique et militaire* 3: 1–56.

————. 1803. "Procès-verbal des conférences de la commission chargée par les différens services publics intéressés à la perfection de la Topographie." *Mémorial topographique et militaire* 5: 1–64.

————. 1829. *Graphic Illustrations of the Colosseum, Regent's Park, in Five Plates, from Drawings by Gandy, Mackenzie, and Other Eminent Artists*. London: Rudolph Ackermann.

————. 1831. *The Portland Directory, Containing Names of the Inhabitants, their Occupations, Places of Business, and Dwelling Houses. With Lists of the Streets, Lanes and Wharves, the Town Officers, Public Offices and Banks*. Portland, ME: S. Colman.

————. ca. 1885. *The National Encyclopaedia: A Dictionary of Universal Knowledge*. London.

————. 1901. "Internationale Einführung gleichmassiger Maasseinheiten und Methoden." In *Verhandlungen des siebenten Internationalen Geographen-Kongresses, Berlin, 1899*, 1: 96–139. Berlin: W. H. Kuhl.

————. 1907. "The Agricultural Hall Show." *The Autocar* (13 April).

————. 1918. *World's Greatest War: Maps of Europe, Italy and the Western Battle Front*. n.p. [USA].

————. 1937. *Bol'shoĭ Sovetskiĭ atlas mira I* [Great Soviet Atlas of the World]. Moscow.

————. 1964. "Definition of Cartography." *Cartographic Journal* 1, no. 1: 17.

————. 2014. "The 500 Year Old Map That Shatters the Official History of the Human Race." *Mind Unleashed* (25 October). http://themindunleashed.org/2014/10/500-year-old-map-shatters-official-history-human-race.html. Accessed 24 February 2014.

————. 2015a. *Map: Exploring the World*. London: Phaidon.

————. 2015b. *The History of the World in Maps: The Rise and Fall of Empires, Countries and Cities*. Glasgow: Times Books.

————. 2015c. Online sales blurb for Antonis Antoniou, Robert Klanten, and Sven Ehmann, eds., *Mind the Map: Illustrated Maps and Cartography* (Berlin: Gestalten). http://usshop.gestalten.com/mind-the-map.html. Accessed 10 September 2015.

————. n.d. "Antarctica and . . . Piri Reis Map." Biblioteca Pleyades. http://www.bibliotecapleyades.net/mapas_pirireis/esp_mapaspirireis04.htm.

Antle, A., and B. Klinkenberg. 1999. "Shifting Paradigms: From Cartographic Communication to Scientific Visualization." *Geomatica* 53, no. 2: 149–55.

Armoghate, Jean-Robert. 2001. "*Un seul poids, une seule mesure*: Le concept de mesure universelle." *Dix-septième siècle* 213: 631–40.

Ash, Eric H. 2007. "Navigation Techniques and Practice in the Renaissance." In Woodward, ed. (2007b, 509–27).

Aumen, William C. 1970. "A New Map Form: Numbers." *Internationales Jahrbuch für Kartographie* 10: 80–84.

Avery, Bruce. 1995. "The Subject of Imperial Geography." In *Prosthetic Territories: Politics and Hypertechnologies*, edited by Gabriel Brahm, Jr. and Mark Driscoll, 55–70. Boulder, CO: Westview Press.

Azócar Fernández, Pablo Iván. 2012. "Paradigmatic Tendencies in Cartography: A Synthesis of the Scientific-Empirical, Critical and Post-Representational Perspectives." Ph.D. dissertation. Dresden University of Technology.

Azócar Fernández, Pablo Iván, and Manfred Ferdinand Buchroithner. 2014. *Paradigms in Cartography: An Epistemological Review of the 20th and 21st Centuries*. Heidelberg: Springer.

Bagrow, Leo. 1951. *Die Geschichte der Kartographie*. Berlin: Safari-Verlag.

———. 1985. *The History of Cartography*. Translated by D. L. Paisley. Edited by R. A. Skelton. 2nd ed. Chicago: Precedent Publishing.

Barber, E. J. W. 2010. "Yet More Evidence from Çatalhöyük." *American Journal of Archaeology* 114, no. 2: 343–45.

Barber, Peter. 2003. "King George III's Topographical Collection: A Georgian View of Britain and the World." In *Enlightenment: Discovering the World in the Eighteenth Century*, edited by Kim Sloan, 158–65. Washington, DC: Smithsonian Books for the British Museum.

———. 2005a. "George III and His Geographical Collection." In *The Wisdom of George the Third: Papers from a Symposium at The Queen's Gallery, Buckingham Palace, June 2004*, edited by Jonathan Marsden, 263–90. London: Royal Collection Publications.

———, ed. 2005b. *The Map Book*. London: Weidenfeld & Nicolson, 2005.

———. 2007. "Mapmaking in England, ca. 1470–1650." In Woodward, ed. (2007b, 1589–669).

Barber, Peter, and Tom Harper. 2010. *Magnificent Maps: Power, Propaganda and Art*. London: British Library.

Barbié du Bocage, Alexandre, P. Amédée Joubert, and Conrad Malte-Brun. 1826. "Rapport des commissaires nommés par la commission centrale de la Société de Géographie pour examiner les résultats du voyage de M. Pacho dans la Cyrénaïque." *Nouvelles annales des voyages, de la géographie et de l'histoire* 30: 96–117.

Barclay, Sheena. 2013. "The Meaning of Cartography: Perspectives from Collins Bartholomew and the *Times Atlas*." *Cartographic Journal* 50, no. 2: 121–27.

Barford, Megan. 2016. "Naval Hydrography, Charismatic Bureaucracy, and the British Military State, 1825–1855." Ph.D. dissertation. University of Cambridge.

Barney, Timothy. 2015. *Mapping the Cold War: Cartography and the Framing of America's International Power*. Chapel Hill: University of North Carolina Press.

Bartholomew, John G. 1902. "The Philosophy of Map-Making and the Evolution of a Great German Atlas." *Scottish Geographical Magazine* 18, no. 1: 34–39.

Basaraner, Melih. 2016. "Revisiting Cartography: Towards Identifying and Developing a Modern and Comprehensive Framework." *Geocarto International* 31, no. 1: 71–91.

Bateson, Gregory. 1972. *Steps to an Ecology of Mind: Collected Essays in Anthropology, Psychiatry, Evolution, and Epistemology*. San Francisco: Chandler.

Baynton-Williams, Ashley. 2015. *The Curious Map Book*. Chicago: University of Chicago Press.

Bendall, A. Sarah. 2019. "Triangulation Surveying." In Edney and Pedley, eds. (2019, forthcoming).

Bennett, J. A. 1991. "Geometry and Surveying in Early-Seventeenth-Century England." *Annals of Science* 48: 345–54.

Bernal, Martin. 1987. *The Fabrication of Ancient Greece, 1785–1985*. Vol. 1 of *Black Athena: The Afroasiatic Roots of Classical Civilization*. London: Free Association Books.

Bernstein, David. 2007. "'We are not now as we once were': Iowa Indians' Political and Economic Adaptations during U.S. Incorporation." *Ethnohistory* 54, no. 4: 605–37.

Berthaut, Henri Marie Auguste. 1898–99. *La Carte de France, 1750–1898: Étude historique*. 2 vols. Paris: Imprimerie du service géographique de l'armée.

Bertuch, Friedrich Justin [attrib.]. 1807. "Prospectus einer topographisch-militarischen Charte von Teutschland in 204 Blättern, unternommen von dem Geographischen Institut zu Weimar." *Allgemeine geographische Ephemeriden* 22, no. 1: 116–22.

Besse, Jean-Marc. 2003. *Face au monde: Atlas, jardins, géoramas*. Paris: Desclée de Brouwer.

Besse, Jean-Marc, and Gilles A. Tiberghien, eds. 2017. *Opérations cartographiques*. Paris: Actes Sud/ENSP.

Beyersdorff, Margot. 2007. "Covering the Earth: Mapping the Walkabout in Andean Pueblos de Indios." *Latin American Research Review* 42, no. 3: 129–60.

Bierut, Michael. 2018. "The Genius of the London Tube Map." *TED: Small Thing Big Idea* (16 January). https://www.facebook.com/SmallThingBigIdea/videos/1328242957321729/.

Bigg, Charlotte. 2007. "The Panorama, or La nature à coup d'œil." In *Observing Nature—Representing Experience: The Osmotic Dynamics of Romanticism, 1800–1850*, edited by Erna Florentini, 73–95. Berlin: Dietrich Reimer Verlag.

Biggs, Michael. 1999. "Putting the State on the Map: Cartography, Territory, and European State Formation." *Comparative Studies in Society and History* 41, no. 2: 374–405.

Black, Jeannette D. 1968. "The Blathwayt Atlas: Maps Used by British Colonial Administrators in the Time of Charles II." *Imago Mundi* 22: 20–29.

———. 1970–75. *The Blathwayt Atlas: A Collection of Forty-Eight Manuscript and Printed Maps of the Seventeenth Century Relating to the British Overseas Empire in That Era, Brought Together about 1683 for the Use of the Lords of Trade and Plantations by William Blathwayt, Secretary*. 2 vols. Providence, RI: Brown University Press.

Black, Jeremy. 1997. *Maps and Politics*. London: Reaktion Books.

Blackstone, William. 1758. *An Analysis of the Laws of England. The Third Edition; to which is Prefixed an Introductory Discourse on the Study of the Law*. Oxford: Clarendon Press.

Blais, Hélène, and Isabelle Laboulais-Lesage, eds. 2006. *Géographies plurielles: Les sciences géographiques au moment de l'émergence des sciences humaines, 1750–1850*. Paris: L'Harmattan.

Blakemore, Michael J. 1981. "Map Image and Map." In *The Dictionary of Human Geography*, edited by R. J. Johnston et al., 199. Oxford: Blackwell Reference.

Blakemore, Michael J., and J. B. Harley. 1980. "Concepts in the History of Cartography: A Review and Perspective." Monograph 26. *Cartographica* 17, no. 4.

Blanding, Michael. 2014. *The Map Thief: The Gripping Story of an Esteemed Rare-Map Dealer Who Made Millions Stealing Priceless Maps*. New York: Gotham Books.

Blaut, James M. 1993. *The Colonizer's Model of the World: Geographical Diffusionism and Eurocentric History*. New York: Guilford Press.

Blaut, James M., David Stea, Christopher Spencer, and Mark Blades. 2003. "Mapping as a Cultural and Cognitive Universal." *Annals of the Association of American Geographers* 93, no. 1 (2003): 165–85.

Board, Christopher. 1967. "Maps as Models." In *Models in Geography*, edited by Richard J. Chorley and Peter Haggett, 671–725. London: Methuen.

———. 2015. "Definitions of Map." In Monmonier, ed. (2015, 798–801).

Boeke, Kees. 1957. *Cosmic View: The Universe in 40 Jumps*. New York: John Day.

Boggs, Samuel W., and Dorothy Cornwall Lewis. 1945. *The Classification and Cataloging of Maps and Atlases*. New York: Special Libraries Association.

Bom HGz, G. D. 1962 [1885]. *Bijdragen tot eene geschiedenis van het geslacht 'Van Keulen' als boekhandelaars, uitgravers, kaart- en Instrumentmakers in Nederland: Eene Biblio-Cartographische Studie*. Amsterdam: Meridian Publishing. Originally published Amsterdam, 1885.

Borges, Jorge Luis. 1964 [1946]. "On Rigor in Science." Translated by Mildred and Harold Boyer. In Borges, *Dreamtigers*, 90. Austin: University of Texas Press. Originally published as "Del rigor en la ciencia," *Los Anales de Buenos Aires* 1, no. 3 (1946): 53; reprinted in *Historia universal de la infamia*, 2nd ed. (Buenos Aires: Emecé, 1954), 131; and in *El Hacedor* (Buenos Aires: Emecé, 1960), 103.

———. 1981. "Partial Enchantments of the *Quixote*." Translated by Emir Rodríguez Monegal and Alastair Reid. In *Borges: A Reader*, 232–35. New York: E. P. Dutton.

Bosse, David. 1995. "Osgood Carleton, Mathematical Practitioner of Boston." *Proceedings of the Massachusetts Historical Society* 107: 141–64.

Bosteels, Bruno. 1996. "A Misreading of Maps: The Politics of Cartography in Marxism and Poststructuralism." In *Signs of Change: Premodern > Modern > Postmodern*, edited by Stephen Barker, 109–38 and 364–82. Albany: State University of New York Press.

Bourdieu, Pierre. 1994. "Rethinking the State: Genesis and Structure of the Bureaucratic Field." *Sociological Theory* 12, no. 1: 1–18.

Boyce, Nell Greenfield. 2014. "There She Blew! Volcanic Evidence of the World's First Map." National Public Radio, 9 January. http://www.npr.org/2014/01/09/260918293/.

Branch, Jordan. 2014. *The Cartographic State: Maps, Territory, and the Origins of Sovereignty*. Cambridge: Cambridge University Press.

Branco, Rui Miguel Carvalhinho. 2005. "The Cornerstones of Modern Government: Maps, Weights and Measures and Census in Liberal Portugal (19th Century)." Ph.D. dissertation. European University Institute.

Brandenberger, René, ed. 2002. *Hans Conrad Escher von der Linth, 1767–1823. Die Ersten Panoramen der Alpen: Zeichnungen, Ansichten, Panoramen und Karten / The First Panoramas of the Alps: Drawings, Views, Panoramas and Maps*. Translated by Hans A. Smith. Glarus, Switzerland: Baeschlin for the Linth-Escher-Stiftung, Mollis.

Bret, Patrice. 2008. "Le moment révolutionnaire: Du terrain à la commission topographique de 1802." In *Les usages des cartes (XVIIᵉ–XIXᵉ siècle): Pour une approche pragmatique des productions cartographiques*, edited by Isabelle Laboulais, 81–97. Strasbourg: Presses Universitaires de Strasbourg.

———. 2019. "Projections Used for Topographical Maps." In Edney and Pedley, eds. (2019, forthcoming).

Brewer, John. 1990. *The Sinews of Power: War, Money and the English State, 1688–1783*. Cambridge, MA: Harvard University Press.

Broman, Thomas. 1998. "The Habermasian Public Sphere and 'Science *in* the Enlightenment.'" *History of Science* 36: 123–49.

Brooke-Hitching, Edward. 2016. *The Phantom Atlas: The Greatest Myths, Lies and Blunders on Maps*. London: Simon & Schuster.

Brotton, Jerry. 2012. *A History of the World in Twelve Maps*. London: Allen Lane.

———. 2014. *Great Maps: The World's Masterpieces Explored and Explained*. London: DK for the Smithsonian Institution.

Brown, Lloyd A. 1949. *The Story of Maps*. Boston: Little, Brown.

Brown, Ralph H. 1941. "The American Geographies of Jedidiah Morse." *Annals of the Association of American Geographers* 31, no. 3: 144–217.

Brownstein, Daniel. 2013. "Up in the Air." Musings on Maps, 26 January. https://dabrownstein.com/category/tableaux-comparatif-et-compare/.

Brückner, Martin. 2008. "Beautiful Symmetry: John Melish, Material Culture, and Map Interpretation." *Portolan* 73: 28–35.

———. 2015. "Maps, Pictures, and the Cartoral Arts in America." *American Art* 29, no. 2: 2–9.

———. 2017. *The Social Life of Maps in America, 1750–1860*. Chapel Hill: University of North Carolina Press for the Omohundro Institute for Early American History and Culture.

Bryars, Tim, and Tom Harper. 2014. *A History of the Twentieth Century in 100 Maps*. Chicago: University of Chicago Press for the British Library.

Buchroithner, Manfred F., and René Pfahlbusch. 2017. "Geodetic Grids in Authoritative Maps: New Findings about the Origin of the UTM Grid." *Cartography and Geographic Information Science* 44, no. 3: 186–200.

Bugge, Thomas. 1779. *Beskrivelse over den Opmaalings Maade, som er brugt ved de Danske geografiske karter; med tilføiet trigonometrisk karte over Siæland, og de der henhørende triangle, beregnede longituder og latitude, samt astronomiske observationer*. Copenhagen: Gyldendals Forlag.

Buisseret, David, ed. 1992. *Monarchs, Ministers, and Maps: The Emergence of Cartography as a Tool of Government in Early Modern Europe*. Chicago: University of Chicago Press.

Bulson, Eric. 2007. *Novels, Maps, Modernity: The Spatial Imagination, 1850–2000*. New York: Routledge.

Burgoyne, Margaret. 1851. "The Balloon: An 'Excursion Trip,' But Not by Railway, by a Lady." *Bentley's Miscellany* 30: 528–34.

Burkeman, Oliver. 2017. "Consumed by Anxiety? Give It a Day or Two." *Guardian* (15 September 2017). www.theguardian.com/lifeandstyle/2017/sep/15/consumed-by-anxiety-give-it-day-or-two. Last accessed, 22 September 2017.

Burroughs, Charles. 1995. "The 'Last Judgment' of Michelangelo: Pictorial Space, Sacred Topography, and the Social World." *Artibus et Historiae* 16, no. 32: 55–89.

Butlin, Robin A. 2009. *Geographies of Empire: European Empires and Colonies, c. 1880–1960*. Cambridge: Cambridge University Press.

Buttenfield, Barbara P. 2015. "Uncertainty and Reliability." In Monmonier, ed. (2015, 1642–44).

Byerly, Alison. 2007. "'A prodigious map beneath his feet': Virtual Travel and the Panoramic Perspective." *Nineteenth-Century Contexts* 29, nos. 2–3: 151–68.

———. 2013. *Are We There Yet? Virtual Travel and Victorian Realism*. Ann Arbor: University of Michigan Press.

Byrd, Max. 2009. "This Is Not a Map." *Wilson Quarterly* 33, no. 3: 26–32.

Cain, Mead. 1994. "The Maps of the Society for the Diffusion of Useful Knowledge: A Publishing History." *Imago Mundi* 46: 151–67.

Calhoun, Craig, ed. 1992. *Habermas and the Public Sphere*. Cambridge, MA: MIT Press.

Cañizares-Esguerra, Jorge. 2017. "On Ignored Global 'Scientific Revolutions.'" *Journal of Early Modern History* 21: 420–32.

Carlson, Julia S. 2010a. *Commentary on 'Written with a Slate-Pencil, on a Stone, on the Side of the Mountain of Black-Comb' by William Wordsworth*. Madison, WI: Silver Buckle Press, University of Wisconsin–Madison, for the History of Cartography Project. https://geography.wisc.edu/histcart/literary-selections/.

———. 2010b. "Topographical Measures: Wordsworth's and Crosthwaite's Lines on the Lake District." *Romanticism* 16, no. 1: 72–93.

Carroll, Lewis [pseud. Charles L. Dodgson]. 1876. *The Hunting of the Snark: An Agony in Eight Fits*. London: Macmillan.

———. 1893. *Sylvie and Bruno Concluded*. London: Macmillan.

Carter, Paul. 1987. *The Road to Botany Bay: An Essay in Spatial History*. London: Faber & Faber.

244

Cartwright, William. 2015. "Rethinking the Definition of the Word 'Map': An Evaluation of Beck's Representation of the London Underground through a Qualitative Expert Survey." *International Journal of Digital Earth* 8, no. 7: 522–37.

Casey, Edward S. 2002. *Representing Place: Landscape Painting and Maps.* Minneapolis: University of Minnesota Press.

Cassini, Jacques. 1720. *De la grandeur et de la figure de la terre.* In *Suite des mémoires de l'Académie royale des sciences, année MDCCXVIII.* 2 vols. Paris: Imprimerie royale.

Cassini de Thury, César François. [1744]. "La description géométrique de la France." Bibliothèque nationale de France, Département des cartes et plans, GE EE-1322 (RES). Untitled manuscript map of the triangulation stored separately as GE C-22286 (RES). Both available through gallica.bnf.fr.

———. 1749. "Sur la description géométrique de la France." In *Mémoires de l'Académie royale des sciences, année 1745,* 553–60. Paris: Imprimerie royale.

Casti, Emanuela. 2000. *Reality as Representation: The Semiotics of Cartography and the Generation of Meaning.* Translated by Jeremy Scott. Bergamo, Italy: Bergamo University Press.

———. 2018. "Bedolina: Map or Tridimensional Model?" *Cartographica* 53, no. 1: 15–35.

Castner, Henry W. 1980. "Special Purpose Mapping in 18th Century Russia: A Search for the Beginnings of Thematic Mapping." *American Cartographer* 7, no. 2: 163–75.

Cattaneo, Angelo. 2009. "Orb and Sceptre: Cosmography and World Cartography in Portugal and Italian Cities in the Fifteenth Century." *Archives internationales d'histoire des sciences* 59, no. 163: 531–53.

Cavelti Hammer, Madlena. 2019. "Heights and Distances, Geometric Determination of." In Edney and Pedley, eds. (2019, forthcoming).

Cavelti Hammer, Madlena, Hans-Uli Feldmann, and Markus Oehrli, eds. 1997. *Farbe, Licht und Schatten: Die Entwicklung der Reliefkartographie seit 1660: Begleitheft zur Sonderausstellung vom 5. April bis 3. August 1997 im Schweizerischen Alpinen Museum Bern und im September/Oktober 1997 im Bundesamt für Eich- und Vermessungswesen in Wien.* Cartographica Helvetica, Sonderheft 13. Murten, Switzerland: Verlag Cartographica Helvetica.

Cep, Casey N. 2014. "The Allure of the Map." *New Yorker* (22 January).

Certeau, Michel de. 1984. *The Practice of Everyday Life.* Translated by Steven F. Rendall. Berkeley: University of California Press.

Cesarz, Gary L. 2012. "Riddles and Resolutions: Infinity, Community, and the Absolute in Royce's Later Philosophy." In *Josiah Royce for the Twenty-First Century: Historical, Ethical, and Religious Interpretations,* edited by Kelly A. Parker and Krzysztof Piotr Skowroński, 63–81. Lanham, MD: Lexington Books.

Chamberlin, Wellman. 1947. *The Round Earth on Flat Paper: Map Projections Used by Cartographers.* Washington, DC: National Geographic Society.

Chang, Kang-tsung. 1980. "Circle Size Judgement and Map Design." *American Cartographer* 7: 155–62.

Chapuis, Olivier. 1999. *À la mer comme au ciel. Beautemps-Beaupré et la naissance de l'hydrographie moderne, 1700–1850: L'émergence de la précision en navigation et dans la cartographie marine.* Paris: Presses de l'Université de Paris–Sorbonne.

———. 2019. "Dépôt des cartes et plans de la Marine (Depository of Maps and Plans of the Navy; France)." In Edney and Pedley, eds. (2019, forthcoming).

Chardon, Roland E. 1988. *Notes and Bibliography on Geographic Linear Measures.* 2 vols. Baton Rouge, LA: Self-published.

Charlesworth, Michael. 2008. *Landscape and Vision in Nineteenth-Century Britain and France.* Aldershot, Surrey: Ashgate.

245

————. 2019. "Packe, Christopher." In Edney and Pedley, eds. (2019, forthcoming).

Chester, Lucy P. 2009. *Borders and Conflict in South Asia: The Radcliffe Boundary Commission and the Partition of Punjab*. Manchester: Manchester University Press.

Chura, Patrick. 2010. *Thoreau the Land Surveyor*. Gainesville: University Press of Florida.

————. 2015. "'Demon est deus inversus': Literary Cartography in Melville's 'The Encantadas.'" *49th Parallel* 35: 47–76.

Chwast, Seymour, Steven Heller, and Martin Venezky. 2004. *The Push Pin Graphic: A Quarter Century of Innovative Design and Illustration*. San Francisco: Chronicle Books.

Clark, John O. E., ed. 2005. *100 Maps: The Science, Art and Politics of Cartography throughout History*. New York: Sterling Publishing.

Clarke, Keith C. 2013. "What Is the World's Oldest Map?" *Cartographic Journal* 50, no. 2: 136–43.

Clarke, Susanna. 2004. *Jonathan Strange & Mr. Norrell*. London: Bloomsbury.

Comment, Bernard. 1999. *The Panorama*. London: Reaktion Books.

Comstock, C. B. 1876. "Notes on European Surveys." In *Annual Report of Major C. B. Comstock, Corps of Engineers, for the Fiscal Year Ending June 30, 1876*, 126–217. Vol. 2.3 of *Report of the Secretary of War*. Washington, DC: Government Printing Office.

Conforti, Joseph A., ed. 2005. *Creating Portland: History and Place in Northern New England*. Hanover, NH: University Press of New England for the University of New Hampshire Press.

Conley, Tom. 2007. *Cartographic Cinema*. Minneapolis: University of Minnesota Press.

Cons, Jason. 2005. "What's the Good of Mercators? Cartography and the Political Ecology of Place." *Graduate Journal of Social Science* 2, no. 1: 7–36.

Cook, Andrew S. 2006. "Surveying the Seas: Establishing the Sea Route to the East Indies." In *Cartographies of Travel and Navigation*, edited by James R. Akerman, 69–96. Chicago: University of Chicago Press.

Cortesão, Armando. 1935. *Cartografia e cartógrafos portugueses dos séculos XV e XVI*. 2 vols. Lisbon: Edição da "Seara nova."

————. 1969–71. *History of Portuguese Cartography*. *Agrupamento de Estudos de Cartografia Antiga, 6 and 8*. 2 vols. Lisbon: Junta de investigacões do ultramar.

Cosgrove, Denis. 1997. "Contested Global Visions: One-World, Whole-Earth, and the Apollo Space Photographs." *Annals of the Association of American Geographers* 84, no. 2: 270–94.

————. 1999. "Introduction." In *Mappings*, edited by Denis Cosgrove, 1–23. London: Reaktion Books.

————. 2001. *Apollo's Eye: A Cartographic Genealogy of the Earth in the Western Imagination*. Baltimore: Johns Hopkins University Press.

————. 2007a. "Epistemology, Geography, and Cartography: Matthew Edney on Brian Harley's Cartographic Theories." *Annals of the Association of American Geographers* 97, no. 1: 202–9.

————. 2007b. "Mapping the World." In *Maps: Finding Our Place in the World*, edited by James R. Akerman and Robert W. Karrow, Jr., 65–115. Chicago: University of Chicago Press.

————. 2008. "Cultural Cartography: Maps and Mapping in Cultural Geography." *Annales de géographie* 117, nos. 660–61: 159–78.

Cosgrove, Denis, and Veronica Della Dora. 2005. "Mapping Global War: Los Angeles, the Pacific, and Charles Owens's Pictorial Cartography." *Annals of the Association of American Geographers* 95, no. 2: 373–90.

Cosgrove, Denis, and William L. Fox. 2010. *Photography and Flight*. London: Reaktion Books.

Cowell, Gillian, and Gert Biesta. 2016. "From Mapreading to Mapmaking: Civic Learning as Orientation, Disorientation and Reorientation." *Policy Futures in Education* 14, no. 4: 431–51.

Crampton, Jeremy W. 1990. "An Elusive Reference: The 1:1 Map Story." *Cartographic Perspectives* 8: 26–27.

———. 1994. "Cartography's Defining Moment: The Peters Projection Controversy, 1974–1990." *Cartographica* 31, no. 4: 16–32.

———. 2001. "Maps as Social Constructions: Power, Communication and Visualization." *Progress in Human Geography* 25, no. 2: 235–52.

———. 2010. *Mapping: A Critical Introduction to Cartography and GIS*. Oxford: Wiley-Blackwell.

Crane, Nicholas. 2003. *Mercator: The Man Who Mapped the Planet*. New York: Henry Holt & Co.

Crary, Jonathan. 1990. *Techniques of the Observer: On Vision and Modernity in the Nineteenth Century*. Cambridge, MA: MIT Press.

Crespo Sanz, Antonio, and Alberto Fernández Wyttenbach. 2011. "¿Cartografía antigua o Cartografía histórica? / Old Cartography or Historical Cartography?" *Estudios geográficos* 72, no. 271: 403–20.

Cresswell, Donald H. 2000. "Colony to Commonwealth: The Eighteenth Century." In *Virginia in Maps: Four Centuries of Settlement, Growth, and Development*, edited by Richard W. Stephenson and Marianne M. McKee, 46–117. Richmond, VA: Library of Virginia.

Crone, G. R. 1953. *Maps and Their Makers: An Introduction to the History of Cartography*. London: Hutchinson.

Crowley, John E. 2011. *Imperial Landscapes: Britain's Global Visual Culture*. New Haven, CT: Yale University Press for the Paul Mellon Centre for Studies in British Art.

Cumming, William P. 1980. "The Colonial Charting of the Massachusetts Coast." In *Seafaring in Colonial Massachusetts: A Conference Held by the Colonial Society of Massachusetts, November 21 and 22, 1975*, edited by Philip Chadwick Foster Smith, 67–118. Boston: Colonial Society of Massachusetts.

———. 1998. *The Southeast in Early Maps*. Edited by Louis De Vorsey, Jr. 3rd ed. Chapel Hill: University of North Carolina Press.

Curtis, Jacqueline W. 2016. "Transcribing from the Mind to the Map: Tracing the Evolution of a Concept." *Geographical Review* 106, no. 3: 338–59.

Dahlberg, Richard E. 1984. "The Public Land Survey: The American Rural Cadastre." *Computers, Environment and Urban Systems* 9, nos. 2–3: 145–53.

Dallet, Gabriel. 1893. "La construction d'une carte." *Annales de géographie* 2: 11–26, 137–50.

Daly, Charles P. 1879. "The Early History of Cartography, or What We Know of Maps and Map-Making before the Time of Mercator." *Journal of the American Geographical Society of New York* 11: 1–40.

Dando, Christina Elizabeth. 2007. "Riding the Wheel: Selling American Women Mobility and Geographic Knowledge." *ACME: An International Journal for Critical Geographies* 6, no. 2: 174–210. http://www.acme-journal.org/.

———. 2017. *Women and Cartography in the Progressive Era*. London: Routledge.

d'Anville, Jean Baptiste Bourguignon. 1777. *Considérations générales sur l'étude et les connoissances que demande la composition des ouvrages de géographie*. Paris: Imprimerie de Lambert.

Darkes, Gilles. 2017. "An Introduction to Map Design." In *The Routledge Handbook of Mapping and Cartography*, edited by Alexander J. Kent and Peter Vujakovic, 287–98. London: Routledge.

Darnton, Robert. 1982. "What Is the History of Books?" *Daedalus: Proceedings of the American Academy of Arts and Sciences* III, no. 3: 65–83.

d'Avezac de Castera-Macaya, Marie Armand Pascal. 1863. *Coup d'oeil historique sur la projection des cartes de géographie*. Paris: E. Martinet. Also published in the *Bulletin de la Société de géographie*, ser. 5, 5 (1863): 257–361 and 438–85.

Deane, Charles. 1887. "Notes on Hubbard's Map of New England." *Proceedings of the Massachusetts Historical Society*, 2nd ser., 4: 13–21.

Defoe, Daniel. 1704. *The Storm*. London.

De Keyzer, Maïka, Iason Jongepier, and Tim Soens. 2014. "Consuming Maps and Producing Space: Explaining Regional Variations in the Reception and Agency of Mapmaking in the Low Countries during the Medieval and Early Modern Periods." *Continuity and Change* 29, no. 2: 209–40.

Dekker, Elly. 2002. "The Doctrine of the Sphere: A Forgotten Chapter in the History of Globes." *Globe Studies: The Journal of the International Coronelli Society*, nos. 49–50: 25–44.

Delaney, John. 2011. "Philippe Vandermaelen (1795–1869): *Atlas universel* (1827)." Princeton, NJ: Princeton University Library. http://libweb5.princeton.edu/visual_materials/maps/websites/vandermaelen/home.htm.

———. 2012. *First X, Then Y, Now Z: An Introduction to Landmark Thematic Maps*. Princeton, NJ: Princeton University Library.

Delano Smith, Catherine. 1987. "Cartography in the Prehistoric Period in the Old World: Europe, the Middle East, and North Africa." In Harley and Woodward, eds. (1987, 54–101).

———. 2001. "The Grip of the Enlightenment: The Separation of Past and Present." In *Plantejaments i Objectius d'una Historia Universal de la Cartografia/Approaches and Challenges in a Worldwide History of Cartography*, edited by David Woodward, Catherine Delano Smith, and Cordell D. K. Yee, 281–97. Vol. II of *Cicle de conferències sobre Història de la Cartografia*. Barcelona: Institut Cartogràfic de Catalunya.

———. 2006. "Milieus of Mobility: Itineraries, Route Maps, and Road Maps." In *Cartographies of Travel and Navigation*, edited by James R. Akerman, 16–68. Chicago: University of Chicago Press.

Delano Smith, Catherine, and Roger J. P. Kain. 1999. *English Maps: A History*. Toronto: University of Toronto Press for the British Library.

Del Casino, Vincent J., Jr., and Stephen P. Hanna. 2006. "Beyond the 'Binaries': A Methodological Intervention for Interrogating Maps as Representational Practices." *ACME: An International Journal for Critical Geographies* 4, no. 1: 34–56. http://www.acme-journal.org/.

Delisle, Guillaume. 1728. "Examen et comparaison de la grandeur de Paris, de Londres, et de quelques autres villes du monde, anciennes et modernes." In *Histoire de l'Académie royale des sciences. Année M.DCCXXV. Avec les Mémoires de mathématique & de physique, pour la même année, Tirés des Registres de cette Académie*, 480–57. Paris.

Della Dora, Veronica. 2007. "Putting the World into a Box: A Geography of Nineteenth-Century 'Travelling Landscapes.'" *Geografiska Annaler Series B—Human Geography* 89B, no. 4: 287–306.

Demeritt, David. 2001. "Scientific Forest Conservation and the Statistical Picturing of Nature's Limits in the Progressive-Era United States." *Environment and Planning D: Society & Space* 19, no. 4: 431–59.

Dempsey, Charles. 2001. *Inventing the Renaissance Putto*. Chapel Hill: University of North Carolina Press.

Dernay, Eugene. 1945. *Longitudes and Latitudes in the United States*. Tempe, AZ: American Federation of Astrologers.

DeRogatis, Amy. 2005. "What Would Jesus Do? Sexuality and Salvation in Protestant Evangelical Sex Manuals, 1950s to the Present." *Church History* 74, no. 1: 97–137.

Desdevises du Dézert, Théophile-Alphonse. 1880. "Le mouvement géographique en France et à l'étranger et les sociétés de géographie." *Bulletin de la société normande de géographie* 2: 5–24.

Desimini, Jill, and Charles Waldheim. 2016. *Cartographic Grounds: Projecting the Landscape Imaginary*. New York: Princeton Architectural Press.

Dickson, Peter W. 2007. *The Magellan Myth: Reflections on Columbus, Vespucci and the Waldseemueller Map of 1507*. Mount Vernon, OH: Printing Arts Press.

Di Fiore, Laura. 2017. "The Production of Borders in Nineteenth-Century Europe: Between Institutional Boundaries and Transnational Practices of Space." *European Review of History* 24, no. 1: 36–57.

Dobbin, Claire. 2012. *London Underground Maps: Art, Design and Cartography*. Farnham, Surrey: Lund Humphries for the London Transport Museum.

Dodge, Martin, Rob Kitchin, and Chris Perkins. 2009a. "Mapping Modes, Methods and Moments: A Manifesto for Map Studies." In Dodge, Kitchin, and Perkins, eds. (2009b, 220–43).

———, eds. 2009b. *Rethinking Maps: New Frontiers in Cartographic Theory*. London: Routledge.

———, eds. 2011. *The Map Reader: Theories of Mapping Practice and Cartographic Representation*. Hoboken, NJ: Wiley-Blackwell.

Doherty, Peter. 2017. "Mappaemundi, Maps and the Romantic Aesthetic in Children's Books." *Children's Literature in Education* 48, no. 1: 89–102.

Doiron, Claude J., Jr. 1972. "A Survey of Scale Concepts in Geographic Research." Ph.D. dissertation. University of Denver.

Dorling, Daniel. 2013. "Cartography: Making Sense of Our Worlds." *Cartographic Journal* 50, no. 2: 152–54.

———. 2015. "Cartogram." In Monmonier, ed. (2015, 193–97).

Dorling, Daniel, and David Fairbairn. 1997. *Mapping: Ways of Representing the World*. Harlow, Essex: Addison-Wesley Longman.

Dorrian, Mark, and Frédéric Pousin, eds. 2013. *Seeing from Above: The Aerial View in Visual Culture*. London: I. B. Tauris.

Downs, Roger M. 2015. "Cognition and Cartography." In Monmonier, ed. (2015, 1065–70).

Driver, Felix. 2001. *Geography Militant: Cultures of Exploration and Empire*. Oxford: Blackwell.

Dufour, Auguste-Henri. [1864]. *Atlas universel, physique, historique et politique de géographie ancienne et moderne*.

Du Mont, John S. 1978. *American Engraved Powder Horns: The Golden Age, 1755/1783*. Canaan, NH: Phoenix Publishing.

Dunlop, Catherine Tatiana. 2015. *Cartophilia: Maps and the Search for Identity in the French-German Borderland*. Chicago: University of Chicago Press.

Dutka, Jacques. 1995. "On Gauss' Priority in the Discovery of the Method of Least Squares." *Archive for History of Exact Sciences* 49, no. 4: 355–70.

Dyce, Matt. 2013. "Canada between the Photograph and the Map: Aerial Photography, Geographical Vision and the State." *Journal of Historical Geography* 39: 69–84.

Dym, Jordana. 2015. "'Mapitas,' Geografías Visualizadas and the Editorial Piedra Santa: A Mission to Democratize Cartographic Literacy in Guatemala." *Journal of Latin American Geography* 14, no. 3: 245–72.

Dym, Jordana, and Karl Offen, eds. 2011. *Mapping Latin America: A Cartographic Reader.* Chicago: University of Chicago Press.

Eames, Charles, and Ray Eames. 1968. "A Rough Sketch for a Proposed Film Dealing with the Powers of Ten and the Relative Size of Things in the Universe." Film, 8 minutes. Distributed by IBM.

———. 1977. "Powers of Ten: A Film Dealing with the Relative Size of Things in the Universe and the Effect of Adding Another Zero." Film, 9 minutes. Distributed by IBM.

Eckert, Max. 1907. "Die Kartographie als Wissenschaft." *Zeitschrift der Gesellschaft für Erdkunde zu Berlin* 1907, no. 8: 539–55.

———. 1908. "On the Nature of Maps and Map Logic." *Bulletin of the American Geographical Society* 40, no. 6: 344–51.

———. 1921–25. *Die Kartenwissenschaft: Forschungen und Grundlagen zu einer Kartographie als Wissenschaft.* 2 vols. Berlin: Walter de Gruyter.

———. 1939. *Kartographie: Ihre Aufgaben und Bedeutung für die Kultur der Gegenwart.* Berlin: Walter de Gruyter.

Eco, Umberto. 1985. "Map of the Empire." *Literary Review* 28, no. 2: 233–38.

———. 1994. "On the Impossibility of Drawing a Map of the Empire on a Scale of 1 to 1." In Umberto Eco, *How to Travel with a Salmon and Other Essays,* 95–106. New York: Harvest.

Edelson, S. Max. 2017. *The New Map of Empire: How Britain Imagined America before Independence.* Cambridge, MA: Harvard University Press.

Edney, Matthew H. 1986. "Politics, Science, and Government Mapping Policy in the United States, 1800–1925." *American Cartographer* 13, no. 4: 295–306.

———. 1991. "The Atlas of India, 1823–1947: The Natural History of a Topographic Map Series." *Cartographica* 28, no. 4: 59–91.

———. 1993. "Cartography without 'Progress': Reinterpreting the Nature and Historical Development of Mapmaking." *Cartographica* 30, nos. 2–3: 54–68.

———. 1994a. "Cartographic Culture and Nationalism in the Early United States: Benjamin Vaughan and the Choice for a Prime Meridian, 1811." *Journal of Historical Geography* 20, no. 4: 384–95.

———. 1994b. "Mathematical Cosmography and the Social Ideology of British Cartography, 1780–1820." *Imago Mundi* 46: 101–16.

———. 1997. *Mapping an Empire: The Geographical Construction of British India, 1765–1843.* Chicago: University of Chicago Press.

———. 2003a. "Bringing India to Hand: Mapping Empires, Denying Space." In *The Global Eighteenth Century,* edited by Felicity Nussbaum, 65–78. Baltimore: Johns Hopkins University Press.

———. 2003b. "New England Mapped: The Creation of a Colonial Territory." In *La cartografia europea tra primo Rinascimento e fine dell'Illuminismo: Atti del Convegno internazionale "The Making of European Cartography," Firenze BNCF-IUE, 13–15 dicembre 2001,* edited by Diogo Ramada Curto, Angelo Cattaneo, and André Ferrand de Almeida, 155–76. Florence: Leo S. Olschki Editore.

———. 2005. "The Origins and Development of J. B. Harley's Cartographic Theories." Monograph 54. *Cartographica* 40, nos. 1–2.

———. 2007a. "Mapping Empires, Mapping Bodies: Reflections on the Use and Abuse of Cartography." *Treballs de la Societat Catalana de Geografia* 63: 83–104.

———. 2007b. "Mapping Parts of the World." In *Maps: Finding Our Place in the World,* edited by James R. Akerman and Robert W. Karrow, Jr., 117–57. Chicago: University of Chicago Press.

———. 2007c. "Printed But Not Published: Limited-Circulation Maps of Territorial Disputes in Eighteenth-Century New England." In *Mappæ Antiquæ: Liber Amicorum Günter Schilder. Vriendenboek ter gelegenheid van zijn 65ste verjaardag*, edited by Paula van Gestel-van het Schip et al., 147–58. 't Goy-Houten, Netherlands: HES & De Graaf Publishers.

———. 2008a. "John Mitchell's Map of North America (1755): A Study of the Use and Publication of Official Maps in Eighteenth-Century Britain." *Imago Mundi* 60, no. 1: 63–85.

———. 2008b. "Putting 'Cartography' into the History of Cartography: Arthur H. Robinson, David Woodward, and the Creation of a Discipline." In *A Reader in Critical Geographies*, edited by Salvatore Engel-Di Mauro and Harald Bauder, 711–28. Praxis (e) Press. www.praxis-epress.org.

———. 2009. "The Irony of Imperial Mapping." In *The Imperial Map: Cartography and the Mastery of Empire*, edited by James R. Akerman, 11–45. Chicago: University of Chicago Press.

———. 2010. "Simon de Passe's Cartographic Portrait of Captain John Smith and a New England (1616/7)." *Word & Image* 26, no. 2: 186–213.

———. 2011a. "A Cautionary Historiography of 'John Smith's *New England*.'" *Cartographica* 46, no. 1: 1–27.

———. 2011b. "Knowledge and Cartography in the Early Atlantic." In *The Oxford Handbook of the Atlantic World, 1450–1850*, edited by Nicholas Canny and Philip Morgan, 87–112. Oxford: Oxford University Press.

———. 2011c. "Progress and the Nature of 'Cartography.'" In *Classics in Cartography: Reflections on Influential Articles from Cartographica*, edited by Martin Dodge, 331–42. Hoboken, NJ: Wiley-Blackwell.

———. 2012a. "Cartography's 'Scientific Reformation' and the Study of Topographical Mapping in the Modern Era." In *History of Cartography: International Symposium of the ICA Commission, 2010*, edited by Elri Liebenberg and Imre Josef Demhardt, 287–303. Heidelberg: Springer for the International Cartographic Association.

———. 2012b. "Field/Map: An Historiographic Review and Reconsideration." In *Scientists and Scholars in the Field: Studies in the History of Fieldwork and Expeditions*, edited by Kristian H. Nielsen, Michael Harbsmeier, and Christopher J. Ries, 431–56. Aarhus, Denmark: Aarhus University Press.

———. 2014a. "Academic Cartography, Internal Map History, and Critical Studies of Mapping Processes." *Imago Mundi* 66 supplement: 83–106.

———. 2014b. "A Content Analysis of *Imago Mundi*, 1935–2010." *Imago Mundi* 66 supplement: 107–31.

———. 2015a. "Cartography and Its Discontents." *Cartographica* 50, no. 1: 9–13.

———. 2015b. "Histories of Cartography." In Monmonier, ed. (2015, 607–14).

———. 2015c. "Modes of Cartographic Practice." In Monmonier, ed. (2015, 978–80).

———. 2016. "Of Lectures, Libraries, and Maps: The Nebenzahl Lectures and the Study of Map History." Keynote lecture, 19th Nebenzahl Lectures, "Maps, Their Collecting and Study: A Fifty Year Retrospective," organized by James R. Akerman. Newberry Library, Chicago. 27 October. Script available at https://www.mappingasprocess.net/blog/2018/1/27/of-maps-libraries-and-lectures-the-nebenzahl-lectures-and-the-study-of-map-history.

———. 2017a. "Map History: Discourse and Process." In *The Routledge Handbook of Map-*

ping and Cartography, edited by Alexander J. Kent and Peter Vujakovic, 68–79. London: Routledge.

———. 2017b. "Mapping, Survey, and Science." In *The Routledge Handbook of Mapping and Cartography*, edited by Alexander J. Kent and Peter Vujakovic, 145–58. London: Routledge.

———. 2017c. "References to the Fore! Local and National Mapping Traditions in the Printed Maps of Antebellum Portland, Maine." Osher Map Library and Smith Center for Cartographic Education, University of Southern Maine. 1 July. www.oshermaps.org /special-map-exhibition/references-to-the-fore.

———. 2017d. "The Rise of Systematic, Territorial Surveys." In *The Routledge Handbook of Mapping and Cartography*, edited by Alexander J. Kent and Peter Vujakovic, 159–72. London: Routledge.

———. 2018. "A Self-Explanatory Map? Come for the Satire, Stay for the Fun." Mapping as Process (16 February). www.mappingasprocess.net/blog/2018/2/16/a-self -explanatory-map-come-for-the-satire-stay-for-the-fun.

———. 2019a. "Geodetic Surveying in the Enlightenment." In Edney and Pedley, eds. (2019, forthcoming).

———. 2019b. "Geographical Mapping in the Enlightenment." In Edney and Pedley, eds. (2019, forthcoming).

———. 2019c. "History and Cartography." In Edney and Pedley, eds. (2019, forthcoming).

———. 2019d. "Meridians, Local and Prime." In Edney and Pedley, eds. (2019, forthcoming).

———. 2019e. "Modes of Cartographic Practice." In Edney and Pedley, eds. (2019, forthcoming).

———. forthcoming. "Modes of Cartographic Practice." In Kain, ed. (forthcoming).

Edney, Matthew H., and Susan Cimburek. 2004. "Telling the Traumatic Truth: William Hubbard's *Narrative* of King Philip's War and His 'Map of New-England.'" *William & Mary Quarterly*, 3rd ser., 61, no. 2: 317–48.

Edney, Matthew H., and Mary S. Pedley, eds. 2019. *Cartography in the European Enlightenment*. Vol. 4 of *The History of Cartography*. Chicago: University of Chicago Press. Forthcoming.

Edwards, Jess. 2003. "How to Read an Early Modern Map: Between the Particular and the General, the Material and the Abstract, Words and Mathematics." *Early Modern Literary Studies* 9, no. 1: art. 6. http://extra.shu.ac.uk/emls/09–1/edwamaps.html.

———. 2006. *Writing, Geometry and Space in Seventeenth-Century England and America: Circles in the Sand*. London: Routledge.

Ehrenberg, Ralph E. 2006. "'Up in the air in more ways than one': The Emergence of Aeronautical Charts in the United States." In *Cartographies of Travel and Navigation*, edited by James R. Akerman, 207–59. Chicago: University of Chicago Press.

Ehrensvärd, Ulla. 2006. *The History of the Nordic Map: From Myths to Reality*. Helsinki, Finland: John Nurminen Foundation.

Eisele, Carolyn. 1963. "Charles S. Peirce and the Problem of Map Projection." *Proceedings of the American Philosophical Society* 107, no. 4: 299–307.

Elden, Stuart. 2013. *The Birth of Territory*. Chicago: University of Chicago Press.

Ellis, Patrick. 2018. "The Panstereorama: City Models in the Balloon Era." *Imago Mundi* 70, no. 1: 79–93.

Ette, Ottmar. 2010. "'Everything is interrelated, even the errors in the system': Alexander von Humboldt and Globalization." *Atlantic Studies* 7, no. 2: 113–26.

Evans, Richard J. 2013. *Altered Pasts: Counterfactuals in History*. Waltham, MA: Brandeis University Press.

Fall, Juliet J. 2006. "Embodied Geographies, Naturalised Boundaries, and Uncritical Geopolitics in *La Frontière Invisible.*" *Environment and Planning D: Society & Space* 24, no. 5: 653–69.

Farinelli, Franco. 2009. *De la raison cartographique.* Paris: CTHS.

———. 2015. "Subject, Space, Object: The Birth of Modernity." In *Mathematizing Space: The Objects of Geometry from Antiquity to the Early Modern Age*, edited by Vincenzo De Risi, 143–55. Basel, Switzerland: Birkhauser.

Fasman, Jon. 2005. *The Geographer's Library.* New York: Penguin.

Fernández-Armesto, Felipe. 2007. "Maps and Exploration in the Sixteenth and Early Seventeenth Centuries." In Woodward, ed. (2007b, 738–70).

Field, Kenneth, and William Cartwright. 2014. "Becksploitation: The Over-use of a Cartographic Icon." *Cartographic Journal* 51, no. 4: 343–59.

Fine, Cordelia. 2017. *Testosterone Rex: Myths of Sex, Science, and Society.* New York: W. W. Norton.

Fisher, Irving, and O. M. Miller. 1944. *World Maps and Globes.* New York: Essential Books.

Fite, Emerson D., and Archibald Freeman. 1926. *A Book of Old Maps Delineating American History from the Earliest Days Down to the Close of the Revolutionary War.* Cambridge, MA: Harvard University Press.

Flem-Ath, Rand, and Rose Flem-Ath. 2012. *Atlantis beneath the Ice: The Fate of the Lost Continent.* 2nd ed. Rochester, VT: Bear & Co.

Fletcher, David H. 1995. *The Emergence of Estate Maps: Christ Church, Oxford, 1600 to 1840. Christ Church Papers, 4.* Oxford: Clarendon Press.

Foliard, Daniel. 2017. *Dislocating the Orient: British Maps and the Making of the Middle East, 1854–1921.* Chicago: University of Chicago Press.

Ford, Lily. 2016a. "'For the sake of the prospect': Experiencing the World from Above in the Late 18th Century." *Public Domain Review* (20 July). http://publicdomainreview .org/2016/07/20/for-the-sake-of-the-prospect-experiencing-the-world-from-above -in-the-late-18th-century/.

Ford, Lily. 2016b. "'Unlimiting the bounds': The Panorama and the Balloon View." *Public Domain Review* (3 August). https://publicdomainreview.org/2016/08/03/unlimiting -the-bounds-the-panorama-and-the-balloon-view/.

Fordham, H. G. 1914. *Studies in Carto-Bibliography, British and French, and in the Bibliography of Itineraries and Road-Books.* Oxford: Clarendon Press.

Foucault, Michel. 1970. *The Order of Things: An Archaeology of the Human Sciences.* New York: Random House.

———. 1972. *The Archaeology of Knowledge and the Discourse on Language.* Translated by A. M. Sheridan Smith. New York: Pantheon Books.

———. 1977. *Discipline and Punish: The Birth of the Prison.* Translated by Alan Sheridan. New York: Random House.

———. 1983. *This Is Not a Pipe.* Translated by James Harkness. Berkeley: University of California Press.

François, Jean. 1652. *La science de la géographie, divisée en trois parties.* Rennes.

Fraser, Robert. 2008. *Book History through Postcolonial Eyes: Rewriting the Script.* London: Routledge.

Freitag, Ulrich. 1962. "Der Kartenmaßstab: Betrachtungen über den Maßstabsbegriff in der Kartographie." *Kartographische Nachrichten* 12: 134–46.

Fremlin, Gerald, and Arthur H. Robinson. 1998. "Maps as Mediated Seeing." Monograph 51. *Cartographica* 35, nos. 1–2.

Friendly, Michael. 2005. "Milestones in the History of Data Visualization: A Case Study in Statistical Historiography." In *Classification: The Ubiquitous Challenge*, edited by C. Weihs and W. Gaul, 34–52. New York: Springer.

————. 2008a. "A Brief History of Data Visualization." In *Handbook of Computational Statistics: Data Visualization*, edited by C. Chen, W. Härdle, and A. Unwin, 1–34. Heidelberg: Springer-Verlag.

————. 2008b. "The Golden Age of Statistical Graphics." *Statistical Science* 23, no. 4: 502–35.

Fulton, Robert. 2017. "Crafting a Site of State Information Management: The French Case of the Dépôt de la Guerre." *French Historical Studies* 40, no. 2: 215–40.

Gaiman, Neil. 2006. *Fragile Things: Short Fictions and Wonders*. New York: Harper.

————. 2012. "The Mapmaker." Author's narration and commentary, 3 October. https://www.youtube.com/watch?v=BmLWtwflMgc.

Galle, Andreas. 1924. "Über die geodätischen Arbeiten von Gauss." In Carl Friedrich Gauss, *Werke*, 12 vols, vol. 11, part 2, separately paginated section. Göttingen: Königliche Gesellschaft der Wissenschaften.

Garfield, Simon. 2013. *On the Map: A Mind-Expanding Exploration of the Way the World Looks*. New York: Gotham.

Garland, Ken. 1994. *Mr Beck's Underground Map*. Harrow Weald, Middlesex: Capital Transport.

Gartner, William Gustav. 2011. "An Image to Carry the World within It: Performance Cartography and the Skidi Star Chart." In *Early American Cartographies*, edited by Martin Brückner, 169–247. Chapel Hill: University of North Carolina Press for the Omohundro Institute of Early American History and Culture.

Gaspar, Joaquim Alves. 2013. "From the Portolan Chart to the Latitude Chart: The Silent Cartographic Revolution." *Comité français de cartographie*, no. 216: 67–77.

————. 2016. "Revisiting the Mercator World Map of 1569: An Assessment of Navigational Accuracy." *Journal of Navigation* 69, no. 6: 1183–96.

Gatterer, Johann Cristoph. 1775. *Abriß der Geographie*. Göttingen, Germany: Johann Christian Dietrich.

Gauss, Carl Friedrich. 1847. "Untersuchung über Gegenstände der höhere Geodäsie; zweite Abhandlung." *Abhandlungen der Mathematischen Classe der Königlichen Gesellschaft der Wissenschaften zu Göttingen* 3: 3–43.

Geddes, Patrick. 1902. "Note on a Draft Plan for an Institute of Geography." *Scottish Geographical Magazine* 18: 141–44.

Gehring, Ulrike, and Peter Weibel, eds. 2014. *Mapping Spaces: Networks of Knowledge in 17th Century Landscape Painting*. Munich, Germany: Hirmer for ZKM Karlsruhe.

Gerardy, Theo. 1977. "Die Anfänge von Gauss' geodätischer Tätigkeit." *Zeitschrift für Vermessungswesen* 102, no. 1: 1–20.

Gilbert, Pamela K. 2004. *Mapping the Victorian Social Body*. Albany: State University of New York Press.

Glennie, Paul, and Nigel Thrift. 2009. *Shaping the Day: A History of Timekeeping in England and Wales, 1300–1800*. Oxford: Oxford University Press.

Godlewska, Anne. 1989. "Traditions, Crisis, and New Paradigms in the Rise of the Modern French Discipline of Geography, 1760–1850." *Annals of the Association of American Geographers* 79, no. 2: 192–213.

————. 1995. "Jomard: The Geographic Imagination and the First Great Facsimile Atlases." In *Editing Early and Historical Atlases: Papers Given at the Twenty-Ninth Annual Conference on Editorial Problems, University of Toronto, 5–6 November 1993*, edited by Joan Winearls, 109–35. Toronto: University of Toronto Press.

———. 1997. "The Idea of the Map." In *Ten Geographic Ideas That Changed the World*, edited by Susan Hanson, 17–39. New Brunswick, NJ: Rutgers University Press.

———. 1999. *Geography Unbound: French Geographic Science from Cassini to Humboldt.* Chicago: University of Chicago Press.

Golinski, Jan. 2012. "Is It Time to Forget Science? Reflections on Singular Science and Its History." *Osiris* 27, no. 1: 19–36.

Gombrich, E. H. 1975. "Mirror and Map: Theories of Pictorial Representation." *Philosophical Transactions of the Royal Society of London, B: Biological Sciences* 270, no. 903: 119–49.

Goodchild, Michael F. 2015a. "Accuracy in Mapping." In Monmonier, ed. (2015, 13–16).

———. 2015b. "Scale." In Monmonier, ed. (2015, 1383–89).

Goode, J. Paul. 1927. "The Map as a Record of Progress in Geography." *Annals of the Association of American Geographers* 17, no. 1: 1–14.

Gould, Peter, and Rodney White. 1974. *Mental Maps.* Harmondsworth, Middlesex: Penguin Books.

Grafton, Anthony. 2007. *What Was History? The Art of History in Early Modern Europe.* Cambridge: Cambridge University Press.

Green, Samuel A. 1905. "John Foster, the Earliest Engraver in New England." *Proceedings of the Massachusetts Historical Society*, 2nd ser., 19: 51–60.

Greenleaf, Moses. 1829. *Atlas Accompanying Greenleaf's Map and Statistical Survey of Maine.* Portland, ME: Shirley & Hyde.

Gregory, J. W. 1917. "The Evolution of the Map of the World." *Scottish Geographical Magazine* 33, no. 2: 49–65.

Griffin, Dori. 2013. *Mapping Wonderlands: Illustrated Cartography of Arizona, 1912–1962.* Tucson: University of Arizona Press.

———. 2017. "Beautiful Geography: The Pictorial Maps of Ruth Taylor White." *Imago Mundi* 69, no. 2: 233–47.

Guiso, Maria Antonietta, and Nicoletta Muratore. 1992. *Ad usum navigantium: Carte nautiche manoscritte di Gerard van Keulen, 1709–1713.* Rome: Istituto poligrafico e Zecca della stato for the Biblioteca Angelica, 1992.

Guptill, Stephen C. 2015. "Electronic Cartography and the Concept of Digital Map." In Monmonier, ed. (2015, 356–59).

Gutowski, Liz. 1980. "*Coitus topographicus.*" In *Push Pin Graphic* 83 ("Couples"): n.p.

Haasbroek, N. D. 1968. *Gemma Frisius, Tycho Brahe and Snellius and Their Triangulations.* Delft, Netherlands: Rijkscommissie voor Geodesie.

Habermas, Jürgen. 1989 [1962]. *The Structural Transformation of the Public Sphere: An Inquiry into a Category of Bourgeois Society.* Translated by Thomas Burger and Frederick Lawrence. Cambridge, MA: MIT Press. Originally published as *Strukturwandel der Öffentlicheit* (Darmstadt: Hermann Luchterhand Verlag, 1962).

Hadlaw, J. 2003. "The London Underground Map: Imagining Modern Time and Space." *Design Issues* 19, no. 1: 25–35.

Haffner, Jeanne. 2013. *The View from Above: The Science of Social Space.* Cambridge, MA: MIT Press.

Hall, David D. 1996. *Cultures of Print: Essays in the History of the Book.* Amherst: University of Massachusetts Press.

———. 2000. "Readers and Writers in Early New England." In *The Colonial Book in the Atlantic World*, edited by Hugh Amory and David D. Hall, 117–51. Vol. 1 of *A History of the Book in America.* Cambridge: Cambridge University Press for the American Antiquarian Society.

255

Hall, Debbie, ed. 2016. *Treasures from the Map Room: A Journey through the Bodleian Collections*. Oxford: Bodleian Libraries, University of Oxford.

Hallisey, Elaine J. 2005. "Cartographic Visualization: An Assessment and Epistemological Review." *Professional Geographer* 57, no. 3: 350–64.

Hane, Joshua G. 1996. "The London Underground Map of 1933." M.S. thesis. University of Wisconsin–Madison.

Hankins, Thomas L. 1999. "Blood, Dirt, and Nomograms." *Isis* 90, no. 1: 50–80.

Hanna, Stephen P. 2012. "Cartographic Memories of Slavery and Freedom: Examining John Washington's Map and Mapping of Fredericksburg, Virginia." *Cartographica* 47, no. 1: 29–49.

Hanna, Stephen P., and Vincent J. Del Casino, Jr., eds. 2003. *Mapping Tourism*. Minneapolis: University of Minnesota Press.

Hannah, Matthew G. 2000. *Governmentality and the Mastery of Territory in Nineteenth-Century America*. Cambridge: Cambridge University Press.

Hansen, Jason. 2015. *Mapping the Germans: Statistical Science, Cartography, and the Visualization of the German Nation, 1848–1914*. Oxford: Oxford University Press.

Hansen, Paul. 2013. *Commentary on 'West with the Night' by Beryl Markham*. Madison, WI: Silver Buckle Press, University of Wisconsin–Madison, for the History of Cartography Project. https://geography.wisc.edu/histcart/literary-selections/.

Hapgood, Charles H. 1966. *Maps of the Ancient Sea Kings: Evidence of Advanced Civilization in the Ice Age*. Philadelphia: Chilton.

Haraway, Donna. 1988. "Situated Knowledges: The Science Question in Feminism and the Privilege of Partial Perspective." *Feminist Studies* 14, no. 3: 575–99.

Hargrave, Kiran Millwood. 2016. *The Cartographer's Daughter*. New York: Knopf.

Harley, J. B. 1987. "The Map and the Development of the History of Cartography." In Harley and Woodward, eds. (1987, 1–42).

———. 1989. "Deconstructing the Map." *Cartographica* 26, no. 2: 1–20.

———. 1991. "Can There Be a Cartographic Ethics?" *Cartographic Perspectives* 10: 9–16.

Harley, J. B., and David Woodward, eds. 1987. *Cartography in Prehistoric, Ancient, and Medieval Europe and the Mediterranean*. Vol. 1 of *The History of Cartography*. Chicago: University of Chicago Press.

———, eds. 1992. *Cartography in the Traditional Islamic and South Asian Societies*. Vol. 2.1 of *The History of Cartography*. Chicago: University of Chicago Press.

———, eds. 1994. *Cartography in the Traditional East and Southeast Asian Societies*. Vol. 2.2 of *The History of Cartography*. Chicago: University of Chicago Press.

Harmon, Katharine A. 2004. *You Are Here: Personal Geographies and Other Maps of the Imagination*. New York: Princeton Architectural Press.

———. 2016. *You Are Here NYC: Mapping the Soul of the City*. New York: Princeton Architectural Press.

Harper, Tom, ed. 2016. *Maps and the 20th Century: Drawing the Line*. London: British Library.

Harrington, Bates. 1879. *How 'Tis Done: A Thorough Ventilation of the Numerous Schemes Conducted by Wandering Canvassers, together with the Various Advertising Dodges for the Swindling of the Public*. Chicago: Fidelity Publishing.

Harrison, Richard Edes. 1944. *Look at the World: The Fortune Atlas for World Strategy*. New York: Alfred A. Knopf.

———. 1958. "Why Our Maps Aren't Good Enough." *Harper's Magazine* (July 1958).

Harrisse, Henry. 1892. *The Discovery of North America: A Critical, Documentary, and Historic Investigation, with an Essay on the Early Cartography of the New World*. London.

Harvey, David. 1989. *The Condition of Postmodernity: An Enquiry into the Origins of Cultural Change*. Oxford: Basil Blackwell.

Harvey, Miles. 2000. *The Island of Lost Maps: A True Story of Cartographic Crime*. New York: Random House.

Harvey, P. D. A. 1980. *The History of Topographical Maps: Symbols, Pictures and Surveys*. London: Thames & Hudson.

———. 1987. "Local and Regional Cartography in Medieval Europe." In Harley and Woodward, eds. (1987, 464–501).

———. 1993. "Estate Surveyors and the Spread of the Scale-Map in England, 1550–1580." *Landscape History* 15: 37–50.

Harwood, Jeremy. 2006. *To the Ends of the Earth: 100 Maps That Changed the World*. Cincinnati, OH: F & W Publications.

Harzinski, Kris. 2010. *From Here to There: A Curious Collection from the Hand Drawn Map Association*. New York: Princeton Architectural Press.

Heffernan, Michael. 2005. "Edme Mentelle's Geographies and the French Revolution." In *Geography and Revolution*, edited by David N. Livingstone and Charles W. J. Withers, 273–303. Chicago: University of Chicago Press.

———. 2014. "A Paper City: On History, Maps, and Map Collections in 18th and 19th Century Paris." *Imago Mundi* 66 supplement: 5–20.

Heidenreich, Conrad E. 1975. "Measures of Distance Employed on 17th and Early 18th Century Maps of Canada." *Canadian Cartographer* 12, no. 2: 121–37.

Heilbron, J. L. 2000. "Domesticating Science in the Eighteenth Century." In *Science and the Visual Image in the Enlightenment*, edited by William R. Shea, 1–24. Canton, MA: Science History Publications.

Heiser, Willem J. 2003. "Early Statistical Modelling of Latent Quantities: The History of Distance Measurement by Triangulation." In *New Developments in Psychometrics: Proceedings of the International Meeting of the Psychometric-Society, IMPS2001, Osaka, Japan, July 15–19, 2001*, edited by Haruo Yanai et al., 33–44. Tokyo: Springer.

Hennerdal, Pontus. 2015. "Beyond the Periphery: Child and Adult Understanding of World Map Continuity." *Annals of the Association of American Geographers* 105, no. 4: 773–90.

Herb, Guntram Henrik. 1997. *Under the Map of Germany: Nationalism and Propaganda, 1918–1945*. London: Routledge.

Herodotus. 2008. *The Histories*. Translated by Robin Waterfield. Oxford: Oxford University Press.

Hessler, John W., and Chet Van Duzer. 2012. *Seeing the World Anew: The Radical Vision of Martin Waldseemüller's 1507 & 1516 World Maps*. Delray Beach, FL: Levenger Press for the Library of Congress.

Higton, Hester. 2001. "Does Using an Instrument Make You Mathematical? Mathematical Practitioners of the 17th Century." *Endeavour* 25: 18–22.

Hill, Gillian. 1978. *Cartographical Curiosities*. London: British Library.

Hodgkiss, Alan G. 1981. *Understanding Maps: A Systematic Enquiry of Their Use and Development*. Folkestone, Kent: Dawson.

Hoffman, Moshe, Uri Gneezy, and John A. List. 2011. "Nurture Affects Gender Differences in Spatial Abilities." *Proceedings of the National Academy of Sciences of the United States of America* 108, no. 36: 14786–88.

Holman, Richard B. 1960. "John Foster's Woodcut Map of New England." *Printing and Graphic Arts* 8: 53–93.

———. 1970. "Seventeenth-Century American Prints." In *Prints in and of America to 1850*, edited by John D. Morse, 23–52. Charlottesville: University Press of Virginia for the Henry Francis du Pont Winterthur Museum.

Holmes, Nigel. 1991. *Pictorial Maps*. New York: Watson-Guptill Publications.

Holtorf, Christian. 2017. "Zur Wissengeschichte von Geografie und Kartografie: Einleitung." *Berichte zur Wissenschaftsgeschichte* 40, no. 1: 7–16.

Holtwijk, Theo H. B. M., and Earle G. Shettleworth, Jr., eds. 1999. *Bold Vision: The Development of the Parks of Portland, Maine*. Portland: Greater Portland Landmarks.

Holwell, John. 1678. *A Sure Guide to the Practical Surveyor in Two Parts*. London: by W. Godbid for Christopher Hussey.

Hooper, Wynnard. 1883. "The Theory and Practice of Statistics." *Journal of the Statistical Society of London* 46, no. 3: 461–516.

Hornsby, Stephen J. 2011. *Surveyors of Empire: Samuel Holland, J. F. W. Des Barres, and the Making of The Atlantic Neptune*. Montreal: McGill-Queen's University Press.

———. 2017. *Picturing America: The Golden Age of Pictorial Maps*. Chicago: University of Chicago Press.

Hornsey, Richard. 2012. "Listening to the Tube Map: Rhythm and the Historiography of Urban Map Use." *Environment and Planning D: Society & Space* 30, no. 4: 675–93.

———. 2016. "The Cultural Uses of the A–Z London Street Atlas: Navigational Performance and the Imagining of Urban Form." *Cultural Geographies* 23, no. 2: 265–80.

Houston, Kerr. 2005. "'Siam not so small!' Maps, History, and Gender in *The King and I*." *Camera Obscura* 20, no. 2: 72–117.

Hubbard, William. 1677. *A Narrative of the Troubles with the Indians in New-England, from the first planting thereof in the year 1607, to this present year 1677. But chiefly of the late Troubles in the two last years, 1675 and 1676. To which is added a Discourse about the Warre with the Pequods In the year 1637*. Boston: Printed by John Foster.

Hudson, Brian. 1977. "The New Geography and the New Imperialism: 1870–1918." *Antipode* 9, no. 2: 12–19.

Huguenin, Marcel. 1948. *Historique de la cartographie de la nouvelle carte de France. Publications techniques de l'Institut géographique national*. Paris: Imprimerie de l'Institut géographique nationale pour le Ministère des travaux publics et des transports.

Hull, John T. 1885. *The Siege and Capture of Fort Loyall, Destruction of Falmouth, May 20, 1690 (O.S.)*. Portland, ME: Owen, Strout & Co., for the City Council of Portland.

Humboldt, Alexander von. 1836–39. *Examen critique de l'histoire de la géographie du nouveau continent et des progrès de l'astronomie nautique aux quinzième et seizième siècles*. 5 vols. Paris: Librairie de Gide.

Humboldt, Alexander von, and Aimé Bonpland. 2009 [1805]. *Essay on the Geography of Plants*. Translated by Stephen T. Jackson and Sylvie Romanowski. Chicago: University of Chicago Press. Originally published as *Essai sur la géographie des plantes* (Paris: Chez Levrault, Schoell et Cie., 1805).

Hutchinson, Dave. 2004. "On the Windsor Branch." In *As The Crow Flies*, 149–59. Wigan, Lancashire: BeWrite Books. Reprinted as *Europe in Autumn* (London: Solaris, 2014), part 2, chap. 2.2.

Huynh, Niem Tu, Sean Doherty, and Bob Sharpe. 2010. "Gender Differences in the Sketch Map Creation Process." *Journal of Maps* 2010: 270–88.

Imhof, Eduard. 1964. "Beiträge zur Geschichte der topographische Kartographie." *Internationales Jahrbuch für Kartographie* 4: 129–53.

———. 1982 [1965]. *Cartographic Relief Presentation*. Translated by H. J. Steward. New York: Walter de Gruyter. Originally published as *Kartographische Geländedarstellung* (Berlin: Walter de Gruyter, 1965).

Isidore of Seville. 1472. *Liber etymologiarum*. Augsburg: Günther Zainer.

Jacob, Christian. 1999. "Mapping in the Mind: The Earth from Ancient Alexandria." In *Mappings*, edited by Denis Cosgrove, 24–49. London: Reaktion Books.

———. 2006. *The Sovereign Map: Theoretical Approaches in Cartography throughout History*. Translated by Tom Conley. Edited by Edward H. Dahl. Chicago: University of Chicago Press.

Jameson, Frederic. 1991. "The Cultural Logic of Late Capitalism." In *Postmodernism, or, The Cultural Logic of Late Capitalism*, 44–54. Durham, NC: Duke University Press.

Javorsky, Irene. 1990. "Pariser und Londoner Georamen des 19. Jahrhunderts." *Der Globusfreund*, nos. 38–39: 179–92.

Jay, Martin. 1993. "Scopic Regimes of Modernity." In *Force Fields: Between Intellectual History and Cultural Critique*, 114–33. New York: Routledge.

Jervis, Thomas Best. 1836a. *The Expediency and Facility of Establishing the Metrological and Monetary Systems throughout India, on a Scientific and Permanent Basis . . . with Respect to Such as Subsist at Present, or Have Hitherto Subsisted in All Past Ages throughout the World*. Bombay: American Mission.

———. 1836b. *Records of Ancient Science, Exemplified and Authenticated in the Primitive Universal Standard of Weights and Measures*. Calcutta.

Johnson, Alexander James Cook. 2017. *The First Mapping of America: The General Survey of British North America*. London: I. B. Tauris.

Johnson, Stephen. 2016. *The Ghost Map: The Story of London's Most Terrifying Epidemic— and How It Changed Science, Cities, and the Modern World*. New York: Riverhead Books.

Jolly, David C. 1986. "Was Antarctica Mapped by the Ancients?" *Skeptical Inquirer* 11, no. 1: 32–43.

Jomard, Edme François. 1809. "Mémoire sur le système métrique des anciens Egyptiens contenant des recherches sur leurs connoissances géométriques et sur les mesures des autres peuples de l'antiquité." In *Description de l'Egypte: Antiquités Mémoires I*, 495–802. Paris: Imprimerie impériale.

Jones, Bennett Melvill, and John Crisp Griffiths. 1925. *Aerial Surveying by Rapid Methods*. Cambridge: Cambridge University Press.

Kain, Roger J. P., ed. forthcoming. *Cartography in the Nineteenth Century*. Vol. 5 of *The History of Cartography*. Chicago: University of Chicago Press.

Kain, Roger J. P., and Richard R. Oliver. 2015. *British Town Maps: A History*. London: British Library.

Kaplan, Caren. 2018. *Aerial Aftermaths: Wartime from Above*. Durham, NC: Duke University Press.

Kasson, John F. 1990. *Rudeness and Civility: Manners in Nineteenth-Century Urban America*. New York: Hill & Wang.

Keates, J. S. 1982. *Understanding Maps*. New York: John Wiley & Sons.

Keillor, Garrison. 1985. *Lake Wobegon Days*. New York: Viking.

Keltie, J. Scott. 1886. "Geographical Education: Report to the Council of the Royal Geographical Society." *Supplementary Papers of the Royal Geographical Society* 1, no. 4: 439–594.

Kennelly, Arthur E. 1928. *Vestiges of Pre-Metric Weights and Measures Persisting in Metric-System Europe, 1926–1927*. New York: Macmillan for the Bureau of International Research of Harvard University and Radcliffe College.

Kent, Alexander J. 2017. "Trust Me, I'm a Cartographer: Post-Truth and the Problem of Acritical Cartography." *Cartographic Journal* 54, no. 3: 193–95.

King, Geoff. 1996. *Mapping Reality: An Exploration of Cultural Cartographies*. New York: St. Martin's Press.

259

Kingston, Ralph. 2006. "How Old Is the Word Cartography." MapHist listserv, 3 March. Available through archived threads at http://www.maphist.nl. Last accessed 2 October 2011.

———. 2011. "The French Revolution and the Materiality of the Modern Archive." *Libraries & the Cultural Record* 46, no. 1: 1–25.

———. 2014. "Trading Places: Accumulation as Mediation in French Ministry Map Depots, 1798–1810." *History of Science* 52, no. 3: 247–76.

Kishimoto, Haruko. 1968. *Cartometric Measurements*. Zurich.

Kitchin, Rob. 1994. "Cognitive Maps: What Are They and Why Study Them?" *Journal of Environmental Psychology* 14: 1–19.

Kitchin, Rob, and Martin Dodge. 2007. "Rethinking Maps." *Progress in Human Geography* 31, no. 3: 331–44.

Kitchin, Rob, Justin Gleeson, and Martin Dodge. 2013. "Unfolding Mapping Practices: A New Epistemology for Cartography." *Transactions of the Institute of British Geographers* 38, no. 3: 480–96.

Kitchin, Rob, Chris Perkins, and Martin Dodge. 2009. "Thinking about Maps." In Dodge, Kitchin, and Perkins, eds. (2009b, 1–25).

Klein, Hans. 1983. "Maßstäbe und Maßstabsberechnungen alter Karten." In *Kartenhistorisches Colloquium Bayreuth '82: Vorträge und Berichte*, edited by Wolfgang Scharfe, Hans Volet, and Erwin Herrmann, 71–77. Berlin: Dietrich Reimer Verlag for the Arbeitskreis "Geschichte der Kartographie" of the Deutsche Gesellschaft für Kartographie und Historische Verein für Oberfranken.

Klein, Judy L. 2001. "Reflections from the Age of Economic Measurement." *History of Political Economy* 33 supplement: 111–36.

Klöti, Thomas. 1994. *Johann Friedrich von Ryhiner, 1732–1803: Berner Staatsmann, Geograph, Kartenbibliograph und Verkehrspolitiker*. Bern, Switzerland: Geographische Gesellschaft Bern.

Koláčný, Anton. 1969. "Cartographic Information: A Fundamental Concept and Term in Modern Cartography." *Cartographic Journal* 6: 47–49.

Koller, Christophe, and Patrick Jucker-Kupper, eds. 2009. *Karten, Kartographie und Geschichte: Von der Visualisierung der Macht zur Macht der Visualisierung / Cartes, cartographie et histoire: De la visualization du pouvoir au pouvoir de la visualisation*. Bern, Switzerland: Verein "Geschichte und Infomatik" / Association "Histoire et Informatique."

Konvitz, Josef W. 1987. *Cartography in France, 1660–1848: Science, Engineering, and Statecraft*. Chicago: University of Chicago Press.

Korzybski, Alfred. 1933. *Science and Sanity: An Introduction to Non-Aristotelian Systems and General Semantics*. Lakeville, CT: International Non-Aristotelian Library Publishing Company.

Kotarbińska, Janina. 1957. "Pojęcie znaku [The notion of the sign]." *Studia Logica* 6: 37–143.

Kraak, Menno-Jan, and Sara Irina Fabrikant. 2017. "Of Maps, Cartogaphy and the Geography of the International Cartographic Association." *International Journal of Cartography 3 supplement*: 9–31.

Kretschmer, Ingrid. 1986a. "Maßstab." In Kretschmer, Dörflinger, and Wawrik, eds. (1986, 2: 469–71).

———. 1986b. "Maßstabsangabe." In Kretschmer, Dörflinger, and Wawrik, eds. (1986, 2: 471–75).

———. 2015a. "Eckert, Max." In Monmonier, ed. (2015, 338–40).

———. 2015b. "Peters Projection." In Monmonier, ed. (2015, 1099–101).

Kretschmer, Ingrid, Johannes Dörflinger, and Franz Wawrik, eds. 1986. *Lexikon zur Ge-*

schichte der Kartographie von den Anfängen bis zum ersten Weltkrieg. 2 vols. Vol. C of *Die Kartographie und ihre Randgebiete: Enzyklopädie*. Vienna: Franz Deuticke.

Krygier, John B. 1995. "Cartography as an Art and a Science?" *Cartographic Journal* 32: 3–10.

———. 2008. "Cartocacoethes: Why the World's Oldest Map Isn't a Map." Making Maps: DIY Cartography, 13 October. http://makingmaps.net/2008/10/13/cartocacoethes-why -the-worlds-oldest-map-isnt-a-map/.

———. 2013. "Making Maps That Don't Look Like Maps: Applied Counter-Cartocacoethes for Spies/1915." Making Maps: DIY Cartography, 2 July. http://makingmaps.net/2013 /07/02/making-maps-that-dont-look-like-maps-applied-counter-cartocacoethes-for -spies-1915/.

Kuhn, Thomas S. 1970. *The Structure of Scientific Revolutions*. 2nd ed. Chicago: University of Chicago Press.

Kumler, Mark P., and Barbara P. Buttenfield. 1996. "Gender Differences in Map Reading Abilities: What Do We Know? What Can We Do?" In *Cartographic Design: Theoretical and Practical Perspectives*, edited by Clifford H. Wood and C. Peter Keller, 125–36. New York: John Wiley & Sons.

Kupčík, Ivan. 2011. *Alte Landkarten von der Antike bis zum Ende des 19. Jahrhunderts: Ein Handbuch zur Geschichte der Kartographie*. 8th ed. Stuttgart: Franz Steiner.

Laboulais, Isabelle. 2019. "Académie des sciences (Academy of Sciences; France)." In Edney and Pedley, ed. (2019, forthcoming).

La Chapelle, Jean Baptiste de. 1755. "Échelle." In *Encyclopédie, ou, Dictionnaire raisonné des sciences, des arts et des métiers*, edited by Denis Diderot and Jean Le Rond d'Alembert, 17 text vols. and 11 plate vols., 5: 248–49. Paris: Briasson, David, Le Breton & Durand.

Lakoff, George. 1987. *Women, Fire, and Dangerous Things: What Categories Reveal about the Mind*. Chicago: University of Chicago Press.

Lambert, J. H. 2011 [1772]. *Notes and Comments on the Composition of Terrestrial and Celestial Maps*. Translated by Waldo R. Tobler. Redlands, CA: Esri Press. Originally published as "Anmerkungen und Zusätze zur Entwerfung der Land- und Himmelscharten," in *Beyträge zum Gebrauche der Mathematik und deren Anwendung* (Berlin: Verlag der Buch-handlung der Realschule, 1772), 3: 105–99.

Lane, Christopher W. 1986. "Whose Map Is It Anyway?" *Map Collector* 36: 16–20.

La Renaudière, Philippe François de. 1828. *Histoire abrégée de l'origine et des progrès de la géo-graphie*. Paris: Decourchant.

Larsgaard, Mary Lynette. 1984. *Topographic Mapping of the Americas, Australia, and New Zealand*. Littleton, CO: Libraries Unlimited.

Latour, Bruno. 1987. *Science in Action: How to Follow Scientists and Engineers through Society*. Cambridge, MA: Harvard University Press.

———. 1990. "Drawing Things Together." In *Representation in Scientific Practice*, edited by Michael Lynch and Steve Woolgar, 19–68. Cambridge, MA: MIT Press.

———. 2005. *Reassembling the Social: An Introduction to Actor-Network-Theory*. Oxford: Ox-ford University Press.

Laussedat, Aimé. 1891. "Notice sur l'histoire des applications de la perspective à la topogra-phie et à la cartographie." *Paris-Photographe* 5: 1–22.

Lem, Stanislaw. 1985. *The Cyberiad: Fables for the Cybernetic Age*. Translated by Michael Kan-del. San Diego: Harcourt Brace Jovanovich.

Lennon, Florence Becker. 1962. *The Life of Lewis Carroll: Victoria through the Looking-Glass*. New York: Collier.

Lennox, Jeffers L. 2017. *Homelands and Empires: Indigenous Spaces, Imperial Fictions, and*

Competition for Territory in Northeastern North America, 1690–1763. Toronto: University of Toronto Press.

Lepetit, Bernard. 1993. "Architecture, géographie, histoire: Usages de l'échelle." *Genèses: Sciences sociales et histoire* 13: 118–38.

———. 1995. "De l'échelle en histoire." In *Jeux d'échelle. La micro-analyse à l'expérience*, edited by Jacques Revel, 71–95. Paris: Gallimard–Le Seuil.

Lester, Toby. 2009. *The Fourth Part of the World: The Race to the Ends of the Earth, and the Epic Story of the Map That Gave America Its Name.* New York: Free Press.

Levinsky, Allen. 2007. *A Short History of Portland.* Beverly, MA: Commonwealth Editions.

Lewes, Darby. 2000. *Nudes from Nowhere: Utopian Sexual Landscapes.* Lanham, MD: Rowman & Littlefield.

Lewes, George Henry. 1853. *Comte's Philosophy of the Sciences: Being an Exposition of the Principles of the 'Cours de philosophie positive' of Auguste Comte.* London: Henry G. Bohn.

Lewis, G. Malcolm. 1987. "The Origins of Cartography." In Harley and Woodward, eds. (1987, 50–53).

Lewis, Martin, and Kären E. Wigen. 1997. *The Myth of Continents: A Critique of Metageography.* Berkeley: University of California Press.

Liberman, Mark. 2013. "(Not) Trusting Data." Language Log, University of Pennsylvania, 4 August. http://languagelog.ldc.upenn.edu/nll/?p=5590.

———. 2014. "Holacracy." Language Log, University of Pennsylvania, 4 March. http://languagelog.ldc.upenn.edu/nll/?p=10840.

Licka, J. L. 1880. "Zur Geschichte der Horizontallinien oder Isohypsen." *Zeitschrift für Vermessungswesen* 9: 37–50.

Lieber, Francis [Franz]. 1834. *A Constitution and Plan of Education for Girard College for Orphans, with an Introductory Report, Laid before the Board of Trustees.* Philadelphia: Carey, Lea & Blanchard.

Lightman, Bernard. 2012. "Spectacle in Leicester Square: James Wyld's Great Globe, 1851–61." In *Popular Exhibitions, Science and Showmanship, 1840–1910*, edited by Joe Kember, John Plunkett, and Jill A. Sullivan, 19–39. London: Pickering & Chatto.

Lindgren, Ute. 2007. "Land Surveys, Instruments, and Practitioners in the Renaissance." In Woodward, ed. (2007b, 477–508).

Livingstone, David N. 1984. "Natural Theology and Neo-Lamarckism: The Changing Context of Nineteenth-Century Geography in the United States and Great Britain." *Annals of the Association of American Geographers* 74, no. 1: 9–28.

———. 1992. *The Geographical Tradition: Episodes in the History of a Contested Enterprise.* Oxford: Blackwell.

———. 2010. "Cultural Politics and the Racial Cartographics of Human Origins." *Transactions of the Institute of British Geographers* 35, no. 2: 204–21.

Lloyd, Robert, and Patricia Gilmartin. 1987. "The South Carolina Coastline on Historical Maps: A Cartometric Analysis." *Cartographic Journal* 24, no. 1: 19–26.

Lois, Carla. 2015. "El mapa, los mapas: Propuestas metolodógicas para abordar la pluralidad y la inestabilidad de la imagen cartográfica." *Geograficando* 11, no. 1: separately paginated.

Long, John H. 2015. "London Underground Map." In Monmonier, ed. (2015, 788–92).

Lüdde, Johann Gottfried. 1849. *Die Geschichte der Methodologie der Erdkunde. In ihrer ersten Grundlage, vermittelst einer historisch-kritischen Zusammenstellung der Literatur der Methodologie der Erdkund.* Leipzig: J. C. Hinrichs.

Ludden, David. 1998. Review of M. H. Edney, *Mapping an Empire: The Geographical Construction of British India, 1765–1843* (Chicago: University of Chicago Press, 1997). *Journal of Interdisciplinary History* 29, no. 2: 332–33.

Lukens, R. R. 1931. "Captain John Smith's Map." *Military Engineer* 23, no. 131: 435–38.

MacDonald, J. Fred. 1979. *Don't Touch That Dial: Radio Programming in American Life from 1920 to 1960*. Chicago: Nelson-Hall.

MacEachren, Alan M. 1979. "The Evolution of Thematic Cartography: A Research Methodology and Historical Review." *Canadian Cartographer* 16, no. 1: 17–33.

———. 1995. *How Maps Work: Representation, Visualization, and Design*. New York: Guilford Press.

Maeer, Alistair S., and Ashley Baynton-Williams. 2019. "*English Pilot, The*." In Edney and Pedley, eds. (2019, forthcoming).

Mah, Harold. 2000. "Phantasies of the Public Sphere: Rethinking the Habermas of Historians." *Journal of Modern History* 72: 153–82.

Maling, D. H. 1973. *Coordinate Systems and Map Projections*. London: George Philip & Son.

———. 1989. *Measurements from Maps: Principles and Methods of Cartometry*. Oxford: Pergamon Press.

———. 1991. "The Origins of That Definition." *Cartographic Journal* 28, no. 2: 221–23.

Malte-Brun, Conrad. 1808. "Carte topographique et militaire de l'Allemagne, en deux cent quatrefeuilles, publiée par l'Institut géographique de Weimar." *Annales des voyages, de la géographie et de l'histoire* 5: 264–68.

———. 1810–29. *Précis de la géographie universelle, ou description de toutes les parties du monde, sur un plan nouveau d'après les grandes divisions naturelles du globe*. 8 vols. Paris: Fr. Buisson and Aimé-André.

———. 1824–29. *Universal Geography, or a Description of All Parts of the World, on a New Plan, According to the Great Natural Divisions of the Globe*. 7 vols. Boston: Wells and Lilly.

———. 1827–32. *Universal Geography: or a Description of All the Parts of the World on a New Plan, According to the Great Natural Divisions of the Globe*. 6 vols. Philadelphia: A. Finley, J. Laval & S. F. Bradford.

Mapp, Paul W. 2011. *The Elusive West and the Contest for Empire, 1713–1763*. Chapel Hill: University of North Carolina Press.

Markham, Beryl. 1942. *West with the Night*. Boston: Houghton Mifflin.

Marschner, F. J. 1944. "Structural Properties of Medium- and Small-Scale Maps." *Annals of the Association of American Geographers* 34, no. 1: 1–46.

Mastronunzio, Marco, and Elena Dai Prà. 2016. "Editing Historical Maps: Comparative Cartography Using Maps as Tools." *e-Perimetron* 11, no. 4: 183–95. *www.e-perimetron.org*.

Mayhew, Henry. 1852. "'In the clouds'; or Some Account of a Balloon Trip with Mr. Green." *Illustrated London News* (18 September): 224.

Mayhew, Robert J. 2003. "'Geography is the eye of history': The Theory and Practice of a Commonplace, 1500–1950." In William Camden, *Britannia (1610)*, v–xi. Bristol: Thoemmes Press.

McClintock, Anne. 1995. *Imperial Leather: Race, Gender and Sexuality in the Colonial Contest*. London: Routledge.

McCorkle, Barbara B. 2001. *New England in Early Printed Maps, 1513 to 1800: An Illustrated Carto-Bibliography*. Providence, RI: John Carter Brown Library.

McDermott, Paul D. 1975. "What Is a Map?" In *Map Librarianship: Readings*, edited by Roman Drazniowsky, 88–97. Metuchen, NJ: Scarecrow Press.

McHaffie, Patrick, Sona Karentz Andrews, and Michael Dobson. 1990. "Ethical Problems in Cartography: A Roundtable Commentary." *Cartographic Perspectives* 7: 3–13.

McKenzie, D. F. 1999. *Bibliography and the Sociology of Texts*. 2nd ed. Cambridge: Cambridge University Press.

McManis, Douglas R. 1972. *European Impressions of the New England Coast, 1497–1620*. Uni-

versity of Chicago, Department of Geography Research Paper, 139. Chicago: University of Chicago Press.

McMaster, Robert B., and Susanna A. McMaster. 2015. "Academic Cartography in Canada and the United States." In Monmonier, ed. (2015, 1–6).

Meece, Stephanie. 2006. "A Bird's Eye View—of a Leopard's Spots: The Çatalhöyük 'Map' and the Development of Cartographic Representation in Prehistory." *Anatolian Studies* 56: 1–16.

Mellaart, James. 1964. "Excavations at Çatal Hüyük, 1963: Third Preliminary Report." *Anatolian Studies* 14: 39–119.

Melton, James Van Horn. 2001. *The Rise of the Public in Enlightenment Europe.* Cambridge: Cambridge University Press.

Merrill, Samuel. 2013. "The London Underground Diagram: Between Palimpsest and Canon." *London Journal* 38, no. 3: 245–64.

Meynen, Emil, ed. 1973. *Multilingual Dictionary of Technical Terms in Cartography / Dictionnaire multilingue de termes techniques cartographiques.* Wiesbaden: Franz Steiner Verlag for the International Cartographic Association, Commission II.

Michelson, Bruce. 1995. *Mark Twain on the Loose: A Comic Writer and the American Self.* Amherst: University of Massachusetts Press.

Mikhailov, Nikolai. 1949. *Across the Map of the U.S.S.R.* Moscow: Foreign Languages Publishing House.

Millard, A. R. 1987. "Cartography in the Ancient Near East." In Harley and Woodward, eds. (1987, 107–16).

Miller, B. A., and R. J. Schaetzl. 2014. "The Historical Role of Base Maps in Soil Geography." *Geoderma* 230: 329–39.

Mills, Sara. 2004. *Discourse.* 2nd ed. London: Routledge.

Mitchell, Peta. 2008. *Cartographic Strategies of Postmodernity: The Figure of the Map in Contemporary Theory and Fiction.* London: Routledge.

Mitchell, Rose, and Andrew Janes. 2014. *Maps, Their Untold Stories: Map Treasures from The National Archives.* London: Bloomsbury for the National Archives.

Mitchell, W. J. T. 1994. "Imperial Landscape." In *Landscape and Power*, edited by W. J. T. Mitchell, 5–34. Chicago: University of Chicago Press.

Moellering, Harold. 1984. "Real Maps, Virtual Maps and Interactive Cartography." In *Spatial Statistics and Models*, edited by Gary L. Gaile and Cort J. Willmot, 109–32. Boston: D. Reidel Publishing.

———. 2012. "The International Cartographic Association Research Agenda: Review, Perspectives, Comments and Recommendations." *Cartography and Geographic Information Science* 39, no. 1: 61–68.

Monmonier, Mark. 1985. *Technological Transition in Cartography.* Madison: University of Wisconsin Press.

———. 1991. *How to Lie with Maps.* Chicago: University of Chicago Press.

———. 2013. "History, Jargon, Privacy and Multiple Vulnerabilities." *Cartographic Journal* 50, no. 2: 171–74.

———, ed. 2015. *Cartography in the Twentieth Century.* Vol. 6 of *The History of Cartography.* Chicago: University of Chicago Press.

———. 2017. *Patents and Cartographic Inventions: A New Perspective for Map History.* Cham, Switzerland: Palgrave Macmillan.

Monmonier, Mark, Adrienne Lee Atterberry, Kayla Fermin, Gabrielle E. Marzolf, and Madeleine Hamlin. 2018. *A Directory of Cartographic Inventors.* Syracuse, NY: Bar Scale Press.

Monsaingnon, Guillaume. 2017. "Réincarnations cartographiques." In Besse and Tiber-ghien, eds. (2017, 118–31).

Montello, Daniel R. 1993. "Scale and Multiple Psychologies of Space." In *Spatial Informa-tion Theory: A Theoretical Basis for GIS. European Conference, COSIT '93, Marciana Marina, Elba Island, Italy, September 19–22, 1993, Proceedings*, edited by Andre U Frank and Irene Campari, 312–21. Berlin: Springer.

Morrison, Joel L. 1976. "The Science of Cartography and Its Essential Processes." *Interna-tionales Jahrbuch für Kartographie* 16: 84–97.

———. 2015. "Coordinate Systems." In Monmonier, ed. (2015, 278–84).

Morrison, Joel L., and Michael Wintle. 2019. "Projections Used for Geographical Maps." In Edney and Pedley, eds. (2019, forthcoming).

Morrison, Philip, Phylis Morrison, and the Office of Charles and Ray Eames. 1982. *Powers of Ten: About the Relative Size of Things in the Universe. Scientific American Library*. San Francisco: W. H. Freeman.

Morse, Jedidiah. 1793. *The American Universal Geography, or, A View of the Present State of All the Empires, Kingdoms, States, and Republics in the Known World, and of the United States of America in Particular*. 2 vols. Boston.

Moser, Molly. 2017. "Carson City Cartograph Housed at Bancroft Library." *Nevada Appeal*, 28 January. http://www.nevadaappeal.com/news/local/carson-city-cartograph-housed -at-bancroft-library/.

Muehrcke, Phillip C. 1981. "Maps in Geography." *Cartographica* 18, no. 2: 1–41.

Muehrcke, Phillip C., and Juliana O. Muehrcke. 1974. "Maps in Literature." *Geographical Review* 64, no. 3: 317–38.

Munroe, Randall. 2016. "Map Age Guide." *xkcd* (1 June). xkcd.com/1688/.

———. 2017. "Most-Used Word in Each State, Based on Something Something Search Data." *xkcd* (2 June). xkcd.com/1845/.

Murphy, David Thomas. 1997. *The Heroic Earth: Geopolitical Thought in Weimar Germany, 1918–1933*. Kent, OH: Kent State University Press.

Murphy, J. 1979. "Measures of Map Accuracy Assessment and Some Early Ulster Maps." *Irish Geography* 11: 88–101.

Nadal, Francesc, and Luis Urteaga. 1990. *Cartography and State: National Topographic Maps and Territorial Statistics in the Nineteenth Century*. Geo critica: Cuadernos criticos de geografia humana, 88, English parallel series 2. Barcelona: Catedra de geografia humana, Facultad de geografia e historia, Universitat de Barcelona.

Neeley, Kathryn A. 2001. *Mary Somerville: Science, Illumination, and the Female Mind*. Cam-bridge: Cambridge University Press.

Neumann, Jan. 1994. "The Topological Information Content of a Map: An Attempt at a Rehabilitation of Information Theory in Cartography." *Cartographica* 31, no. 1: 26–34.

Nicolai, Roel. 2015. "The Premedieval Origin of Portolan Charts: New Geodetic Evidence." *Isis* 106, no. 3: 517–43.

Nicolar, Joseph [pseud. "Young Sebbatis"]. 2004 [ca. 1887]. "The Scribe of the Penobscots Sends Us His Weekly Message: Some of the Names That the Indian Has Bestowed— Quaint and Old—Our Indian Correspondent Continues the Legends of His Race." In *Dawnland Voices: An Anthology of Indigenous Writing from New England*, edited by Siobhan Senier, 204–6. Lincoln: University of Nebraska Press, 2014. Originally pub-lished in *Old Town Herald* (ca. 1887).

Nigg, Joe. 2013. *Sea Monsters: A Voyage around the World's Most Beguiling Marine Map*. Chi-cago: University of Chicago Press.

Nikolow, Sybilla. 2001. "A. F. W. Crome's Measurements of the 'Strength of the State': Statistical Representations in Central Europe around 1800." *History of Political Economy* 33 supplement: 23–56.

Nordenskiöld, Adolf Erik. 1889. *Facsimile-Atlas to the Early History of Cartography with Reproductions of the Most Important Maps Printed in the XV and XVI Centuries.* Translated by Johan Adolf Ekelöf and Clements R. Markham. Stockholm.

Oettermann, Stephan. 1997. *The Panorama: History of a Mass Medium.* Translated by Deborah Lucas Schneider. New York: Zone Books.

Ogilby, John. 1675. *Britannia, Volume the First, or, an Illustration of the Kingdom of England and Dominion of Wales by a Geographical and Historical Description of the Principal Roads Thereof, Actually Admeasured and Delineated in a Century of Whole-Sheet Copper-Sculps.* London: John Ogilby.

O'Gorman, Francis. 2003. "'To see the finger of God in the dimensions of the pyramid': A New Context for Ruskin's 'The Ethics of the Dust' (1866)." *Modern Language Review* 98, no. 3: 563–73.

Oleksijczuk, Denise Blake. 2011. *The First Panoramas: Visions of British Imperialism.* Minneapolis: University of Minnesota Press.

Oliver, Richard. 2014. *The Ordnance Survey in the Nineteenth Century: Maps, Money and the Growth of Government.* London: Charles Close Society.

Olson, David R. 1994. *The World on Paper: The Conceptual and Cognitive Implications of Writing and Reading.* Cambridge: Cambridge University Press.

Olsson, Gunnar. 2007. *Abysmal: A Critique of Cartographic Reason.* Chicago: University of Chicago Press.

Openshaw, Stan. 1991. "A View on the GIS Crisis in Geography, or, Using GIS to Put Humpty-Dumpty Back Together Again." *Environment and Planning A* 23, no. 5: 621–28.

Ormeling, Ferjan. 2007. "The Development of Cartography Manuals in Western Europe: Henri Zondervan." In *Ormeling's Cartography: Presented to Ferjan Ormeling on the Occasion of His 65th Birthday and His Retirement as Professor of Cartography,* edited by Elger Heere and Martijn Storms, 183–89. Utrecht: Faculteit Geowetenschappen Universiteit Utrecht for the Koninklijk Nederlands Aardrijkskundig Genootschap.

———. 2015. "Academic Cartography in Europe." In Monmonier, ed. (2015, 6–13).

Ortiz-Ospina, Esteban. 2018. "Economists Are Unfairly Maligned—But They Are Often Pretty Prejudiced Themselves." *The Conversation* (14 March). http://theconversation.com/economists-are-unfairly-maligned-but-they-are-often-pretty-prejudiced-themselves-91456.

Ozouf-Marignier, Marie-Vic. 1992. *La formation des départements: La représentation du territoire français à la fin du 18e siècle.* Paris: Éditions de l'École des hautes études en sciences sociales.

Pacho, Jean Raymond. 1827. *Relation d'un voyage dans la Marmarique, la Cyrénaïque, et les oasis d'Audjelah et de Maradèh, accompagnée de cartes géographiques et topographiques.* Paris: Firmin Didot Père & Fils.

Packe, Christopher. 1737. *A Dissertation upon the Surface of the Earth, As Delineated in a Specimen of a Philosophico-Chorographical Chart of East-Kent, Herewith Humbly Presented to, and Read before the Royal Society, Nov. 25. 1736. In a Letter to Cromwel Mortimer, M.D. F.R.S. Secr.* London: J. Roberts.

———. 1743. *Αγκογραφια, sive Convallium Descriptio. In Which Are Briefly But Fully Expounded the Origine, Course and Insertion; Extent, Elevation and Congruity of All the Valleys and Hills, Brooks and Rivers, (as an Explanation of a New Philosophico-Chorographical Chart) of East-Kent.* Canterbury, Kent: J. Abree for the Author.

Padrón, Ricardo. 2007. "Mapping Imaginary Worlds." In *Maps: Finding Our Place in the World*, edited by James R. Akerman and Robert W. Karrow, Jr., 255–87. Chicago: University of Chicago Press.

Palsky, Gilles. 1999. "Borges, Carrol et la carte au 1/1." *Cybergeo: European Journal of Geography*. Cartographie, Imagerie, SIG, doc. 106. http://cybergeo.revues.org/5233.

———. 2002. "Emmanuel de Martonne and the Ethnographical Cartography of Central Europe (1917–1920)." *Imago Mundi* 54: 111–19.

Pápay, Gyula. 2017. "Max Eckert and the Foundations of Modern Cartographic Praxis." In *The Routledge Handbook of Mapping and Cartography*, edited by Alexander J. Kent and Peter Vujakovic, 9–28. London: Routledge.

Pearce, Margaret Wickens. 1998. "Native Mapping in Southern New England Indian Deeds." In *Cartographic Encounters: Perspectives on Native American Mapmaking and Map Use*, edited by G. Malcolm Lewis, 157–86. Chicago: University of Chicago Press.

Pearson, Alastair, and Michael Heffernan. 2015. "Globalizing Cartography? The International Map of the World, the International Geographical Union, and the United Nations." *Imago Mundi* 67, no. 1: 58–80.

Pearson, Alistair, D. R. F. Taylor, K. D. Kline, and Mike Heffernan. 2006. "Cartographic Ideals and Geopolitical Realities: International Maps of the World from the 1890s to the Present." *Canadian Geographer* 50, no. 2: 149–76.

Peary, Robert Edwin. 1941 [1919]. *Flat Globe of the Earth*. New York: Wm. H. Wise. 1st ed., 1919.

Pedersen, Kurt Møller, and Peter de Clercq. 2010. "Thomas Bugge, 1740–1815." In Thomas Bugge, *An Observer of Observatories: The Journal of Thomas Bugge's Tour of Germany, Holland and England in 1777*, ix–xix. Århus: Aarhus Universitetsforlag.

Pedley, Mary Sponberg. 1992. *Bel et Utile: The Work of the Robert de Vaugondy Family of Mapmakers*. Tring, Hertfordshire: Map Collector Publications.

Peirce, Charles Saunders. 1931–58. *Collected Papers*. Edited by Charles Hartshorne and Paul Weiss. 8 vols. Cambridge, MA: Harvard University Press.

Pelletier, Monique. 2013. *Les cartes de Cassini: La science au service de l'état et des provinces*. Paris: Éditions du CTHS.

———. 2019. "*Carte de France*." In Edney and Pedley, eds. (2019, forthcoming).

Petchenik, Barbara Bartz. 1977. "Cognition in Cartography." In *The Nature of Cartographic Communication*, edited by Leonard Guelke, 117–28. Cartographica Monograph, 19. Toronto: B. V. Gutsell.

———. 1979. "From Place to Space: The Psychological Achievement of Thematic Mapping." *American Cartographer* 6, no. 1: 5–12.

Peters, Arno. 1983. *Die neue Kartographie / The New Cartography*. New York: Friendship Press.

Philip, George. 1895–96. "The Enlargement of the Geographical Horizon, as Illustrated in the History of Cartography, down to the End of the Age of Discovery." *Proceedings of the Literary and Philosophical Society of Liverpool* 50: 313–39.

Pickles, John. 1992. "Texts, Hermeneutics and Propaganda Maps." In *Writing Worlds: Discourse, Text and Metaphor in the Representation of Landscape*, edited by Trevor J. Barnes and James S. Duncan, 193–230. London: Routledge.

———. 2004. *A History of Spaces: Cartographic Reason, Mapping and the Geo-Coded World*. London: Routledge.

Pietkiewicz, Stanisław. 1968. "The Evolution of the Map Definition during the Last Hundred Years." In *Actes du XIe congrès international d'histoire des sciences, Varsovie-Cracovie, 24–31 août 1965*, edited by Bogdan Suchodolski, 272–75. Warsaw.

Piper, Karen. 2002. *Cartographic Fictions: Maps, Race, and Identity*. New Brunswick, NJ: Rutgers University Press.

Poeppel, David. 2012. "The Maps Problem and the Mapping Problem: Two Challenges for a Cognitive Neuroscience of Speech and Language." *Cognitive Neuropsychology* 29, nos. 1–2: 34–55.

Pogo, Alexander. 1935. "Gemma Frisius, His Method of Determining Differences of Longitude by Transporting Timepieces (1530), and His Treatise on Triangulation (1533)." *Isis* 22, no. 2: 469–506.

Porter, Roy. 1980. "The Terraqueous Globe." In *The Ferment of Knowledge: Studies in the Historiography of Eighteenth-Century Science*, edited by G. S. Rousseau and Roy Porter, 285–324. Cambridge: Cambridge University Press.

Porter, Theodore M. 1986. *The Rise of Statistical Thinking, 1820–1900*. Princeton, NJ: Princeton University Press.

Portlock, J. E. 1855. "Continuation of Memoir of the Late Major-General Colby, Royal Engineers, L.L.D., F.R.S.L. and E., M.R.I.A., F.G.S., &c., &c., with a Sketch of the Origin and Progress of the British Trigonometrical Survey." *Papers on Subjects Connected with the Duties of the Corps of Royal Engineers*, new ser. 4: i–xxxii.

———. 1858. *Papers on Geometrical Drawing, on the Arms in Use and Permanent Fortification, on the Attack and Defence of Fortresses, on Military Mining, and on the Defence of Coasts.* London: George Edward Eyre & William Spottiswoode.

Potter, Jefferson R. 1891. "History of Methods of Instruction in Geography." *Pedagogical Seminary* 1, no. 3: 415–24.

Potter, Simon R. 2001. "The Elusive Concept of 'Map': Semantic Insights into the Cartographic Heritage of Japan." *Geographical Review of Japan*, ser. B 74, no. 1: 1–14.

Pratt, Mary Louise. 1992. *Imperial Eyes: Travel Writing and Transculturation*. London: Routledge.

Proclus. 1970. *A Commentary on the First Book of Euclid's Elements*. Translated by Glenn R. Morrow. Princeton, NJ: Princeton University Press.

Propen, Amy D. 2009. "Cartographic Representation and the Construction of Lived Worlds: Understanding Cartographic Practice as Embodied Knowledge." In Dodge, Kitchin, and Perkins, eds. (2009b, 113–30).

Ptolemy, Claudius. 1991. *The Geography*. Translated by Edward Luther Stevenson. New York: Dover. Originally published 1932 (New York: New York Public Library).

———. 2000. *Ptolemy's Geography: An Annotated Translation of the Theoretical Chapters*. Translated and edited by J. Lennart Berggren and Alexander Jones. Princeton, NJ: Princeton University Press.

Quam, Louis O. 1943. "The Use of Maps in Propaganda." *Journal of Geography* 42: 21–32.

Raisz, Erwin. 1938. *General Cartography*. 1st ed. New York: McGraw-Hill.

Rajanikant (pseud. Pant, Rajani Kant). 1974. "The Origin of Geometry in Ancient India." *Jijnasa* 1: 51–61.

Rakosi, Carl. 1931. "Before You." *Poetry* 37, no. 5: 240–41.

Ramaswamy, Sumathi. 2004. *The Lost Land of Lemuria: Fabulous Geographies, Catastrophic Histories*. Berkeley: University of California Press.

———. 2010. *The Goddess and the Nation: Mapping Mother India*. Durham, NC: Duke University Press.

———. 2017. *Terrestrial Lessons: The Conquest of the World as Globe*. Chicago: University of Chicago Press.

Rankin, William. 2014. "The Geography of Radionavigation and the Politics of Intangible Artifacts." *Technology and Culture* 55, no. 3: 622–74.

————. 2016. *After the Map: Cartography, Navigation, and the Transformation of Territory in the Twentieth Century*. Chicago: University of Chicago Press.

————. 2017. "Zombie Projects, Negative Networks, and Multigenerational Science: The Temporality of the International Map of the World." *Social Studies of Science* 47, no. 3: 353–75.

Ravenhill, William. 1976. "'As to its position in respect to the Heavens.'" *Imago Mundi* 28: 79–93.

————. 1983. "Christopher Saxton's Surveying: An Enigma." In *English Map-Making, 1500–1650: Historical Essays*, edited by Sarah Tyacke, 112–19. London: British Library.

Ravenstein, Ernest George, Charles F. Close, and Alexander Ross Clarke. 1911. "Map." In *Encyclopaedia Britannica*, 17: 629–63. 11th ed. Cambridge: Cambridge University Press.

Rees, Ronald. 1980. "Historical Links between Cartography and Art." *Geographical Review* 70, no. 1: 60–78.

Reeves, Edward A. 1910. *Maps and Map-Making: Three Lectures Delivered under the Auspices of the Royal Geographical Society*. London: Royal Geographical Society.

Reitinger, Franz. 1999. "Mapping Relationships: Allegory, Gender and the Cartographical Image in Eighteenth-Century France and England." *Imago Mundi* 51: 106–29.

————. 2008. *Kleiner Atlas Amerikanischer Überempfindlichkeiten*. Klagenfurt: Ritter.

Relaño, Francesc. 2001. *The Shaping of Africa: Cosmographic Discourse and Cartographic Science in Late Medieval and Early Modern Europe*. Burlington, VT: Ashgate.

Rice, Matthew T. 2015. "Copyright Traps." In Monmonier, ed. (2015, 284–85).

Richards, Penny L. 2004. "'Could I but mark out my own map of life': Educated Women Embracing Cartography in the Nineteenth-Century Antebellum South." *Cartographica* 39, no. 3: 1–17.

Robert de Vaugondy, Didier. 1755. *Essai sur l'histoire de la géographie, ou sur son origine, ses progrès et son état actuel*. Paris: Antoine Boudet.

Roberts, Maxwell J. 2005. *Underground Maps after Beck: The Story of the London Underground Map in the Hands of Henry Beck's Successors*. Harrow, Middlesex: Capital Transport.

Robinson, Arthur H. 1960. *Elements of Cartography*. 2nd ed. New York: John Wiley & Sons.

————. 1965. "The Potential Contribution of Cartography in Liberal Education." In *Geography in Undergraduate Liberal Education: A Report of the Geography in Liberal Education Project*, 34–47. Washington, DC: Association of American Geographers.

————. 1971. "The Genealogy of the Isopleth." *Cartographic Journal* 8, no. 1: 49–53.

————. 1979. "The Image and the Map: The Cartographic Problem." In *Congress Proceedings: 22nd International Geographical Congress / Actes du Congrès: 22e Congrès international de géographie, Montreal, August 10–17, 1972*, edited by J. Keith Fraser, 50–61. Ottawa: Canadian Committee for Geography.

————. 1982. *Early Thematic Mapping in the History of Cartography*. Chicago: University of Chicago Press.

Robinson, Arthur H., Joel L. Morrison, Phillip C. Muehrcke, A. Jon Kimerling, and Stephen C. Guptill. 1995. *Elements of Cartography*. 6th ed. New York: John Wiley & Sons.

Robinson, Arthur H., and Barbara Bartz Petchenik. 1976. *The Nature of Maps: Essays toward Understanding Maps and Mapping*. Chicago: University of Chicago Press.

Robinson, Arthur H., and Helen M. Wallis. 1967. "Humboldt's Map of Isothermal Lines: A Milestone in Thematic Cartography." *Cartographic Journal* 4, no. 2: 119–23.

Rochberg, Francesca. 2012. "The Expression of Terrestrial and Celestial Order in Ancient Mesopotamia." In *Ancient Perspectives: Maps and Their Place in Mesopotamia, Egypt, Greece, and Rome*, edited by Richard J. A. Talbert, 9–46. Chicago: University of Chicago Press.

Ross, Sydney. 1962. "Scientist: The Story of a Word." *Annals of Science* 18, no. 2: 65–85.

Rossetto, Tania. 2014. "Theorizing Maps with Literature." *Progress in Human Geography* 38, no. 4: 513–30.

———. 2015. "Semantic Ruminations on 'Post-Representational Cartography.'" *International Journal of Cartography* 2, no. 1: 151–67.

Rossi, Massimo. 2007. *L'officina della Kriegskarte: Anton von Zach e le cartografie delle stati veneti, 1796–1805*. Pieve di Soligno (Treviso): Edizioni Fondazione Benetton / Grafiche V. Bernardi.

Roszak, Theodore. 1972. *Where the Wasteland Ends: Politics and Transcendence in Postindustrial Society*. Garden City, NY: Doubleday.

Rouse, Joseph. 1987. *Knowledge and Power: Toward a Political Philosophy of Science*. Ithaca, NY: Cornell University Press.

Rousseau, Jean-Jacques. 1762. *Émile, ou de l'education*. Paris.

Royce, Josiah. 1899–1901. "Supplementary Essay: The One, the Many, and the Infinite." In *The World and the Individual*, 1: 471–588. London: Macmillan.

Ruge, Sophus. 1883. "Map." In *The Encyclopaedia Britannica*, 15: 515–23. 9th ed. New York: Charles Scribner's Sons.

Rundstrom, Robert A. 1989. "A Critical Appraisal of 'Applied' Cartography." In *Applied Geography: Issues, Questions, and Concerns*, edited by Martin S. Kenzer, 175–91. Dordrecht: Kluwer Academic.

———. 1991. "Mapping, Postmodernism, Indigenous People and the Changing Direction of North American Cartography." *Cartographica* 28, no. 2: 1–12.

Ryan, James R. 1998. *Picturing Empire: Photography and the Visualization of the British Empire*. Chicago: University of Chicago Press.

Ryden, Kent C. 2001. *Landscape with Figures: Nature and Culture in New England*. Iowa City: University of Iowa Press.

Sahlins, Peter. 1989. *Boundaries: The Making of France and Spain in the Pyrenees*. Berkeley: University of California Press.

Saint-Exupéry, Antoine de. 1931. *Vol de nuit*. Paris.

Sandler, Christian. 1905. *Die Reformation der Kartographie um 1700*. Munich: R. Oldenbourg.

Sandman, Alison. 2007. "Spanish Nautical Cartography in the Renaissance." In Woodward, ed. (2007b, 1095–142).

———. 2008. "Controlling Knowledge: Navigation, Cartography, and Secrecy in the Early Modern Spanish Atlantic." In *Science and Empire in the Atlantic World*, edited by James Delbourgo and Nicholas Dew, 31–51. New York: Routledge.

———. 2019. "Longitude and Latitude." In Edney and Pedley, eds. (2019, forthcoming).

Santarém, Manuel Francisco de Barros e Sousa, Visconde de. 1906. *Algumas cartas ineditas do Visconde de Santarem*. Edited by Vicente de Almeida de Eça. Lisbon.

Sartre, Jean-Paul. 1964. *Nausea*. Translated by Lloyd Alexander. New York: New Directions.

Schaffer, Simon. 1997. "Metrology, Metrication, and Victorian Values." In *Victorian Science in Context*, edited by Bernard Lightman, 438–74. Chicago: University of Chicago Press.

Scharfe, Wolfgang. 1972. *Abriss der Kartographie Brandenburgs, 1771–1821*. Berlin: Walter de Gruyter.

———. 1986. "Max Eckert's *Kartenwissenschaft*: The Turning Point in German Cartography." *Imago Mundi* 38: 61–66.

———. 1997. "Approaches to the History of Cartography in German-Speaking Countries." In *La cartografia dels països de parla Alemanya: Alemanya, Àustria i Suïssa*, edited by Wolfgang Scharfe, Ingrid Kretschmer, and Hans-Uli Feldmann, 19–41. Vol. 6 of *Cicle de conferències sobre Història de la Cartografia*. Barcelona: Institut Cartogràfic de Catalunya.

Scherer, Heinrich, S.J. 1710. *Critica quadripartita, in qua plura recens inventa, et emendata circa geographiae artificium, historiam, technicam, et astrologiam scitu dignissima explicantur.* Munich: Johann Kaspar Bencard.

Schilder, Günter. 2017. *Early Dutch Maritime Cartography: The North Netherland School of Cartography (c. 1580–c. 1620).* Leiden: HES & De Graaf.

Schilder, Günter, and Marco van Egmond. 2007. "Maritime Cartography in the Low Countries during the Renaissance." In Woodward, ed. (2007, 1384–432).

Schlögel, Karl. 2016. *In Space We Read Time: On the History of Civilization and Geopolitics.* Translated by Gerrit Jackson. New York: Bard Graduate Center.

Schmitt, Axel K., Martin Danišík, Erkan Aydar, et al. 2014. "Identifying the Volcanic Eruption Depicted in a Neolithic Painting at Çatalhöyük, Central Anatolia, Turkey." *PLoS ONE* 9, no. 1: e84711.

Schuiten, François, and Benoît Peeters. 2002. *Cities of the Fantastic: The Invisible Frontier.* Translated by Joe Johnson. 2 vols. New York: Nantier, Beall, Minoustchine.

Schulten, Susan. 1998. "Richard Edes Harrison and the Challenge to American Cartography." *Imago Mundi* 50 (1998): 174–88.

———. 2007. "Emma Willard and the Graphic Foundations of American History." *Journal of Historical Geography* 33, no. 3: 542–64.

———. 2012. *Mapping the Nation: History and Cartography in Nineteenth Century America.* Chicago: University of Chicago Press, 2012.

———. 2017. "Map Drawing, Graphic Literacy, and Pedagogy in the Early Republic." *History of Education Quarterly* 57, no. 2 (2017): 185–220.

Schwartz, Joan M., and James R. Ryan, eds. 2003. *Picturing Place: Photography and the Geographical Imagination.* London: I. B. Tauris.

Schwartz, Seymour I., and Ralph E. Ehrenberg. 1980. *The Mapping of America.* New York: Harry N. Abrams.

Seed, Patricia. 2014. *The Oxford Map Companion: One Hundred Sources in World History.* New York: Oxford University Press.

———. 2015. "Unpasted: A Guide to Surviving Prints of Mercator's Nautical Chart of 1569." In *A World of Innovations: Cartography in the Time of Gerhard Mercator,* edited by Gerhard Holzer et al., 146–57. Newcastle upon Tyne: Cambridge Scholars.

Self, Will. 2013. "Will Self Reads 'On Exactitude in Science' by Jorge Luis Borges." *The Guardian Books Podcast.* 4 January. https://www.theguardian.com/books/audio/2013/jan/04/will-self-jorge-luis-borges.

Shapin, Steven. 1996. *The Scientific Revolution.* Chicago: University of Chicago Press.

Sheynin, Oscar. 1994. "C. F. Gauss and Geodetic Observations." *Archive for History of Exact Sciences* 46, no. 3: 253–83.

———. 1999. "Gauss and the Method of the Least Squares." *Jahrbücher für Nationalökonomie und Statistik / Journal of Economics and Statistics* 219, nos. 3–4: 458–67.

———. 2001. "Gauss, Bessel and the Adjustment of Triangulation." *Historia Scientiarum: International Journal of the History of Science Society of Japan* 11: 168–75.

Shirley, Rodney W. 2001. *The Mapping of the World: Early Printed World Maps, 1472–1700.* 2nd ed. Riverside, CT: Early World Press.

Shweder, Richard A. 1985. "Has Piaget Been Upstaged? A Reply to Hallpike." *American Anthropologist,* new ser. 87, no. 1: 138–44.

Sieg, Wilfried, and Dirk Schlimm. 2005. "Dedekind's Analysis of Number: Systems and Axioms." *Synthese* 147, no. 1: 121–70.

Silbernagel, Janet. 1997. "Scale Perception: From Cartography to Ecology." *Bulletin of the Ecological Society of America* 78, no. 2: 166–69.

Silverberg, Joel S. 2015. "The Rise of 'the Mathematicals': Placing Maths into the Hands of Practitioners: The Invention and Popularization of Sectors and Scales." In *Research in History and Philosophy of Mathematics: The CSHPM 2014 Annual Meeting in St. Catharines, Ontario,* edited by Maria Zack, 23–50. Heidelberg: Springer International.

Silvestre, Marguerite. 2016. *Philippe Vandermaelen, Mercator de la jeune Belgique: Histoire de l'Établissement géographique de Bruxelles et de son fondateur.* Vol. 7 of *Inventaire raisonné des collections cartographiques Vandermaelen conservées à la Bibliothèque royale de Belgique.* Brussels: Bibliothèque royale de Belgique.

Singer, Thea. 2016. "Researchers Reveal Inconsistent Borders in Online Maps." news@ Northeastern, 19 May. http://www.northeastern.edu/news/2016/05/researchers-reveal -inconsistent-borders-in-online-maps/.

Sismondo, Sergio, and Nicholas Chrisman. 2001. "Deflationary Metaphysics and the Natures of Maps." *Philosophy of Science* 68, no. 3 supplement (Proceedings of the 2000 Biennial Meeting of the Philosophy of Science Association): S38–S49.

Sitwell, O. F. G. 1993. *Four Centuries of Special Geography: An Annotated Guide to Books That Purport to Describe All the Countries in the World Published in English before 1888, with a Critical Introduction.* Vancouver: University of British Columbia Press.

Sivin, Nathan, and Gari Ledyard. 1994. "Chinese Maps in Political Culture." In Harley and Woodward, eds. (1994, 23–31).

Skelton, R. A. 1958. "Cartography." In *The Industrial Revolution, c. 1750 to c. 1850,* edited by Charles Singer, 596–628. Vol. 4 of *The History of Technology.* Oxford: Oxford University Press.

———. 1965. *Looking at an Early Map. The Annual Public Lecture on Books and Bibliography, University of Kansas, October 1962. University of Kansas Publications, Library Series 17.* Lawrence: University of Kansas Press.

———. 1966. "Introduction." In Georg Braun and Frans Hogenberg, *Civitates Orbis Terrarum: "The Towns of the World," 1572–1618,* edited by R. A. Skelton and A. O. Vietor, 1: vii–xlvi. Cleveland: World Publishing.

———. 1972. *Maps: A Historical Survey of Their Study and Collecting.* Chicago: University of Chicago Press.

Skurnik, Johanna. 2017. *Making Geographies: The Circulation of British Geographical Knowledge of Australia, 1829–1863.* Annales Universitatis Turkensis, ser. B, 444. Turku: University of Turku.

Sloan, Kim. 2007. *A New World: England's First View of America.* Chapel Hill: University of North Carolina Press.

Slocum, Terry A., and Fritz C. Kessler. 2015. "Thematic Mapping." In Monmonier, ed. (2015, 1500–1524).

Smith, Edgar Crosby. 1902. "Bibliography of the Maps of Maine." In *Moses Greenleaf: Maine's First Map-Maker,* 139–65. Bangor, ME: For the De Burians.

Smith, James L. 2014. "Europe's Confused Transmutation: The Realignment of Moral Cartography in Juan de la Cosa's Mappa Mundi (1500)." *European Review of History* 21, no. 6: 799–816.

Smits, Jan. 1996. "Mathematical Data for Bibliographic Descriptions of Cartographic Materials and Spatial Data." Koninklijke Bibliotheek, 15 February 1996; updated 15 February 2013. http://archive.is/SzXdv#selection-9.0-25.17.

Snyder, John P. 1993. *Flattening the Earth: Two Thousand Years of Map Projections.* Chicago: University of Chicago Press.

Sobel, Dava. 1995. *Longitude: The True Story of a Lone Genius Who Solved the Greatest Scientific Problem of His Time.* New York: Walker & Co.

Sorum, Eve. 2009. "'The place on the map': Geography and Meter in Hardy's Elegies." *Modernism/Modernity* 16, no. 3: 543–64.

Speier, Hans. 1941. "Magic Geography." *Social Research* 8: 310–30.

Stavenhagen, W. 1904. *Skizze der Entwicklung und des Standes des Kartenwesens des ausserdeutschen Europa. Petermanns Mitteilungen*, Ergänzungsheft 148. Gotha: Justus Perthes.

Stevens, Henry, and Roland Tree. 1985 [1951]. "Comparative Cartography." In *The Mapping of America*, edited by R. V. Tooley, 41–107. London: Holland Press. Originally published in *Essays Honoring Lawrence C. Wroth* (Portland, ME: Anthoensen Press, 1951), 305–64.

Stevenson, Robert Louis. 1894. "My First Book: Treasure Island." *The Idler Magazine: An Illustrated Monthly* (August).

Steward, H. J. 1974. *Cartographic Generalisation: Some Concepts and Explanation.* Cartographica Monograph, 10. Toronto: Department of Geography, York University.

Stewart, John Q. 1943. "The Use and Abuse of Map Projections." *Geographical Review* 33, no. 4: 589–604.

Stigler, Stephen M. 1986. *The History of Statistics: The Measurement of Uncertainty before 1900.* Cambridge, MA: Harvard University Press.

Strachey, Henry. 1848. "Notes of the Construction of the Map of the British Himálayan Frontier in Kumaon and Gurhwál." *Journal of the Royal Asiatic Society of Bengal*, new ser. 19: 532–38.

Sullivan, Garrett A., Jr. 1998. *The Drama of Landscape: Land, Property, and Social Relations on the Early Modern Stage.* Stanford, CA: Stanford University Press.

Sullivan, Robert. 1883. *Geography Generalised: Or, an Introduction to the Study of Geography on the Principles of Classification and Comparison.* Edited by Samuel Haughton. 65th ed. Dublin: Sullivan Brothers.

Tagg, John. 1988. *The Burden of Representation: Essays on Photographies and Histories.* Minneapolis: University of Minnesota Press.

Taliaferro, Henry G. 2013. "Fry and Jefferson Revisited." *Journal of Early Southern Decorative Arts* 34. http://www.mesdajournal.org/2013/fry–jefferson–revisited/.

Taylor, Andrew. 2004. *The World of Gerard Mercator: The Mapmaker Who Revolutionized Cartography.* New York: Walker & Co.

Taylor, T. Griffith. 1928. *Australia in Its Physiographic and Economic Aspects.* Oxford: Clarendon Press. 1st ed., 1911.

Tebel, René. 2019. "Josephinische Landesaufnahme (Josephine Survey; Austrian Monarchy)." In Edney and Pedley, eds. (2019, forthcoming).

Tessicini, Dario. 2011. "Definitions of Cosmography and Geography in the Wake of Ptolemy's *Geography*." In *Ptolemy's Geography in the Renaissance*, edited by Zur Shalev and Charles Burnett, 31–50. London: Warburg Institute.

Thébaud-Sorger, Marie. 2013. "Thomas Baldwin's Airopaidia, or the Aerial View in Colour." In Dorrian and Pousin, eds. (2013, 46–65).

Thomas, Leah M. 2017. "Tactile Semiotics: Design in Eighteenth- and Nineteenth-Century Maps by, and for, the Blind." *IMCos Journal* 150: 21–32.

Thompson, Clive. 2017. "From Ptolemy to GPS, the Brief History of Maps." *Smithsonian* 48, no. 4: 16–22.

Thompson, Edward V. 2010. *Printed Maps of the District and State of Maine, 1793–1860: An Illustrated and Comparative Study.* Bangor, ME: Nimue Books & Prints.

Thompson, F. M. L. 1968. *Chartered Surveyors: The Growth of a Profession.* London: Routledge & Kegan Paul.

Thoreau, Henry D. 1848. "Ktaadn, and the Maine Woods, No. II. Life in the Maine Woods." *Union Magazine of Literature and Art* 3, no. 2: 73–79.

273

Thrift, Nigel. 2007. *Non-Representational Theory: Space, Politics, Affect.* London: Routledge.

Thrower, Norman J. W. 1972. *Maps and Man: An Examination of Cartography in Relation to Culture and Civilization.* Englewood Cliffs, NJ: Prentice-Hall.

———. 1991. "When Mapping Became a Science." *UNESCO Courier* (June 1991): 31–34.

———. 1996. *Maps and Civilization: Cartography in Culture and Society.* 2nd ed. Chicago: University of Chicago Press.

Tobler, Waldo R. 1959. "Automation and Cartography." *Geographical Review* 49, no. 4: 529–34.

———. 1979. "A Transformational View of Cartography." *American Cartographer* 6, no. 2: 101–6.

Tooley, R. V. 1963. *Geographical Oddities, or Curious, Ingenious, and Imaginary Maps and Miscellaneous Plates Published in Atlases.* Vol. 1 of *Map Collectors' Circle.* London: Map Collectors' Circle.

———. 1969. *Collectors' Guide to Maps of the African Continent and Southern Africa.* London: Carta Press.

Török, Zsolt. 2019. "Thematic Mapping in the Enlightenment." In Edney and Pedley, eds. (2019, forthcoming).

Toulmin, Stephen. 1953. *The Philosophy of Science: An Introduction.* New York: Harper & Row.

Travis, Trysh. 2000. "What We Talk about When We Talk about *The New Yorker*." *Book History* 3: 253–85.

Tuan, Yi-Fu. 1977. *Space and Place: The Perspective of Experience.* Minneapolis: University of Minnesota Press.

———. 1979. "Sight and Pictures." *Geographical Review* 69, no. 4: 413–22.

Tucker, Jennifer. 1996. "Voyages of Discovery on Oceans of Air: Scientific Observation and the Image of Science in an Age of 'Balloonacy.'" *Osiris* 11: 144–76.

Turnbull, David. 1993. *Maps Are Territories: Science Is an Atlas: A Portfolio of Exhibits.* Chicago: University of Chicago Press.

———. 1996. "Cartography and Science in Early Modern Europe: Mapping the Construction of Knowledge Spaces." *Imago Mundi* 48: 5–24.

Twain, Mark [pseud. Samuel L. Clemens]. 1870a. "Fortifications of Paris." *Buffalo Express* (17 September): 2.

———. 1870b. "Mark Twain's Map of Paris." *The Galaxy* 10, no. 5 (5 November): 724–25.

———. 1894. *Tom Sawyer Abroad.* New York: Charles L. Webster, 1894.

———. 1995. *1870–1871.* Edited by Victor Fischer, Michael B. Frank, and Lin Salamo. Vol. 4 of *Mark Twain's Letters.* Berkeley: University of California Press.

Tyner, Judith A. 2015a. "Persuasive Cartography." In Monmonier, ed. (2015, 1087–95).

———. 2015b. *Stitching the World: Embroidered Maps and Women's Geographical Education.* London: Ashgate.

———. 2017. "Persuasive Map Design." In *The Routledge Handbook of Mapping and Cartography*, edited by Alexander J. Kent and Peter Vujakovic, 439–49. London: Routledge.

United States Geological Survey. 2002. "USGS Fact Sheet 015–02: Map Scales." February. http://pubs.er.usgs.gov/publication/fs01502.

Unno, Kazutaka. 1994. "Cartography in Japan." In Harley and Woodward, eds. (1994, 346–477).

Valerio, Vladimiro. 2007. "Cartography, Art and Mimesis: The Imitation of Nature in Land Surveying in the Eighteenth and Nineteenth Centuries." In *Observing Nature—Representing Experience: The Osmotic Dynamics of Romanticism, 1800–1850*, edited by Erna Florentini, 57–71. Berlin: Dietrich Reimer.

Van der Heijden, H. A. M. 1998. "Heinrich Bünting's *Itinerarium Sacrae Scripturae*, 1581: A Chapter in the Geography of the Bible." *Quaerendo* 28, no. 1: 49–71.

Van der Krogt, Peter. 2006. "Kartografie of Cartografie?" *Caert-Thresoor* 25, no. 1: 11–12.

———. 2015. "The Origin of the Word 'Cartography.'" *e-Perimetron* 10, no. 3: 124–42. *www.e-perimetron.org*.

———. 2017. "Een kaart met een schaal van 1:53 miljard." Bijzondere Collecties, Allard Pierson Museum, UvA Erfgoed, University of Amsterdam. 5 June. http://www.blogs -uva-erfgoed.nl/een-kaart-met-een-schaal-van-153-miljard/. Accessed 16 July 2017.

Van Duzer, Chet, and Ilya Dines. 2016. *Apocalyptic Cartography: Thematic Maps and the End of the World in a Fifteenth-Century Manuscript*. Leiden: Brill and HES & De Graaf.

Van Sant, Tom. n.d. "The Earth from Space." Tom Van Sant. http://www.tomvansant.com /id4.html. Last accessed 21 September 2017.

Vasiliev, Irina, Scott Freundschuh, David M. Mark, G. D. Theisen, and J. McAvoy. 1990. "What Is a Map?" *Cartographic Journal* 27, no. 2: 119–23.

Verdier, Nicolas. 2015. *La carte avant les cartographes: L'avènement du régime cartographique en France au XVIIIe siècle*. Paris: Publications de la Sorbonne.

Veres, Madalina Valeria. 2015. "Constructing Imperial Spaces: Habsburg Cartography in the Age of Enlightenment." Ph.D. dissertation. University of Pittsburgh.

Verner, Coolie. 1985. "Smith's *Virginia* and Its Derivatives: A Carto-Bibliographical Study of the Diffusion of Geographical Knowledge." in *The Mapping of America*, edited by R. V. Tooley, 135–72. London: Holland Press.

Vertesi, Janet. 2008. "Mind the Gap: The London Underground Map and Users' Representations of Urban Space." *Social Studies of Science* 38, no. 1: 7–33.

Vervust, Soetkin. 2016a. "Count de Ferraris's Maps of the Austrian Netherlands (1770s): Cassini de Thury's Geodetic Contribution." *Imago Mundi* 69, no. 2: 164–82.

———. 2016b. "Deconstructing the Ferraris Maps (1770–1778): A Study of the Map Production Process and Its Implications for Geometric Accuracy." Ph.D. dissertation. Ghent University.

Visvalingam, M. 1989. "Cartography, GIS and Maps in Perspective." *Cartographic Journal* 26, no. 1: 26–32.

Vivan, Itala. 2000. "Geography, Literature, and the African Territory: Some Observations on the Western Map and the Representation of Territory in the South African Literary Imagination." *Research in African Literatures* 31, no. 2: 49–70.

Vries, Dirk de, Günter Schilder, W. F. J. Mörzer Bruyns, P. D. J. van Iterson, and I. Jacobs. 2005. *The Van Keulen Cartography, Amsterdam 1680–1885*. Alphen aan den Rijn, Netherlands: Caneletto/Repro-Holland, 2005.

Wagner, Henry R. 1932. "Biblio-Cartography." *Pacific Historical Review* 1, no. 1: 103–10.

Wagner, Hermann. 1914. "Der Kartenmaßstab: Historisch-kritische Betrachtungen." *Zeitschrift der Gesellschaft für Erdkunde zu Berlin* 1914, nos. 1–2: 1–34, 81–117.

Walckenaer, Charles-Athanase. 1835a. "Cartes géographiques (notices historiques)." In *Encyclopédie des gens du monde, répertoire universel des sciences, des lettres et des arts*, 5.1: 11–18. 22 vols. Paris: Treuttel & Würtz.

———. 1835b. "Découvertes (voyages de)." In *Encyclopédie des gens du monde, répertoire universel des sciences, des lettres et des arts*, 7.2: 637–41. 22 vols. Paris: Treuttel & Würtz.

Waldseemüller, Martin, Matthias Ringmann, Jacob Aeszler, and Georg Übelin. 1513. *Geographie opus nouissima traductione e Grecorum archtypis castigatissime*. Strasbourg: Johannes Schott.

Wallace, Timothy R. 2015. "Map Pin." In Monmonier, ed. (2015, 814–16).

———. 2016. "Cartographic Journalism: Situating Modern News Mapping in a History of Map-User Interaction." Ph.D. dissertation. University of Wisconsin–Madison.

Wallach, Alan. 2005. "Some Further Thoughts on the Panoramic Mode." In *Within the Landscape: Essays on Nineteenth-Century American Art and Culture*, edited by Phillip Earenfight and Nancy Siegal, 99–128. Carlisle, PA: Trout Gallery, Dickinson College.

Wallis, B. C. 1911. "Apparatus for Use in Teaching Geography." *School World: A Monthly Magazine of Educational Work and Progress* 13, no. 152: 288–91.

Wallis, Helen M. 1973. "The Map Collections of the British Museum Library." In *My Head Is a Map: Essays and Memoirs in Honour of R. V. Tooley*, edited by Helen Wallis and Sarah Tyacke, 3–20. London: Francis Edwards and Carta Press.

Wallis, Helen M., and Arthur H. Robinson, eds. 1987. *Cartographical Innovations: An International Handbook of Mapping Terms to 1900*. London: Map Collector Publications for the International Cartographic Association.

Wang, Jessica. 2017. "Reckoning with the Spatial Turn: Cartography, Territoriality, and International History." *Diplomatic History* 41, no. 5: 1010–18.

Warner, Michael. 1990. *The Letters of the Republic: Publication and the Public Sphere in Eighteenth-Century America*. Cambridge, MA: Harvard University Press.

Waselkov, Gregory A. 1998. "Indian Maps of the Colonial Southeast: Archaeological Implications and Prospects." In *Cartographic Encounters: Perspectives on Native American Mapmaking and Map Use*, edited by G. Malcolm Lewis, 205–21. Chicago: University of Chicago Press.

Watson, Ruth. 2008. "Cordiform Maps since the Sixteenth Century: The Legacy of Nineteenth-Century Classificatory Systems." *Imago Mundi* 60, no. 2: 182–94.

———. 2009. "Mapping and Contemporary Art." *Cartographic Journal* 46, no. 4: 293–307.

Weber, Eugen. 1976. *Peasants into Frenchmen: The Modernization of Rural France, 1870–1914*. Stanford, CA: Stanford University Press.

Weibel, Peter. 2014. "Media, Mapping and Painting." In Gehring and Weibel (2014, 440–59).

Wheat, James Clements, and Christian F. Brun. 1978. *Maps and Charts Published in America before 1800: A Bibliography*. Rev. ed. London: Holland Press.

Wheeler, George M. 1885. "Government Land and Marine Surveys (Origin, Organization, Administration, Functions, History, and Progress)." In Wheeler, *Report upon the Third International Geographical Congress and Exhibition at Venice, Italy, 1881*, 76–569. Washington, DC: Government Printing Office.

Wheeler, James O. 1998. "Mapphobia in Geography? 1980–1996." *Urban Geography* 19, no. 1: 1–15.

Whewell, William [attrib.]. 1834. "*On the Connexion of the Physical Sciences* by Mrs. Somerville." *Quarterly Review* 51: 54–68.

Whittington, Karl. 2014. *Body-Worlds: Opicinus de Canistris and the Medieval Cartographic Imagination*. Toronto: Pontifical Institute of Mediaeval Studies, 2014.

Wigen, Kären, Sugimoto Fumiko, and Cary Karacas, eds. 2016. *Cartographic Japan: A History in Maps*. Chicago: University of Chicago Press.

Williams, James L. 2015. *Blazes, Posts, and Stones: A History of Ohio's Original Land Subdivisions*. Columbus, OH: Compass & Chain Publishing.

Williams, Robert Lee. 1959. "Map Projections, Linear Scale, and the Representative Fraction." *Annals of the Association of American Geographers* 49, no. 1: 88.

Wilson, Holly L. 2011. "The Pragmatic Use of Kant's *Physical Geography* Lectures." In *Reading Kant's Geography*, edited by Stuart Elden and Eduardo Mendieta, 161–72. Albany: SUNY Press.

276

Winchester, Simon. 2001. *The Map That Changed the World: William Smith and the Birth of Modern Geology*. New York: HarperCollins.

Winichakul, Thongchai. 1994. *Siam Mapped: A History of the Geo-Body of a Nation*. Honolulu: University of Hawaii Press.

Winlow, Heather. 2006. "Mapping Moral Geographies: W. Z. Ripley's Races of Europe and the U.S." *Annals of the Association of American Geographers* 96, no. 1: 119–41.

———. 2009. "Mapping the Contours of Race: Griffith Taylor's Zones and Strata Theory." *Geographical Research* 47, no. 4: 390–407.

Withers, Charles W. J. 2001. *Geography, Science and National Identity: Scotland since 1520*. Cambridge: Cambridge University Press.

———. 2017. *Zero Degrees: Geographies of the Prime Meridian*. Cambridge, MA: Harvard University Press.

———. 2019a. "Histories of Cartography." In Edney and Pedley, eds. (2019, forthcoming).

———. 2019b. "Metaphor, Maps as." In Edney and Pedley, eds. (2019, forthcoming).

Wolkenhauer, Wilhelm. 1895. *Leitfaden zur Geschichte der Kartographie in tabellarischer Darstellung*. Breslau: Ferdinand Hirt, Königliche Universitäts- und Verlags-Buchhandlung.

Wood, Denis. 1973. "Humanization of Cartography." *Bulletin of the Special Libraries Association, Geography and Map Division* 91: 2–10.

———. 1977. "Now and Then: Comparisons of Ordinary Americans' Symbol Conventions with Those of Past Cartographers." *Prologue* 9: 151–61.

———. 1978. "Introducing the Cartography of Reality." In *Humanistic Geography: Prospects and Problems*, edited by David Ley and Marwyn S. Samuels, 207–19. Chicago: Maaroufa Press.

———. 1992a. "How Maps Work." *Cartographica* 29, nos. 3–4: 66–74.

———. 1992b. *The Power of Maps*. With contributions by John Fels. New York: Guilford Press.

———. 1993. "Maps and Mapmaking." *Cartographica* 30, no. 1: 1–9.

———. 2003. "Cartography Is Dead (Thank God!)." *Cartographic Perspectives* 45: 4–7.

———. 2010. *Rethinking the Power of Maps*. With contributions by John Fels and John B. Krygier. New York: Guilford Press.

———. 2012. Review of Jordana Dym and Karl Offen, eds., *Mapping Latin America: A Cartographic Reader* (Chicago: University of Chicago Press, 2011). *Cartographica* 47, no. 2: 136–38.

Wood, Denis, and John Fels. 1986. "Designs on Signs: Myth and Meaning in Maps." *Cartographica* 23, no. 3: 54–103. Reprinted as Wood with Fels (1992, 95–142).

———. 2008. *The Natures of Maps: Cartographic Constructions of Nature*. Chicago: University of Chicago Press.

Wood, Denis, and John B. Krygier, eds. 2006. Special issue on art and cartography. *Cartographic Perspectives* 53: 4–50, 61–82.

Woodbridge, William C. 1831. *Preparatory Lessons for Beginners or First Steps to Geography*. [New Haven, CT.]

———. 1838. *Rudiments of Geography, on a New Plan, Designed to Assist the Memory by Comparison and Classification*. 19th ed., "with preparatory lessons, a series of questions, &c." Hartford, CT: John Beach.

Woodward, David. 1967. "The Foster Woodcut Map Controversy: A Further Examination of the Evidence." *Imago Mundi* 21: 50–61.

———. 1987. "Medieval *Mappaemundi*." In Harley and Woodward, eds. (1987, 286–370).

———. 1992. "Representations of the World." In *Geography's Inner Worlds: Pervasive Themes in Contemporary American Geography*, edited by Ronald F. Abler, Melvin G. Marcus, and Judy M. Olson, 50–73. New Brunswick, NJ: Rutgers University Press.

———. 2001. "The 'Two Cultures' of Map History—Scientific and Humanistic Traditions: A Plea for Reintegration." In *Plantejaments i Objectius d'una Història Universal de la Cartografia / Approaches and Challenges in a Worldwide History of Cartography*, edited by David Woodward, Catherine Delano Smith, and Cordell D. K. Yee, 49–67. Vol. II of *Cicle de conferències sobre Història de la Cartografia*. Barcelona: Institut Cartogràfic de Catalunya.

———. 2007a. "Cartography and the Renaissance: Continuity and Change." In Woodward, ed. (2007b, 3–24).

———, ed. 2007b. *Cartography in the European Renaissance*. Vol. 3 of *The History of Cartography*. Chicago: University of Chicago Press.

———. 2007c. "Techniques of Map Engraving, Printing, and Coloring in the European Renaissance." In Woodward, ed. (2007b, 591–610).

Woodward, David, and G. Malcolm Lewis, eds. 1998. *Cartography in the Traditional African, American, Arctic, Australian, and Pacific Societies*. Vol. 2.3 of *The History of Cartography*. Chicago: University of Chicago Press.

Worcester, Joseph E. 1849. *A Universal and Critical Dictionary of the English Language*. Boston: Wilkins, Carter & Co.

Wright, J. K. 1942. "Map Makers Are Human." *Geographical Review* 32: 527–44.

——— [attrib.]. 1945. Review of Lawrence C. Wroth, "The Early Cartography of the Pacific," *Papers of the Bibliographical Society of America* 38, no. 2 (1944): 87–268. *Geographical Review* 35, no. 3: 505–6.

———. 1955. "'Crossbreeding' Geographical Quantities." *Geographical Review* 45, no. 1: 52–65.

Wright, J. K., George Kish, and R. A. Skelton. 1969. "Map: History." In *Encyclopaedia Britannica*, 14: 827–38. 14th ed. London: H. H. Benton.

Wu, Shellen. 2014. "The Search for Coal in the Age of Empires: Ferdinand von Richthofen's Odyssey in China, 1860–1920." *American Historical Review* 119, no. 2: 339–63.

Wyckoff, William. 2016. "Cartography and Capitalism: George Clason and the Mapping of Western American Development, 1903–1931." *Journal of Historical Geography* 52: 48–60.

Yee, Cordell D. K. 1994. "Reinterpreting Traditional Chinese Geographical Maps." In Harley and Woodward, eds. (1994, 35–70).

Zähringer, Raphael. 2017. "X Marks the Spot—Not: Pirate Treasure Maps in *Treasure Island* and *Käpt'n Sharky und das Geheimnis der Schatzinsel*." *Children's Literature in Education* 48, no. 1: 6–20.

Zelinsky, Wilbur. 1973. "The First and Last Frontier of Communication: The Map as Mystery." *Bulletin of the Geography and Map Division, Special Libraries Association* 94: 2–8.

Zentai, László. 2015. "Orienteering Map." In Monmonier, ed. (2015, 1043–46).

Zeune, August. 1811. *Goea: Versuch einer wissenschaftlichen Erdbeschreibung*. 2nd ed. Berlin: Julius Eduard Hitzig.

———. 1833. *Allgemeine naturgemässe Erdkunde mit Bezug auf Natur und Völkerleben*. Berlin.

Ziman, John. 1978. "World Maps and Pictures." In *Reliable Knowledge: An Exploration of the Grounds for Belief in Science*, 77–94. Cambridge: Cambridge University Press.

Ziter, Edward. 2003. *The Orient on the Victorian Stage*. Cambridge: Cambridge University Press.

Zupko, Ronald Edward. 1977. *British Weights and Measures: A History from Antiquity to the Seventeenth Century*. Madison: University of Wisconsin Press.

———. 1978. *French Weights and Measures before the Revolution: A Dictionary of Provincial and Local Units*. Bloomington: Indiana University Press.

———. 1985. *A Dictionary of Weights and Measures for the British Isles: The Middle Ages to the Twentieth Century.* Philadelphia: American Philosophical Society.

———. 1990. *Revolution in Measurement: Western European Weights and Measures since the Age of Science.* Philadelphia: American Philosophical Society.

Zynda, Lyle. 2004. "We're All Just Floating in Space." In *Finding Serenity: Anti-Heroes, Lost Shepherds and Space Hookers in Joss Whedon's 'Firefly,'* edited by Jane Espenson, 85–95. Dallas, TX: BenBella Books.

279

Index

Barney, Timothy: on "death" of cartography, 232;
    work by, also cited, 97, 153
Bartholomew, John G.: *Citizen's Atlas of the
    World*, 215; work by, also cited, 229
Bartholomew Archive, 148f
Basaraner, Melih, cited, 24
base maps and mapping, 83, 130, 145
baselines, 91, 197
Batavian Republic, systematic mapping of, 108
Bateson, Gregory, definition of map, 76
battles on maps, 89
Bavaria, systematic mapping of, 108
Baynton-Williams, Ashley, cited, 36, 231
beating the bounds, 38
Beck, Henry, *Railway Map, No. 1*, 218–19, 218f, 226
*Becken-Entwerfung und zwar die Erde von einem
    Luftball aus gesehen* (Zeune), 137f
Bedolina Map (Valcamonica, Italy), 71
Belgium, systematic mapping of, 109
Belgrade (Serbia), 21
Bendall, A. Sarah, cited, 199
Bennett, J. A., cited, 177
Bentham, Jeremy, panopticon, 135, 136f
Berlin (Germany), 214
Bern (Switzerland), 136f, 137f, 147
Bernal, Martin, cited, 129
Bernstein, David, cited, 39
Berthaut, Henri Marie Auguste, cited, 130, 209
Bertuch, Friedrich Justin, cited, 117
Besse, Jean-Marc, cited, 138, 161
Beyersdorff, Margot, cited, 219
"bibliography" (word), 116
Biblioteca nacional de España, 197f
Bibliothèque nationale de France, 35f, 114, 185f
Bibliothèque nationale et universitaire de Stras-
    bourg, 209f
bicycle maps, 159
bicycles, 156
Bierut, Michael, on London Underground map,
    219
Biesta, Gert, on how maps "work," 83
big data, 82n
Bigg, Charlotte, cited, 135
Biggs, Michael: identifies some cartographic
    preconceptions, 52; on inequalities inherent
    in mapping, 25; work by, also cited, 5, 77, 93
Bird's Eye View of India (Illustrated London
    News), 139f
Bismarck, Otto von, on Twain's map of Paris, 11
Black, Jeannette D., cited, 146
Black, Jeremy: on politicization of mapping, 26;
    work by, also cited, 2
Blackstone, William, map metaphor, 178
Blaeu, Joan: *Atlas maior*, 79; town atlas of the
    Netherlands, 188
Blais, Hélène, cited, 111
Blakemore, Michael J., cited, 23, 66, 87, 92, 99,
    117, 125
Blanding, Michael, cited, 92, 231

blank sheets, referred to as *cartes*, 74
Blathwayt, William, 146
Blaut, James M.: on mapping as "a basic, en-
    during human instinct," 66; work by, also
    cited, 121
blind, map for the, 39
"Blue Marble," 140, 141f
Board, Christopher, cited, 18, 56
Board of Ordnance, 108, 113. *See also* Ordnance
    Survey (Great Britain)
Boeke, Kees, *Cosmic View*, 79, 80f
Boggs, Samuel W.: "natural scale indicator," 220;
    work by, also cited, 209n, 220n, 221
Bohnenberger, Johann Gottlieb Friedrich von,
    108
Bom HGz, G. D., cited, 93, 117
Bonaparte, Napoleon, 110, 156, 205
Bonpland, Aimé, cited, 143
book history applied to map studies, 45
"Book of Roger" (al-Idrīsī), 57, 78
border maps, 33t
Borges, Jorge Luis, 18n; cartographic satire, 10, 15,
    18n, 19, 56; "On Rigor in Science," 15
Bosse, David, on Osgood Carleton, 98
Bosteels, Bruno, on Borges, 18n
Boston, MA, 43–44, 89–90, 98; Public Library, 34f
Boston Sunday Post, 160f
botanical mapping, 142–43, 204
boundary maps and mapping (mode), 28, 32, 33t,
    38, 42–43, 45, 100, 110, 178, 181, 186, 188, 223
Bourbon restoration, 129
Bourbourg (France), 188
Bourdieu, Pierre, definition of cartography, 82
Boyce, Nell Greenfield, cited, 70
Branch, Jordan, cited, 19
Branco, Rui Miguel Carvalhinho, cited, 129
Brandenberger, René, cited, 135
Braun, Georg: *Civitates orbis terrarum*, 104; on
    the pleasures of atlases, 104
Bret, Patrice, cited, 202, 206
Brewer, John, cited, 114
Brill, Emily, 179n
Britain as focus of map history, 127
British: army, 60; Cartographic Society, 120;
    Library, 114; mapping of South Asia, 111;
    Museum, 146; North America, "first" map
    of, 89–90; overseas mapping, 101; Parliament,
    114, 209; schools, map use in, 46. *See also*
    Great Britain
Broekburg (now Bourbourg, France), 188
Broman, Thomas, cited, 113
Brooke-Hitching, Edward, cited, 231
Brotton, Jerry: *History of the World in Twelve
    Maps*, 79; work by, also cited, 66, 76, 78, 231,
    235
Brown, Lloyd A., cited, 127
Brown, Ralph H., cited, 28
Brownlow, Kevin, 19n
Brownstein, Daniel, cited, 135

Brückner, Martin: on the "cartoral arts," 219; on "maps are bad" critique, 25; on patterns of production and consumption, 46; work by, also cited, 138, 152

Brun, Christian F., cited, 30

Bry, Theodore de, *Americae pars magis cognita*, 195

Bryars, Tim: on the London Underground map, 218; work by, also cited, 153, 163, 231

Buache, Philippe, 200

Buchroithner, Manfred F., cited, 23, 204

Buffalo Express (Buffalo, NY), 11

Bugge, Thomas, cited, 220

"Bugs Bunny" cartoons, 161

building plans, 33t

Buisseret, David, cited, 5

Bulson, Eric, cited, 39

Bünting, Heinrich, map of world as a cloverleaf, 189

Burdett, P. P.: survey of Cheshire, 196; *This survey of the county palatine of Chester*, 197f

Burgoyne, Margaret, on the view from a balloon, 137–38

Burkeman, Oliver, cited, 80

"burlesque map," 12

Burroughs, Charles, cited, 39

Burton, Robert: *The Anatomy of Melancholy*, 105; on the function of maps, 105

Butlin, Robin A., cited, 152

Buttenfield, Barbara P., cited, 73, 146

Byerly, Alison, cited, 135, 138

by-passes, 83n

Byrd, Max, cited, 19

Byzantine cosmography, 191

cadastral mapping, 33t, 108–10, 123, 147, 226

Cadiz (Spain) as prime meridian, 131

Cage, Nicolas, 219

Cain, Mead, cited, 114

calculation (cartographic preconception), 54

calendar, French revolutionary, 129

Calhoun, Craig, cited, 113

Camp Fire Girls, 161f

Canadian Institute of Geomatics, 73

Cañizares-Esguerra, Jorge, cited, 49

Cape Cod, Smith's explorations in, 50, 75

Cape Horn, 63

Cape of Good Hope, 35f

Cape Town (South Africa), 35f

capitalism, 73, 81–82, 161

Carleton, Osgood, 30f, 98; *The District of Main*, 45

Carlson, Julia S., cited, 78

Carolina map, 92

Carroll, Lewis: cartographic satire, 10, 14–19, 56, 133, 229; *The Hunting of the Snark*, 15–16, 17f; *Sylvie and Bruno Concluded*, 14–15

carta (word), 74, 115

Carta gothica (Magnus), 85

Carta marina (Magnus), 85

carte (word), xiii, 74, 115, 230

Carte de Cassini, 79, 79n, 202. See also *Carte générale et particulière de la France* (Cassini de Thury)

Carte de France (Cassini de Thury), 79, 79n, 202. See also *Carte générale et particulière de la France* (Cassini de Thury)

Carte de France corrigee par ordre du Roy (Picard and de la Hire), 128f

Carte de l'état-major (Dépôt de la Guerre), 109, 130, 208, 209f, 211, 216

Carte de l'Inde (d'Anville), 194f

Carte des provinces de France traversées par la méridie (Cassini), 198f

Carte générale et particulière de la France (Cassini de Thury), 79, 107, 107f, 111, 122, 202–4, 209

Carte qui comprend touts les lieux de la France . . . (Cassini de Thury), 200

Carter, Paul: on "imperial history," 55; on "spatial history," 47; work by, also cited, 57

cartes géographiques, non-normative definition, 3

cartes marines, non-normative definition, 3

Cartesian geometry, xiii, 52, 133, 177, 205, 224

cartobibliographers, 57, 88

cartobibliographies, 60, 92, 116–17

"cartocacoethes," 62

"cartograph" (word), 164

"cartographer" (word), 103, 118

cartographers, modern, and cartographic ideal, 21–24

Cartographer's Daughter, The (Hargrave), 65n

cartographes (word), 118

cartographia (word), 117

cartographic: archive (*see* archive of spatial knowledge [the corpus of maps]); "culture," 48; "diagram," 219; history as a field of study, 125–27; ideal (*see* cartographic ideal); "language," 24, 26, 42, 60, 66, 68, 100–101, 104, 214, 221; "logic," 150; oddities, 99, 162; "reason" (cartographic preconception), 55; revolution, 36, 48; sign plane, 24

cartographic ideal: a belief system, 1; "checklist of wrong convictions," 52–55; damaged by sociocultural critique, 20; defined, 1; emergence and development of, 7, 103–65, 228–31; geometrical essence of, 5; introduced, 1–5; not a logical construct, 164; persistence of, 20; self-effacement and internal complexity of, 24; stages of development, 228–31; supposed timelessness of, 5

cartographie (word), 118–19, 126, 140

cartography: as "applied technology," 82; as the "art and science" of map making, 150; coinage of term, 7, 114–20; as comfort zone, 21; conflated with "the map," 9; core idea seen as indexicality, 25; as crucial in perpetuation of social inequality, 25; "death" of, 231–32; "dialects" of, 100–101, 172; as example of Haeckel's "ontogeny" formulation, 26; an idealization of mapping, 26; as "immoral, anti-egalitarian, and misogynistic," 25; as a "map of mapping," 228; as mathematical process, 164; as a "myth," 50, 103; not a universal phenomenon, 8; origin in Renaissance perspectivism, 76; origins of, 67–71, 76; as paradoxical, 166; as patriarchal and misogynistic, 73; politicization of, 26; as sterile, unintellectual technique, 82n; the "why," "what," and "how" of, 99; the word, 2, 7–8, 20, 82, 101, 103, 114–20, 124–25, 149, 229, 233–35. *See also* map making; mapping

"cartology," 149

cartoons, animated, 161

cartoons, maps in, 86

"cartoral arts" (Brückner), 219

Cartwright, William, cited, 219

Casa de la Contratación, 57

Casey, Edward S., cited, 69

Cassini, Giovanni Domenico, *mappemonde*, 57

Cassini, Jacques, *Carte des provinces de France traversées par la méridie*, 198f

Cassini de Thury, César-François, 79n; *Carte générale et particulière de la France*, 107, 107f, 202–4; *Carte qui comprend touts les lieux de la France . . .* , 200; *Nouvelle carte qui comprend les principaux triangles*, 200, 201f, 202, 202f, 203f

Casti, Emanuela: on map communication, 67; work by, also cited, 224

Castner, Henry W., cited, 145

Çatalhöyük "map," 69, 69f, 86

catalogs, dealers', 44

Cattaneo, Angelo, cited, 191

Catwoman (film), 85

Cavelti Hammer, Madlena, cited, 61, 180

CCTV cameras, 78

"Ceci n'est pas une pipe" (Magritte), 17

celestial maps and mapping (mode), 33t. *See also* cosmographical maps and mapping (mode)

census, national, 114, 142, 203

centimeter, 130, 167, 207t

Central America, Dufour's map of, 212

Central European academics, 149–50

Cep, Casey N., cited, 19

Certeau, Michel de, on "the map," 82

Cesarz, Gary L., cited, 18

Chamberlin, Wellman, cited, 73, 150

Champlain, Lake, 84f

Chang, Kang-tsung, cited, 56

Chanlaire, Pierre Grégoire: *Atlas national de la France*, 211–12; on the scale of his maps, 212, 212n

Chaplin, Charlie, 19n

Chapuis, Olivier, cited, 34, 110–11, 146, 178

Chardon, Roland E., cited, 220

Charlesworth, Michael, cited, 105, 135

"chart" (word), 115

chart and plan, seen as equivalent to map, 22

charta (word), 74, 115

chartographie (word), 117

"chartography" (word), 115, 119

charts: compared with maps and plans, 3; defined, 3, 235

chemistry, 124

Cheshire (England), 135; Burdett's survey of, 196

Chester, Lucy P., cited, 110

chī (unit of measurement), 167

Chicago as depicted in *The Fugitive*, 78

"childlike" maps, 72

Chile, map of, 193f

Chile on Waldseemüller map, 63, 63f

China, 93

China, Ancient, cartographic satire, 15

Chinese conceptions of world, 68

Chinese embassy in Belgrade, 21

chorographical maps and mapping (mode), 33t, 105, 106f, 111, 126, 134, 179–81, 186–88, 191, 193, 207t, 208. *See also* place maps and mapping (mode); regional maps and mapping

Chrisman, Nicholas, cited, 18

chromolithography, 11

chronological conventions used, xiii

chronological fixity (cartographic preconception), 53

chronology: conventions used in this book, xiii; of map production, 69, 76, 105

Chura, Patrick: on literary place mapping, 100; work by, also cited, 181

Chwast, Seymour, cited, 161

Cimburek, Susan, cited, 37, 39, 44, 90, 93

circulation of blood compared to hydrography, 106f

circulation of maps, 27, 32, 33n, 35, 41–42, 47–49, 91, 93–94

Citizen's Atlas of the World (Bartholomew), 215

city directories, maps in, 42

city plans and views. *See* urban maps and mapping

civil mapping, 31

Civil War (U.S.), 42

Civitates orbis terrarum (Braun and Hogenberg), 104

Clark, John O. E., cited, 231

Clarke, Alexander Ross, cited, 209n

Clarke, Keith C.: on Çatalhöyük "map," 70; work by, also cited, 68, 236

Clarke, Susanna, on the difficulty of changing maps, 60

class (social concept), 49, 113, 135; compared to mode of mapping, 31; and map consumption, 47

classroom, maps of the, 153–54, 155f

classrooms, maps in, 46

Clements (William L.) Library, 51f

Cleveland Museum of Art, 181, 184f

clients of map makers, 65

climatic zones, 190

Close, Charles F., cited, 209n

285

INDEX

289

French (*continued*)
113, 129, 205, 211; "school" of cartography, 98; world maps, 192f
Frenchmen in North America, spatial knowledge among, 48
Friendly, Michael, cited, 145, 219
Fromaget, C., 154f
frontier maps, 33t
Fry, Joshua, map of Virginia, 88
Fugitive, The (film), 78
Fulton, Robert, cited, 147
functions of maps, 3
furlong (unit of measurement), 190

Gaiman, Neil, cartographic satire, 15
Galaxy, The, 11
Galeron, Paul, 148f
Galle, Andreas, cited, 109
Gannett, Henry, 130
Garfield, Simon, cited, 231
Garland, Ken, cited, 219
Gartner, William Gustav, on Skiri star chart, 39
Gaspar, Joaquim Alves, cited, 34
Gatterer, Johann Christoph, outline of geographical knowledge, 123
Gauss, Carl Friedrich, 109, 145; on projections, 123, 132
Geddes, Patrick, 147–49; "temple of geography," 148f
Gehring, Ulrike, cited, 5, 134
Gemma Frisius, 196
gender bias, 73, 122, 125. *See also* women
General Topography of North America (Jefferys), 88
generalization, 64, 155, 164, 168, 186–87, 232; cartographic preconception, 53; as the "core process" of cartography, 53, 56; expressed by map scale, 57, 60, 130, 167, 171–72, 187; as primary basis of map, 55–58
geodesists, 109, 132, 199, 225
geodesy, 115n, 179
geodetic maps and mapping (mode), 33t, 101, 108–10, 124, 145, 168, 199. *See also* systematic (territorial) maps and mapping (mode); topographical maps and mapping
geographers: academic, 2, 40; Arab, 78; attitude toward cartography, 82n; Belgian, 212; as cartographers, 216; and cartographic ideal, 2; critical, 193; Danish, 117; Dutch, 131; eighteenth century, 25, 48, 98, 131, 193, 201; English, 131; formal training of, 47; French, 116, 131, 152, 201; German, 215; and graticule, 131, 191, 193–95, 204, 215–16, 222–23, 225, 229–30; humanistic, 81, 218; and map projections, 122; and marine maps, 32, 34, 62, 194; and modes of mapping, 229; nineteenth century, 104, 123–24, 152, 212; and scales, 191, 193–95, 215–16, 222–23, 230; Scottish, 147; seventeenth century, 129; and topographic mapping, 117; twentieth century, 123, 150; in World War II, 150

géographes (word), 118
Geographic Information Systems (GIS), 31, 82n, 231, 236
geographic locations as signs, 24
geographical: archive (*see* archive of spatial knowledge [the corpus of maps]); changes on maps, 87–88; congresses, 214; data, 55–56, 78n, 82n, 88–89, 91, 235; features, 14, 39, 53, 55, 64, 127, 145; index, 44; instruction, maps in, 153–59; knowledge, 44, 68, 88, 123, 127; maps and mapping (mode) (*see* geographical maps and mapping [mode]); regions, 107; societies, 72, 117–18, 120, 123–24, 146, 152, 155, 233; studies, role of visual in, 81; textbooks, 28; truth, 97; writing, 37, 39, 86, 152, 160
geographical maps and mapping (mode), 27, 33t, 44–47, 76, 82, 101, 120, 125–26, 147, 216, 222, 229; of American colonies, 44; compared with marine mapping, 32, 34, 34f, 35, 57, 62, 85, 100, 235; compared with place mapping, 32; compared with urban views, 135; defined, 33t, 37; educational use, 46, 153–56, 154f; in the Enlightenment, 111; as "eye" of history, 105; of France, 202; and geographical writing, 37; and graticule, 104–5, 111, 190f, 191, 195, 229; and history, 44, 121, 126; and map projections, 122–23, 203; and map scale, 78n, 101, 123, 173t, 193–95, 207t, 209, 214–15, 217, 223–24; of New England, 30, 39, 44; and printing, 91; production and consumption, 34f, 46, 91, 114, 194n, 235; of Scandinavia, 85; seen as "charts," 85; of South Asia, 101; space represented on, 83, 225–26; terms used for their makers, 118–19; "top down," 191. *See also* marine maps and mapping; regional maps and mapping; special-purpose maps and mapping; world: maps and mapping (mode)
"geographical pivot of history" (Mackinder), 79
"geographical space" (Montello), 225–26
"geographically abstract" and "geographically concrete maps" (Eckert), 215–16
geography: "age of computation" in, 229; begins with topography, 155; as the "eye" of history, 105; as a field of study, 123–27, 153–55, 155f, 191, 213; historical, ix; history of, 118, 124; human, 40; sociocultural critique in, 19; textbooks, 22, 28; the word, 115, 126
geological maps and mapping, 152, 203, 225
geological metaphor, 144f
Geological Society of London, 123
Geological Survey (U.S.), 130, 167, 174
geologists, 70, 123, 143
"geometrical alterations," 176, 193–94
Geometrical Ascent to the Galleries in the Coloseum, The (Hornor), 136f
géométrie pratique (term), 179
geometry: cartography's idealized, 205–27; development in South Asia, 178; as essence of cartographic ideal, 5; of world/archive, 52.

Hansen, Jason, cited, 64, 145

Hansen, Paul, cited, 95

Hapgood, Charles H., 86; *Maps of the Ancient Sea Kings*, 61–62

Haraway, Donna, on the "god trick," 77

harbor charts, 33t, 35f. *See also* marine maps and mapping

Hargrave, Kiran, 65

Harley, J. B.: criticized by J. H. Andrews, 65; *History of Cartography* (University of Chicago Press), 39; on sociocultural critique, 26; work by, also cited, 2, 18, 23, 87, 92, 96, 99, 114–15, 117, 125, 146, 163

Harmon, Katharine A.: "I map therefore I am," 66; work by, also cited, 19, 163

Harper, Tom: on the London Underground map, 218; work by, also cited, 153, 163, 231

Harrington, Bates, cited, 96

Harriot, Thomas, 90

Harrison, Richard Edes, 97; *Europe from the East*, 97f; work by, also cited, 150

Harrisse, Henry, cited, 127

Harvard University, 21, 98; Map Collection, 155f

Harvey, David: on cognitive maps, 73; conflates viewing a map with viewing the world, 77; on Renaissance perspectivism, 76; work by, also cited, 5, 105

Harvey, Miles, cited, 231

Harvey, P. D. A.: on proportionality, 167; work by, also cited, 168–69, 179

Harwood, Jeremy, cited, 231

Harzinski, Kris, cited, 19, 163

Hasan Dağ (volcano, Turkey), 69–70

Haywood, Jonathan, 183f

Haywood, Samuel, 183f

headland profiles, 35f, 37

heaven and hell, maps to, 160

heavenly bodies, maps of, 33t

Heffernan, Michael, cited, 116, 147, 149, 153

Heidenreich, Conrad E., cited, 220

Heilbron, J. L., cited, 179

Heimatskunde, 68, 154

Heiser, Willem J., cited, 91

hell and heaven, maps to, 160

Hellenistic astrology, 189

Hellenistic era, 127

hemispheric projection, 192f

hemispheric world map, 217f

Hennerdal, Pontus, on projections, 222n

Herb, Guntram Henrik, cited, 98, 150

Herodotus, cited, 178

Herrman, Augustine, 88

Hessler, John W., cited, 63

Higton, Hester, cited, 177

hiking maps, 159

Hill, Gillian, cited, 163

Hire, Philippe de la, 128f

historians of science, 27, 49

historians of the book, 94

historical: geography, ix; mapping, 143, 152; materialism, 73

"historical cartography," 117, 119, 127, 233

"historical map," ambiguity of term, 58

History of Cartography (University of Chicago Press), 32, 36, 39

history of cartography as field of study, 23, 48, 52, 54, 98–99, 101, 119, 121, 125–27, 155

History of Cartography Project, x

history of exploration, 57, 124–26, 152

history of geography, 118, 124

history of mapping (cartographic preconception), 54

history of science, sociocultural critique in, 19

history of the book, 26

History of the World in Twelve Maps (Brotton), 79

Hodgkiss, Alan G., cited, 61

Hoffman, Moshe, cited, 122

Holiday, Henry, 17f

Holman, Richard B., cited, 89

Holmes, Nigel, cited, 160, 163

Holtorf, Christian, cited, 19

Holtwijk, Theo H. B. M., cited, 58

Holwell, John, 185, 185n

Holy Land, 85–86

Holy Roman Empire, 110

hooks, copyright, 96–97

Hooper, Wynnard, cited, 219

Hornor, Thomas, panorama of London, 136f

Hornsby, Stephen J., cited, 20, 110, 162–63

Hornsey, Richard, cited, 156, 219

hospital plans, 84

hot-air balloons, 135–38, 137f

Houston, Kerr, cited, 153

How Maps Work (MacEachren), 2

"how" of cartography, 99

Hubbard, William, 89; *Map of New-England*, 37–39, 38f; *A Narrative of the Troubles with the Indians*, 38–39, 38f, 43–44

Hudson, Brian, cited, 152

Hudson River map, 84f

Huguenin, Marcel, cited, 130, 209

Hull, John T., 58, 59f

human geography, 40

humanistic geographers, 81–82, 218

humanities, sociocultural critique in, 19

humans, interactions with and without maps, 4

Humboldt, Alexander von, 121, 142–43; and map history, 126; work by, also cited, 5

Hunting of the Snark, The (Carroll), 15–16, 17f

Hutchinson, Dave, on a fictional territorial survey, 60

Huynh, Niem Tu, cited, 122

hydrographers, treatment of coastlines by, 62

hydrographic maps and mapping, 33t, 34f, 62, 110–11, 123, 147, 152, 204, 229. *See also* coastal maps and mapping; marine maps and mapping; systematic (territorial) maps and mapping (mode)

nitive development, 72; work by, also cited, 20, 40, 218

Lewis, Martin, cited, 57

li (unit of measurement), 195

Libby, Ebenezer, 29f

Liberman, Mark, cited, 58, 115

librarians, 57, 61, 125, 147, 167, 220, 220n

Library of Congress (U.S.), 12, 12f, 63f, 114, 201f, 211n

Licka, J. L., cited, 145

Lieber, Francis (Franz): on the teaching of map drawing, 119; work by, also cited, 229

lighthouses, 156

Lightman, Bernard, cited, 138

ligne (unit of measurement), 107f, 187, 203, 211–12

Lindgren, Ute, cited, 196

linear measures, 127–32; standardized, 129, 167.
See also individual units of measurement

linear perspective, 93

Lissac, Pierre, 162f

List, John A., cited, 122

listservs, 133

literary maps, 39

literary studies, 16, 18–19, 75, 82, 100, 161

lithographic maps, 17f, 42, 59f, 80f, 97f, 144f, 148f, 151f, 157f, 158f, 162f, 175f, 213f

"lithography" (word), 116

Livingstone, David N., cited, 123, 144, 152

Lloyd, Robert, cited, 62

local places, mapping of, 33t

locational accuracy, and cartographic ideal, 16

locational mapping, 145

"logic" of cartography, 150

logical empiricism, 55

Lohengrin (Wagner), 19n

Lois, Carla, cited, 3

London (England), 88–90, 92, 188; coffeehouses in, 113; commissioners of maps of New England, 42; maps of, 196; panorama of, 135, 136f; as prime meridian, 131; site of publication of Hubbard map, 43–44; University of, 124

London Underground map, 218–19, 218f, 226

Long, John H., cited, 219

longitude, determination at sea, 34–35. *See also* latitude and longitude

longue durée, significance of mapping in, 47

lost property markers, 181

Louis XIV (king of France), 146, 199

Louis XV (king of France), 202

lower classes, consumption of maps by, 49. *See also* class (social concept)

lower photopause, 81

Lüdde, Johann Gottfried, cited, 123

Ludden, David, cited, 101

Lukens, R. R., cited, 91

MacDonald, J. Fred, cited, 163

MacEachren, Alan M., cited, 2, 22, 36, 76, 145

Machigone peninsula, ME, 58

Mackinder, Halford, 79

Maconochie, Alexander, 124

macro-level analysis of mapping, 45

Madawamkeag, ME, 157

Maeer, Alistair S., cited, 36

Magellan, Ferdinand, 63

Magritte, René, "Ceci n'est pas une pipe," 17

Mah, Harold, cited, 113

Maine: Carleton's map of, 30f, 45; Greenleaf map of, 159f; Historical Society, 29f; province of, 28, 30f; Thoreau's travels in, 157; University of Southern (*see* Osher Map Library and Smith Center for Cartographic Education)

Maling, D. H.: comparative table of map scales, 173t; on representative fraction, 168; work by, also cited, 87, 120, 150, 174

Malte-Brun, Conrad, cited, 117–18

"mania, uncontrollable urge," to see maps everywhere, 62

manuscript maps, 29f, 32, 42, 46, 54, 91–94, 111, 113, 154f, 182f, 183f, 184f, 186f, 187f, 192f, 206; cartographic preconception about, 54; use in schools, 46

map (concept, "the map"), 2–4, 6, 10, 21, 55–56, 66, 82, 85, 91, 123, 202, 216, 224, 230, 233–35

"map" (word; i.e. italicized or in quotes), xiii, 1, 3–4, 8, 20, 22, 41, 55, 66–67, 82, 103, 119, 132–33, 138, 158, 218–19, 233–36; derivation of English word, 74

map author, map reader as, 67

map catalogs, 57, 61, 117, 212n, 220, 220n

map collections, 146–49

map compilation techniques, 62

map consumption, 4, 7, 10, 31–32, 33n, 34f, 35, 40, 42, 45–49, 103, 138, 164, 194n. *See also* map use

map design, 149–50

map drawing, 46, 62

map folding, 159

map form, not necessarily the same as map mode, 32

map functions, 3

map historians: adopting sociocultural critique, 2, 7, 26; analysis of geometric accuracy of old maps, 87; attention to "new" data and corrections on maps, 88–91; attention to patterns of map consumption, 46, 49; belief that the default function of maps is navigation, 85–87; concern with technicalities of map making, 2; conflate early with later maps, 58; drawn to regional and nationalist narratives, 57, 99; and an early modern origin for cartography, 104–6; and emerging field of map history, 125–27; and emphasis on analytic mapping, 145; and Hubbard map of New England, 43–45, 89–90; and map authorship, 64–67; and map scale, 167–70, 220; misunderstand the transition from script to print, 91–93; must adjust to different modes of discourse,

concept, 21; for the blind, 39; as "cartographic language," 24; of classrooms, 47; compared to scientific theories, 18; compared with charts and plans, 3; compared with diagrams, 92; as "concentrations," 18; and concept of "authorship," 65; as "condensed history of God's creation," 56; conflated with "cartography," 9; as "depleted homologues," 18; destruction of, 48, 74; as devices with which humans create meaning, 21; do not "explain themselves," 11; drawn in sand, 39; embroidered, 46; as empowering vision, 36, 77–78, 137; evaluation of, 49, 76, 86–87, 89, 157; exist within a web of texts, 40; the "first," 67–69, 92, 113, 127; as functional tools, 164; as "general image of the world or of a region," 235; "general purpose," 76, 83–84, 91, 142n; as "graphic modeling" of world, 60; as "graphic representations," 39; have no easily discerned boundaries, 37; how they "work," 2, 19, 83, 219; as human products, 3, 19; are images, but also objects, 74; as "Immutable mobile," 74; "language" of, 53, 60–61 (see also language); linked to texts, 37; "literal meaning" of, 61; made for power elites, 19; made for social reasons, 19; make the invisible visible, 53; as measured, graphic abstractions, 22; as metaphors, 19; as mimetic, 18, 24, 58, 77; as mirrors of nature, 54, 77; as multi-component "things," 40; must be "correct," 21; nongraphic, nonmaterial, 40; normalization of concept, 1–2; normative (see normative maps); as normative statements of spatial fact, 3–4; not "picture(s)" but "argument(s)," 24; "oldest known," 69f; "openness," 36; perfectibility of, 14; physical forms of, 27; as "pictures" of the world, 58; on powder horns, 84; as products of "cartography," 1; as products of intellect, 64; are read like books, 36; "repackaged" as historical artifacts, 44; as "representing" the world, 58; representational strategies, 27; as "representation[s] of a representation," 76; as "representation[s] of the earth's . . . surface," 22; as scientific instruments, 64; seen as needing no definition, 21; seen as unmediated representations, 58; as semiotic texts, 19, 36, 64; showing distributions of phenomena, 33t; sociocultural critique, 2; spaces depicted by, 27; "special-purpose" (see special-purpose maps and mapping); are stable and fixed, 74; as "standing for" the world, 58; storage of, 48; as "structurally equivalent" to the world, 18; as "system[s] of propositions," 24; as tables or spreadsheets, 56; as texts, 36; as "text[s] representing spatial complexity," 41; thematic (see thematic maps and mapping); as tools of visualization, 14, 64, 100, 105–6, 142, 145, 164, 172, 230; topographical (see topographical maps and mapping); as "totalizing devices," 82; as "a type of congruent diagram," 18; as universal and timeless products, 1; "universal language" of, 60; are used like instruments, 36; and the "view from above," 76; of villages, 47; as "visual representation[s] of an environment," 22; as "works in progress," 40

Maps and Politics (Black), 2
"Maps Showing War Zone in Belgium and France" (*Boston Sunday Post*), 160f
Maraldi, Giovanni Domenico, 200–202, 201f, 202f, 203f
marine archives, 57
marine charts, 32, 34f, 35
marine maps and mapping, 31–32, 33t, 35–37, 35f, 76, 100, 119, 126, 152, 169, 204; compared with place mapping, 36. *See also* coastal maps and mapping; hydrographic maps and mapping; regional maps and mapping; special-purpose maps and mapping; world: maps and mapping (mode)
Markham, Beryl: on African maps, 96; on morality of maps, 95
Marschner, F. J.: on the limitations of maps, 150; work by, also cited, 221
Maryland, Herrman's map of, 88
mass mapping literacy, 151–63
Massachusetts, 181
Massachusetts Bay (province), 43
Massieu, Elise, 154f
Mastronunzio, Marco, cited, 133
materialism, historical, 73
materiality (cartographic preconception), 7, 37, 51, 53, 64, 74–75, 91, 111, 125, 146, 153, 156
"mathematical cosmography," as a mode of mapping, 25
"mathematical practitioners," 177
mathematicians, 123
mathematics, 120–34
May Show (Cleveland Museum of Art), 181, 184f
Mayhew, Henry, on the view from a balloon, 137
Mayhew, Robert J., cited, 105
McClintock, Anne, cited, 160
McCorkle, Barbara B., cited, 30, 92
McDermott, Paul D., cited, 18
McHaffie, Patrick, cited, 96
McKenzie, D. F., on "sociology of texts," 47
McManis, Douglas R., on exploration of New England, 57
McMaster, Robert B., cited, 24
McMaster, Susanna A., cited, 24
measurement: as characteristic of Renaissance, 5; conflated in "surveying and mapping," 94–95; contrasted with metaphor and rhetoric, 163; defining characteristic of *plan*, 235; and dividers on maps, 187; emphasized in education, 47, 154; equated with "science," 150; exemplified by graphic style, 181; as intuitive

297

measurement (*continued*)

human practice, 223; linear, 127–32 (*see also specific units of measurement*); and map resolution, 87, 226; of physical features, 35; as Platonic ideal, 87, 129, 179, 221; rationalization of, 127–32, 205, 213, 223; "scale of" (Doiron), 223; of size of earth, 199; and terms used in mapping, 179; as underlying all maps, 53, 61, 67–68, 70, 78, 95, 99, 103, 119, 129, 133, 138, 155, 188, 217, 223–24. *See also* map scale; proportionality (map-to-world)

mechanical drawing, 207t

medieval maps, 20, 82, 93, 142n, 189

Mediterranean Sea, 126

"medium scale" maps and mapping, 171–73, 217, 225

Meece, Stephanie: on Çatalhöyük "map," 69–71; on prehistoric maps, 68

Mellaart, James, cited, 69, 69f

Mellon (Paul) Collection, 106f, 139f, 140f

Melton, James Van Horn, cited, 113

Mendoza Collection, 197f

mensurare (word), 179

mensuration, 110, 119, 179, 229

mental maps, 53, 66. *See also* cognitive maps and mapping

Mentelle, Edme, 153

menus in restaurants, referred to as *cartes*, 74

Mercator, Gerard, world map (1569), 34, 85

Mercator projection, 34–35, 85, 97, 123, 133, 169, 174, 175f, 176–77, 195, 213

meridians: of longitude, 104, 111, 131, 174, 176, 185, 190f, 191, 192f, 194, 200, 204, 225; prime, 131–32. *See also* graticule; latitude and longitude

Merrill, Samuel, cited, 219

meso-level analysis of mapping, 46

metaphor and mapping, 16, 18–19, 37, 55, 66, 77, 84, 103, 144f, 162–64, 178, 205, 228

metaphysical maps, 33t, 84, 189

"metaphysics of presence" (cartographic preconception), 54

meter (unit of measurement), 69, 77, 109, 129, 138, 207t, 208, 212

metes and bounds surveys, 38, 182f, 183f, 186f

metric system, 127, 129–30, 169, 205–9, 214–15, 220, 223

Metropolitan Museum of Art, 84f

Meynen, Emil, cited, 120, 167, 170

Michelangelo, 39

Michelson, Bruce, cited, 19

Michigan, University of, 51f

micro-level analysis of mapping, 45

Microsoft Word, 31

Middle Ages, 126, 127

middle class as map consumers, 47

Middle East, 101

middle photopause, 81

Migration Zones/Ethnic Strata (Taylor), 144f

Mikhailov, Nikolai, cited, 59

Mi'kmaq Indians, spatial knowledge among, 48

mile, 14, 28, 60, 62, 111, 128, 130, 141f, 187–89, 193f, 195, 209–11, 215. *See also* English mile (unit of measurement); French: mile (unit of measurement); German: mile (unit of measurement); Italian mile (unit of measurement); nautical mile; Russian mile

militarism, 81

military maps and mapping, 21, 31, 45, 152, 156, 158n, 181, 187–88, 207, 207t

Millard, A. R., cited, 169

Millaria Gallica (unit of measurement), 193f

Miller, B. A., cited, 146

Miller, O. M., cited, 221

millimeter, 203, 207t, 208

Mills, Sara, cited, 41

mimetic nature of maps, 18, 24–25, 58, 64, 77

minorities, consumption of maps by, 49

mirror of nature, map as, 54, 77

misogynistic nature of cartography, 25, 73

misogyny, seen as inherent in mapping, 25

Mitchell, Peta, cited, 82

Mitchell, Rose, cited, 231

Mitchell, W. J. T., cited, 134

mobility, mapping for, 54, 156–59

Mode (P. J.) Collection, 17f, 161f, 162f

modern period, defined, xiii

modern nation-state, 25, 33t, 79, 82, 144–45, 156, 203, 231

modernity, 23, 54, 57, 82, 121

modes of mapping, 27–36; change and intersect, 31; defined and listed, 31–32, 33t. *See also* individual modes

Moellering, Harold, cited, 23, 74

Mohammed, Salim, 207t

Mohawk River, NY, 84f

Monde connu des anciens (Dufour), 214f

Monhegan Island, ME, 50

Monmonier, Mark, cited, 21, 32, 36, 150, 153, 159, 163, 236

Monsaingnon, Guillaume, cited, 19n

"monstrosities," medieval maps as, 20

Montanus, Arnoldus, map of Chile, 193f

Montello, Daniel R.: on categories of space, 225–26; work by, also cited, 19

morality (cartographic preconception), 7, 51, 55, 88–89, 95–99, 111, 121

Morrison, Joel L., cited, 57, 122, 134, 191, 204

Morrison, Philip, cited, 79

Morrison, Phylis, cited, 79

Morse, Jedidiah, 28, 30f

Moser, Molly, cited, 219

Mount Vernon, VA, 181, 184f, 187f

mountains, obsession with, 156

movable type, 93

"Mr. Wyld's Model of the Earth," 140f

Muehrcke, Juliana and Philip, on Lewis Carroll, 18–19, 19n

Muehrcke, Phillip, cited, 82, 82n

multiple scales, 193, 194f, 214f

Munroe, Randall: guide to dating world maps, 32; work by, also cited, 230

Muratore, Nicoletta, cited, 93

Murphy, David Thomas, cited, 150

Murphy, J., cited, 87

myth, cartography as, 8, 50, 53, 103

Nadal, Francesc, cited, 109

Naples, systematic mapping of, 108

Napoleon, 110, 156, 205

Napoleonic Wars, 60, 108

Narrative of the Troubles with the Indians, A (Hubbard), 38–39, 43–44

nation-states, 25, 33t, 57, 79, 82, 144–45, 156, 203, 223, 230–31

national: bias in map studies, 45; map collections, 146–49; museums and libraries, 114, 147; standards, 42; surveys, 110, 181, 214, 230. *See also* geodetic maps and mapping (mode); systematic (territorial) maps and mapping (mode); topographical maps and mapping

National Atmospheric and Space Agency (U.S.), 140, 141f

National Geographic (magazine), 72

National Geographic Society, 72

national grid (U.K.), 204

National Library of Scotland, 148f

National Treasure (film), 219

nationalism, 14, 45, 57, 90, 99, 121, 127, 153, 162

Native Americans, spatial knowledge among, 48

NATO, 204

"natural scale," 209, 209n; "Natural Scale Indicator," 220

Nature of Maps (Robinson and Petchenik), 2, 21

nautical mile, 141f, 215

Naval Observatory (Washington, DC), 132

navigation: aids to, 85; charts, 34; instruments, 83, 159; oceanic, 34, 51, 194n, 235; as primary function of maps, 14, 16, 22, 72, 83–86, 95, 152, 159, 164, 231–32

navigators, 3, 38, 47, 85

Nazi cartography, 97–98f, 149

Neeley, Kathryn A., cited, 125

negativity, principle of, 113

Nekola, Peter, cited, 84, 156

Neolithic maps and mapping, 69–70, 86

neologism "cartograph," 219

neologism "cartography," 101, 114–19, 125, 149, 164

Netherlands: Blaeu's town atlas of, 188; as focus of map history, 127; survey of, 197; systematic mapping of, 108–9

network (of latitude and longitude). *See* graticule

Netz (of latitude and longitude). *See* graticule

Neumann, Jan, cited, 67

New England: archives in, 28; cartobibliography, 92; exploration of, 57; maps of, 42–45, 50–51, 51f, 88–90, 188; material life in, 44; readership in, 28

New England (Smith), 51, 51f, 75, 85

New Jersey road map, 158f

New Philosophico-Chorographical Chart of East-Kent, A (Packe), 106f

New York (province), 88

New Yorker, 163

New Zealand, 91

Newberry Library, 94f

Newfoundland, 91

newspaper map, 160f

Nicolai, Roel, on portolan charts, 100

Nicolar, Joseph, verbal map by, 39

Nigg, Joe, cited, 85

Nikolow, Sybilla, cited, 114

Nile River, 178

Nolli, Giovanni Battista, 180f

nombre abstrait (Allent), 207, 207t

nongraphic maps, 60

"nonrepresentational theory," 40

Nordenskiöld, Adolf Erik, cited, 127

normative maps: academic acceptance of, 21–24, 65; attempts at defining, 3, 103, 120; and cartographic "oddities," 162–63, 219; Çatalhöyük "map" as, 70; challenged by actual mapping practices, 167; challenged by satires, 10, 18–19, 229–30; challenged by sociocultural critique, 2–10, 236; as "comfort zone," 21; contrasted to early maps, 231; contrasted with pictorial or metaphorical maps, 162; as distillations of reality, 55–56; geometrical basis of, 87; hegemony and naturalization of, 164; and iconic signs, 134n; and locational indexicality, 232; and "the map," 203; map scale as a core property of, 167, 169, 176; material essence of, 74, 91; metaphorical use of, 103; "must be eliminated," 233, 236; ontology of, 45; and Peirce's mathematical definition, 133; the product of "cartography," 120; public comfort with, 20–21, 231; seen as like scientific laws, 40; seen as self-contained, 37; singularity, universality of, 101; topographical map as default form of, 138

North America: coastal mapping, 110; Dufour's map of, 212

north arrow, 42, 186, 222

North Carolina map, 62

Nouvelle carte qui comprend les principaux triangles (Maraldi and Cassini de Thury), 200, 201f, 202, 203f

Nova Scotia, spatial knowledge in early, 48

novels as literary maps, 39

Numbers (Biblical book), 85

numerical ratio of map scale, 8, 18, 57, 130, 164, 166–77, 188–89, 205–6, 208–18, 210f, 214f, 217f, 220–26, 229–30; defined, 170; validity for various maps, 216. *See also* map scale; representative fraction

numerischer Maßstab (term), 170

Nuova pianta di Roma (Nolli), 180f

299

scale (graphic element) (*continued*)
    169, 186–87, 187f, 188; as used with projective
    geometry, 203. *See also* graduated lines; map
    scale; proportionality (map-to-world)
scalelessness of world/archive, 52
Scarborough, ME, 28, 29f
Schaetzl, R. J., cited, 146
Schaffer, Simon, cited, 129
Scharfe, Wolfgang, cited, 11, 115n, 216
Scherer, Heinrich, 190f
Schilder, Günter, cited, 93
Schlimm, Dirk, cited, 132
Schlögel, Karl, 75
Schmitt, Axel K., cited, 70
schoolroom, maps of the, 153–54, 155f
Schuiten, François, cited, 149
Schulten, Susan, cited, 46, 97, 145, 153
Schwartz, Joan M., cited, 138
Schwartz, Seymour I., cited, 57
science: central to cartographic ideal, 36, 55;
    esoteric, 16; fiction, 80; of map design, 149;
    structural changes in, 123–25
scientific: instruments, maps compared to,
    64; mapping, 31; "revolution," 27, 124, 134;
    theories, compared to maps, 18
"scientist" (word), 124–25
Scotland: National Library of, 148f; systematic
    mapping of, 107, 111
Scottish Geographical Institute, 147
Seckford, Thomas, 185f
sections (vertical cuts), 120, 140f, 167
Seed, Patricia, cited, 85, 231
"seeing the world," 134–46
Seekarten (word), 3
Self, Will, cited, 15
self-containedness (cartographic preconcep-
    tion), 53
self-explanatory (cartographic preconception), 53
Seller, John, 35f
semantic proximity and definition of map, 22
semiotics, 19, 22, 26, 36–37, 41–44, 60, 64, 66–
    67, 69–70, 73, 75, 134n, 145, 169–70, 181, 224,
    234–36
Sepoy Rebellion, 139f
set theory, 5, 132–33
sewage systems, mapping, 209
sex manuals, 160
sexist attitudes in mapping, 64, 71–73
shapes of areas misinterpreted, 61–64
Shapin, Steven: on "diverse array of cultural
    practices," 45; on history of science, 27; work
    by, also cited, 124
Sharif al-Idrīsī, 57, 78
Sharpe, Bob, cited, 122
sheets, standardized, 107
Shettleworth, Earle G., Jr., cited, 58
Sheynin, Oscar, cited, 109
Shirley, Rodney W., cited, 122, 189, 194n
shopping mall plans, 84

shorthand, map as, 60
Shweder, Richard A., cited, 121
Sicily home of al-Idrīsī, 78n
Sieg, Wilfried, cited, 132
Siege and Capture of Fort Loyall, The (Hull), 58, 59f
sign plane, cartographic, 24
Silbernagel, Janet, cited, 174
Silverberg, Joel S., cited, 5, 177
Silvestre, Marguerite, cited, 212
Singer, Thea, 95
"singular logic" of maps, 25
singularity and universality (cartographic pre-
    conception), 31, 55, 99–102, 108, 114, 120, 122,
    125, 127, 131–32, 134, 138, 142, 146, 153, 156, 160
Sinitic languages, 235
Sismondo, Sergio, cited, 18
Sistine Chapel, 39
Sitwell, O. F. G., cited, 28
Sivin, Nathan, cited, 235
Skelton, R. A.: on ease with which map histo-
    rians can be misled, 62; on "mental maps,"
    67; work by, also cited, 71, 90, 93, 105, 111, 125,
    146, 169
sketch maps, 20, 53, 66
Sketch of the Outline and Principal Rivers of India
    (Arrowsmith), 112f
Skiri (Band of Pawnee people), 39
Skurnik, Johanna, cited, 26, 47–48
Sloan, Kim, cited, 90
Slocum, Terry A., cited, 145, 219
small scale mapping. *See* coarser resolution maps
    and mapping
"small scale" vs. "large scale," 8, 171–74, 178, 210–
    11, 230
Smith, Edgar Crosby, cited, 30
Smith, J., on Twain's map of Paris, 11
Smith, James L., on morality of maps, 95
Smith, John, 50–51, 87; *Description of New En-
    gland*, 75; explorations, 75; *New England*, 51,
    51f, 75, 85
Smith Center for Cartographic Education. *See*
    Osher Map Library and Smith Center for
    Cartographic Education
Smithsonian Institution, 76
Smits, Jan, cited, 220
Snel van Royen, Willebrord, 197
Snyder, John P., cited, 122, 168, 191
Sobel, Dava, cited, 231
"social history of culture," 47
social movements, post-WWII, 81
Société de géographie, 124
sociocultural critique, 20, 26; Andrews on, 65;
    contrasted with normative critique, 2–6;
    dismissed by some scholars, 2; enlarges field
    of images accepted as maps, 24; and existen-
    tialism, 20–21; in modern scholarship, 2, 7,
    19, 26; normative maps challenged by, 2–10,
    236; political implications of, 2, 19; sustaining
    cartographic ideal, 25

304

sociological critique, 2–3
sociology, 144; and imperialism, 120–22; of texts, 47
Soens, Tim, cited, 179
"Solitude of Alexander Selkirk, The" (Cowper), 78
Somerville, Mary, 125
Sorum, Eve, cited, 39
South America on Waldseemüller map, 63, 63f, 64
South Asia: British mapping of, 101; geometry
    in, 178; as if seen from a balloon, 138, 139f;
    systematic mapping of, 111
South Carolina, 62
sovereignty (cartographic preconception), 53
Soviet Union, propaganda tract quoted, 58–59
space: Montello's categories of, 225–26; and time,
    rationalization of, 131–34
Spain, systematic mapping of, 107
Spanish geographers, 131
Spanish league, 193f
spatial: ability (cartographic preconception), 53;
    discourses, xiii, 26, 33t, 37, 40–49, 75, 77, 82–83,
    102, 145, 173, 177–78, 181, 189, 222, 225–26, 230,
    233–36, "history" (Carter), 47; knowledge, 4,
    14, 27, 48, 66, 73, 87, 102, 116, 119, 123, 147, 149,
    155, 205, 224, 234; relationships (cartographic
    preconception), 54
special-purpose maps and mapping, 33t, 83–84,
    142n, 145, 156. See also analytical maps and
    mapping (mode); geographical maps and
    mapping (mode); thematic maps and map-
    ping
Speculum orbis terrarum (Jode), 77
Speier, Hans, cited, 150
sphere, "doctrine" of, 191
spherical triangles, 199n
spy thrillers, 78
St. Louis, MO, 131
St. Paul's Cathedral (London), 136f
St. Quentin (Aisne, France), 154f
stability (cartographic preconception), 53
"Standard" Road Map of New Jersey, 158f
standardization: of coins, 128; of linear measures,
    129, 167; of map design, 173; of map scales,
    206–8, 216, 223; of sheet sizes, 107; of survey-
    ing practices, 187; of use of pins in maps, 158
Stanford University, 112f, 207t, 210f, 214f
star charts, 33t
stars (celestial bodies), 189
State Plane Coordinate System, 224–25, 229
states and empires, 120–34. See also nation-states
statistical: maps, 216, 219; techniques, 87, 122,
    145–46, 168
statisticians, 219
Stavenhagen, W., cited, 109
Steinberg, Saul, 163
stereographic projection, 137f, 192f
stereoscope, 219n
sterility of maps (cartographic preconception), 54
Stevens, Henry, cited, 88
Stevenson, Robert Louis, 161

Steward, H. J., cited, 173
Stewart, John Q., cited, 150
Stigler, Stephen M., cited, 109
storage of maps, 48
Strachey, Henry, cited, 204
Strasbourg, Bibliothèque nationale et universi-
    taire de, 209f
"stratigraphy" (word), 115
stratus (word), 115
"Study in Empires, A" (German Library of In-
    formation), 151f
sub-Saharan Africa, 96
Suggested Plan for a National Institute of Geog-
    raphy (Galeron), 148f
Sullivan, Garrett A., Jr., cited, 179
Sullivan, Robert, 154–55
Sumerian plan drawn to scale, 169
Supan, Alexander, on map scales, 214–16
superiority, as a Western myth, 8
surveier (word), 77, 179
surveillance cameras, 78; "surveying and map-
    ping" (the phrase), 94, 154, 172
surveyors: "commanding" and "domineering"
    view of, 77–78, 138; as creator of plans, 3,
    28; desire to improve spatial knowledge,
    87; Dutch, 188; Egyptian, 178; employed
    by Ogilby, 185, 185n; formal training of, 47;
    instruments used by, 185; Libby as, 29f; and
    mapping discourse, 47; map scales employed
    by, 173, 220; professional organizations for,
    146; and property mapping, 181, 186–88;
    referred to as "cartographers," 127; Saxton as,
    91, 181, 185, 185f; systematic, projections used
    by, 123, 203; Thoreau as, 100; Tucker as, 210f;
    use of triangulation, 90, 196, 199; Washing-
    ton as, 182f
surveys and surveying: attributed to ancient
    peoples, 86; author's training in, ix; boundary,
    28, 188; cadastral, 123, 147, 230; cartographic
    preconception, 54; chorographical, 111, 179,
    181, 188, 191, 193; "commanding" and "domi-
    neering" perspective of, 77–78; commercial,
    107; as craft-based activities, 179–80, 187; and
    dating of maps, 75; definitions, 179; errors in,
    60; geodetic, 124, 199; hydrographic, 111, 123,
    147, 203, 229; military, 91, 107, 111, 187–88, 208;
    not differentiated by mapping modes, 100;
    as the only way to map the world, 78–79; or-
    igins of terms, 77–78; prompted by errors in
    maps, 54, 87–89, 111, 199; property, 28, 179, 181,
    186–88; regional, 186, 195; resolution of, 78n,
    88, 100, 104, 172–74, 205–17, 230; of roads, 181;
    as subdivision of cartography, 54, 172; system-
    atic (territorial), 5, 33t, 79, 83, 106–11, 114, 118,
    122–23, 127, 130–31, 134, 199–205, 229; without
    triangulation, 90; trigonometrical (see trian-
    gulation); as underlying all mapping, 61, 76,
    154–55, 226; vignette depicting, 180f. See also
    plane: surveying

topography as the beginning of geographical studies, 155

topologically structured map, 218f

toponyms, 37, 90, 118; forming a map, 39

Török, Zsolt, cited, 142

totalization (cartographic preconception), 54, 82, 135, 236

Toulmin, Stephen, comparing maps to scientific theories, 18

tourist maps, 156

town plan, supposed Neolithic, 69

tracing a map, 157

transatlantic telegraph, 131

transverse cylindrical projection, 200

transverse Mercator projection, 123

trapezoidal projection, 192f

travel books, maps in, 46

Travis, Trysh, cited, 163

Treasure Island (Stevenson), 161

treasure maps, 160–61, 161f

Tree, Roland, cited, 88

triades (Allent's classification), 173t, 206–8, 207t, 211–13, 216, 221, 224–25

triangulation, 79, 90–91, 101, 107–11, 127, 131, 145–46, 181, 196–203, 229; coinage of term, 108–9; map of, in France, 201f, 203f

trigonometric methods, 90–91, 110, 131, 204

trigonometrical geometry, 196–99

tripartite map, 93, 94f

Tripp, B. Ashburton, 181, 184f, 187f

"Triptiks" (AAA), 186

Trolley and Bus Routes and Sight Seeing Places (Philadelphia Rapid Transit Co.), 157f

Tuan, Yi-Fu, cited, 81–82

Tucker, Henry, 210f

Tucker, Jennifer, cited, 137

Tuileries Palace, 212, 212n

Turnbull, David, cited, 18–19, 40, 58

Turteltaub, John, 219

Tuscany, systematic mapping of, 109

Twain, Mark, 40; cartographic satire, 10–13; "Fortifications of Paris," 11, 12f, 40; *Tom Sawyer Abroad*, 12–13

Tyner, Judith A., cited, 46, 97, 153, 167

typus (word), 74, 115

"uncontrollable urge" to see maps everywhere, 62

Union Pacific Railroad, 131

United Kingdom, 154; adoption of metric system, 129; National Grid, 204; Royal Military Academy, 209. *See also* Great Britain

United Nations, 97

United States, 120; Civil War, 42; Coast and Geodetic Survey, 110; Geological Survey, 130, 167, 174; Library of Congress (*see* Library of Congress [U.S.]); lost property markers, 181; National Atmospheric and Space Agency, 140, 141f; popularity of wall maps, 138; Revolutionary War, 108, 113; Survey of the Coast

(*see* Coast and Geodetic Survey [U.S.]); systematic mapping of, 109–10

"universal language" of maps, 60

Universal Transverse Mercator coordinate system, 204

universality and singularity (cartographic preconception), 31, 55, 99–102, 108, 114, 120, 122, 125, 127, 131–32, 134, 138, 142, 146, 153, 156, 160

Universitätsbibliothek (Bern), 136f, 137f

University of London, 124

University of Michigan, 51f

University of Ottawa, 73

University of Southern Maine. *See* Osher Map Library and Smith Center for Cartographic Education

University of Wisconsin–Madison, 198f

Unknown Chaplin, The (film), 19n

Unno, Kazutaka, cited, 32

up-to-dateness (cartographic preconception), 54. *See also* currency of maps

upper class as map consumers, 47

upper photopause, 81

Uppsala as prime meridian, 131

urban maps and mapping, 32, 33t, 42, 134, 169, 207t. *See also* engineering maps and mapping (mode); place maps and mapping (mode); property maps and mapping (mode)

Ursprung of cartography, 71

Urteaga, Luis, cited, 109

use of maps. *See* map use

Valcamonica (Italy), 71

Valerio, Vladimiro, cited, 134

Van der Heijden, H. A. M., cited, 189

Van der Krogt, Peter, 188–89, 189n; work by, also cited, 74, 115–19

Van Duzer, Chet, cited, 63, 142n

Van Egmond, Marco, cited, 93

Van Sant, Tom, 141

Van Sype, Nicola, 194n

Vandermaelen, Philippe: *Atlas universel*, 212, 213f; *Partie des Etats-Unis*, 213f

variable scales on Dufour maps, 212

Vasiliev, Irina, cited, 1, 21–22

Vedic cultures, geometry in, 178

Venetia, systematic mapping of, 108

Venezky, Martin, cited, 161

verbal expression of a map's proportionality to the world, 170, 177, 187–88, 193, 195, 205, 207–11, 217, 222–24

verbal maps, 39, 60

Verdier, Nicolas, cited, 135, 208

Veres, Madalina Valeria, cited, 181

Vermont, 88

Verner, Coolie, on cartographic communication, 91

verst (unit of measurement), 195, 211

Vertesi, Janet, cited, 219

307

Vervust, Soetkin, cited, 181
"very large scale," 172–73, 211
"very small scale," 172–73
Videnskabernes Selskab (Denmark), 108
Vietnam War, 81
"View of the World from 9th Avenue" (Steinberg), 163
views: aerial, 135, 137; on harbor chart, 35f; perspective, 78, 134–35, 137–38, 169, 206; and plans of cities, 33t
Virginia, 181; Fry and Jefferson map, 88; Herrman map of, 88; Smith's explorations in, 75
vision "empowered" by maps, 36, 77–78, 137
"vista space" (Montello), 225
visualization: cartographic preconception, 53; seen as basic to mapping, 64
Visvalingam, M.: on common properties of maps, 120; work by, also cited, 134
Vivan, Itala, cited, 19
Vopel, Caspar, 194n
Voute, Kathleen, 161f
Vries, Dirk de, cited, 93

Wagner, Henry R., cited, 117
Wagner, Hermann, cited, 168–69, 209, 209n, 211n, 212
Wagner, Richard, 19n
Walckenaer, Charles-Athanase, 121; work by, also cited, 126
Walden Pond, 100, 183f
Waldheim, Charles, cited, 224
Waldseemüller, Martin: *Orbis typus universalis*, 34f; *Universalis cosmographia secundum Ptholomæi*, 63, 63f
wall decorations, maps as, 46
wall maps, 36, 39, 46, 59, 69, 83, 85, 138, 152–53, 194n
Wallace, Timothy R., cited, 159
Wallach, Alan, cited, 135
Wallis, B. C., cited, 219n
Wallis, Helen M., cited, 145, 147, 168, 204
Wang, Jessica, on "death" of cartography, 232
Warner, Michael, cited, 93, 113
Warner Brothers cartoons, 161
Waselkov, Gregory A., cited, 39
Washington, DC: conference on the prime meridian, 131; as prime meridian, 132
Washington, George, 181, 184f, 186f; plat of survey by, 182f, 186f
Watson, Ruth, cited, 122, 153
Wawrik, Franz, cited, 108–9
wayfinding (cartographic preconception), 54, 68, 84
web-based mapping, 82, 82n, 86
Weber, Eugen, cited, 129
Weber City (fictitious place), 163
Weibel, Peter: on map making as result of "engineering culture," 105; work by, also cited, 5, 19, 134

West with the Night (Markham), 95
Western cultures, 93–94; cognitive abilities of, 53, 72, 100, 121; contribution of cartography to, 25, 54; exceptionalism of, 93; sketch maps made in, 72; view of non-Western cultures, 71, 73, 121
Western mapping: as defining cartography, 120, 126, 233; geometries of, 176–205; and imperialism, 5, 7, 121; as innately rational, 5, 7–8, 101, 121; modes of, 32, 33t; not a template for earlier or non-Western mapping, 23; and single-point perspective, 138
Westphalia, systematic mapping of, 108
"what" of cartography, 99
Wheat, James Clements, cited, 30
Wheeler, George M., cited, 109–10
Wheeler, James O., cited, 82n
Whewell, William, 124–25
White, Rodney, cited, 66
"White Hills" map, 89–90
white women interpreting indigenous maps, 72
Whittington, Karl, cited, 9
"why" of cartography, 99
Wigen, Kären E., cited, 32, 57
"Wile E. Coyote" cartoons, 161
Willard, Emma, 153
Williams, James L., 56
Williams, Robert Lee, 221
Wilson, Holly L., cited, 123
Winchester, Simon, cited, 142
"Wine Hills" map, 89–90
Winichakul, Thongchai, cited, 110, 155
Winlow, Heather, cited, 144
Wintle, Michael, cited, 122, 191
Withers, Charles W. J., cited, 55, 113, 125, 129, 131, 147
witness trees, 183f
Wolkenhauer, Wilhelm, cited, 229
women: as "colonized other," 121; consumption of maps by, 48; inserting maps in magazines, 72; mapped "as if their bodies were landscapes," 72, 160; as map readers, 28, 48, 53, 72–73, 100, 113, 121–22, 125, 156; spatial abilities of, 53, 72–73, 100, 122; and theorization, 125
Wood, Denis: "cartography is dead," 231–32; on mapping and cognitive development, 72; on mimetic nature of maps, 24–25; operational definition of map, 41n; "paratext," 37; on "veneer of objectivity" of maps, 4; work by, also cited, 2, 5, 61, 68, 121, 141, 153, 218
wood engraving, 140f
Woodbridge, William C., 153, 154, 155f
woodcut maps, 11, 12f, 34f, 38f, 44, 63f, 89–90, 94f, 137f, 155f
Woodward, David: analysis of Hubbard map, 90; *History of Cartography* (University of Chicago Press), 39; work by, also cited, 20, 56, 81, 83, 93, 115, 117, 146, 189, 218
Woolwich (England), 209

Worcester, Joseph E., cited, 117
Wordsworth, William, 78
world: ancient, map of, 214f; map, Dufour's, 212; maps and mapping (mode), 22, 33t, 34f, 62, 79, 122, 153, 154f, 155, 175f, 189–96, 207t, 217f (*see also* cosmographical maps and mapping [mode]; geographical maps and mapping [mode]; marine maps and mapping); "as seen from a balloon," 137f; the word, contrasted with *earth* or *globe*, 189
world/archive, 52, 54–57, 60, 86–87, 99, 149, 230. *See also* archive of spatial knowledge (the corpus of maps)
"world-knowledge" (Kant), 123
World on the Mercator Projection, The (Anonymous), 175f
World War I, 98, 204; newspaper map, 160f
World War II, 97, 204; advances in mapping, 127; and American map making, 150; postwar social movements, 81; widespread map use during, 220
Wright, John K.: on "mental maps," 67; on the need for professionalism, 150; work by, also cited, 71, 98, 126, 229
"writing" (word), 115–16
"Written with a Slate-Pencil, on a Stone" (Wordsworth), 78
Wu, Shellen, cited, 152

Wyckoff, William, cited, 159
Wyld, James, 140f
Wyttenbach, Alberto Fernández, cited, 121

X that "marks the spot," 161, 161f

Yale Center for British Art, 106f, 139f
yard (unit of measurement), 14, 128, 188, 210
"Ye Mappe of Happie Girlhood" (Voute), 161f
Yee, Cordell D. K., cited, 93
Yensen Collection, 160f
York (England), 209, 210f
Young Sebbatis (pseud.), 39

Zähringer, Raphael, cited, 161
Zainer, Günther, 93
Zelinsky, Wilbur: on Chaplin's globe dance, 19n; work by, also cited, 21, 68
Zentai, László, cited, 156
Zeune, August, 137f
Ziman, John, cited, 18
Zitei, Edward, cited, 135
"zombie project," cartography as, 232–33
zones, climatic, 190
zoological mapping, 143
Zoological Society of London, 123
Zupko, Ronald Edward, cited, 128–30, 179
Zynda, Lyle, cited, 21